Advances in Plant Biology

Volume 4

Series Editor

John J. Harada, Davis, CA, USA

For further volumes:
http://www.springer.com/series/8047

Maureen C. McCann · Marcos S. Buckeridge
Nicholas C. Carpita
Editors

Plants and BioEnergy

Springer

Editors
Maureen C. McCann
Department of Biological Sciences
Purdue University
West Lafayette
IN, USA

Nicholas C. Carpita
Department of Botany and Plant Pathology
Purdue University
West Lafayette
IN, USA

Marcos S. Buckeridge
Department of Botany
Institute of Biosciences
University of Campinas
Sao Paolo
Brazil

ISBN 978-1-4614-9328-0 ISBN 978-1-4614-9329-7 (eBook)
DOI 10.1007/978-1-4614-9329-7
Springer New York Heidelberg Dordrecht London

Library of Congress Control Number: 2013953273

Printed on acid-free paper

Springer is part of Springer Science+Business Media (www.springer.com)

Preface

A nation's vision for developing renewable and sustainable energy resources is typically propelled by three drivers—security, cost, and environmental impact. The U.S. currently accounts for one-quarter of the world's total oil consumption. Technology improvements in the recovery of oil from shale, production of hybrid and electric vehicles, and light vehicle fuel efficiencies have reduced but far from eliminated dependence on foreign oil imports. At the same time, Brazil, because of its embrace of ethanol as an alternative liquid fuel in the 1970s, is today energy independent. Issues of energy security are compounded by increased demand from emerging economies and the supply of that demand from politically unstable parts of the world. Economic growth and development worldwide depend increasingly on secure supplies of reliable, affordable, and clean energy. As easily accessible reserves of oil become exhausted, the energy return on energy invested, currently a ratio of 30:1, will decrease, driving up costs for the consumer, not just of liquid transportation fuels, but of all of the oil-based chemicals and materials supplied by the petrochemical industry. Agriculture itself is an oil-intensive enterprise, with about 2 % of total oil consumption used directly in farm vehicles or indirectly for mechanized processing. As carbon dioxide levels reach an unprecedented 400 ppm, there is an unequivocal imperative to mitigate greenhouse gas emissions by decreasing fossil fuel consumption and transition to carbon-neutral or carbon-negative fuels as well as improving efficiency of fuel use.

It was with the urgency conferred by these three drivers that the American Society of Plant Biologists convened the First Pan-American Congress on *Plants and BioEnergy* in June, 2008, in Mérida, Mexico. This congress was designed to initiate Pan-American research collaborations in energy biosciences and to showcase advances in the development of new energy crop plants, their genetic improvement based on new knowledge of plant growth and development, their fit into regional environments, and the development of a sustainable energy agriculture. Subsequent biennial meetings, one in Brazil and another in the US, have served to connect advances in second and third generation of biofuels with the realities of economic success and sustainability. This edition encompasses specific examples of progress to this goal yet keeps in perspective the realities of the economic drivers and pressures that govern the translation of scientific success into a commercial success.

In Part I, **Social and Economic Impacts of a Bioenergy Agriculture,** we begin with Patricia Guardabassi and José Goldemberg's overview of the prospects of global ethanol production in developing countries and the relevant social and economic issues unique to their environments. Jeremy Woods and Nicole Kalas extend this theme, drawing on the lessons of biofuels policy development concerning direct and indirect land use over the last decade to inform energy policies that will drive sustainable land use. Wally Tyner and Farzad Taheripour discuss the uncertainties, policy options, and land-use impacts in moving from ethanol to advanced fuels. Finally, Gal Hochman and David Zilberman discuss the implementation and economics of algal farming and how economic success hinges on high-value bio-products generation.

In Part II, **Biomass Feedstocks**, we explore the breadth of bioenergy crops and the progress that is being made to introduce them into the agricultural landscape, the underlying biology of bioenergy plants, and new ideas to enhance biomass yield and quality for the energy crops of the future. Andrew Jakubowski and Michael Casler show that improved and locally collected ecotypes of switchgrass, big bluestem, and Indiangrass can coexist on the landscape and help to jumpstart sustainably the shift to a bioenergy-based economy. Cynthia Damasceno, Robert Schaffert, and Ismail Dweikat's article considers how to mine the vast genetic diversity in the sorghum genome and its advantages as an annual crop for use in both tropical and temperate biomes. Angela Karp and her colleagues explore the challenges and prospects for integrative approaches to improve woody biomass species, such as willow, as lignocellulosic feedstocks. Two perspectives consider oil production platforms for advanced biofuels, biodiesel, and bio-based products. Umidjon Iskandarov, Hae Jin Kim, and Edgar Cahoon present the advantages of Camelina as an emerging drought-resistant oilseed suitable for marginal lands that are tractable to genetic improvement, and Janaina Meyer and Antonio Salatino, present their ideas for Brazil's contribution to biodiesel with palm. To conclude this section, Ahmed Faik, Nan Jiang, and Mick Held present a thorough update on our present knowledge of the biochemistry of xylan synthesis, with particular emphasis on the unique aspects of synthesis in grasses. Catherine Rayon, Anna Olek, and Nick Carpita close this section with a perspective on the complexities of cellulose biosynthesis that suggests strategies for how cellulose might be designed for improved bioenergy feedstocks.

In Part III, **Biomass Conversion Technologies** we explore the culmination of the technologies that drive the ethanol industry and the promise for the efficient conversion of biomass into energy-dense liquid fuels and high value co-products. Harry Gilbert begins with an extensive review of how novel enzyme repertories are developed for the efficient deconstruction of plant biomass tailored for the bioenergy industry. Adriana Grandis and her colleagues in the Laboratory of Marcos Buckeridge extend this strategy in their perspective on exploiting natural plant cell wall degradation systems to improve bioethanol production. Rebecca Garlock Ong and colleagues in the laboratory of Bruce Dale present a comprehensive review on how knowledge of the fine structure of the plant cell wall informs better design principles for the biorefinery. Ken Reardon gives some guiding principles for the lignocellulosic refineries of the future for how both the economics and the environmental impacts of biofuel production could be improved by developing processes to obtain a wider range of chemicals with higher value from biomass. The edition concludes with two articles that explore the early successes in the direct conversion of lignocellulosic biomass into advanced biofuels. Basudeb Saha, Nathan Mosier, and Mahdi Abu-Omar show the path to catalytic dehydration of lignocellulosic-derived xylose to furfural, while Joe Bozell identifies pathways for the catalytic oxidation of lignin for the production of low molecular weight aromatics.

In this volume, we bring together perspectives from a wide range of disciplines, recognizing that the grand challenge of displacing a century of global dependence on oil requires a new research paradigm, a "Manhattan Project" for the twenty-first century. Production of carbon-neutral reliable, affordable biofuels for a growing population in a manner that does not compromise food and feed production takes a community that is fully engaged, committed, and international in scope. The success of that community will be measured in our contributions to climate security, economic growth, and self-sufficient energy production for our nations.

Maureen C. McCann
Marcos S. Buckeridge
Nicholas C. Carpita

Contents

Part I Economics of Bioenergy

1 The Prospects of First Generation Ethanol in Developing Countries 3
Patricia Guardabassi and José Goldemberg

2 Can Energy Policy Drive Sustainable Land Use? Lessons from Biofuels Policy Development Over the Last Decade 13
Jeremy Woods and Nicole Kalas

3 Advanced Biofuels: Economic Uncertainties, Policy Options, and Land Use Impacts 35
Wallace E. Tyner and Farzad Taheripour

4 Algae Farming and Its Bio-Products 49
Gal Hochman and David Zilberman

Part II Biomass Biology

5 Regional Gene Pools for Restoration, Conservation, and Genetic Improvement of Prairie Grasses 67
Andrew R. Jakubowski and Michael D. Casler

6 Mining Genetic Diversity of Sorghum as a Bioenergy Feedstock 81
Cynthia M. B. Damasceno, Robert E. Schaffert and Ismail Dweikat

7 Genetics, Genomics and Crop Modelling: Integrative Approaches to the Improvement of Biomass Willows 107
Angela Karp, Goetz M. Richter, Ian F. Shield and Steven J. Hanley

8 Camelina: An Emerging Oilseed Platform for Advanced Biofuels and Bio-Based Materials 131
Umidjon Iskandarov, Hae Jin Kim and Edgar B. Cahoon

9 Perspectives in Brazil of the Contribution of Palm Trees to Biodiesel Production 141
Janaina M. Meyer and Antonio Salatino

10 Xylan Biosynthesis in Plants, Simply Complex 153
Ahmed Faik, Nan Jiang and Michael A. Held

11 Towards Redesigning Cellulose Biosynthesis for Improved Bioenergy Feedstocks 183
Catherine Rayon, Anna T. Olek and Nicholas C. Carpita

Part III Biomass Processing

12 Developing Novel Enzyme Repertoires for the Efficient Deconstruction of Plant Biomass Tailored for the Bioenergy Industry 197
Harry J. Gilbert

13 Using Natural Plant Cell Wall Degradation Mechanisms to Improve Second Generation Bioethanol 211
Adriana Grandis, Amanda P. de Souza, Eveline Q. P. Tavares and Marcos S. Buckeridge

14 Linking Plant Biology and Pretreatment: Understanding the Structure and Organization of the Plant Cell Wall and Interactions with Cellulosic Biofuel Production 231
Rebecca Garlock Ong, Shishir P. S. Chundawat, David B. Hodge, Sai Keskar and Bruce E. Dale

15 Lignocellulosic Biorefineries: Concepts and Possibilities 255
Kenneth F. Reardon

16 Catalytic Dehydration of Lignocellulosic Derived Xylose to Furfural 267
Basudeb Saha, Nathan S. Mosier and Mahdi M. Abu-Omar

17 Catalytic Oxidation of Lignin for the Production of Low Molecular Weight Aromatics 277
Joseph J. Bozell

Part I
Economics of Bioenergy

Chapter 1
The Prospects of First Generation Ethanol in Developing Countries

Patricia Guardabassi and José Goldemberg

Abstract There are great perspectives to the development of second-generation technologies to biofuels production, nevertheless its production in large scale is depending on a technological breakthrough to become feasible. The production of ethanol from sugarcane based on first generation technology has evolved in the last decades; however gains of productivity can still be achieved. Latin American and African countries have suitable conditions to the growth of sugarcane. Many of these countries are highly dependent on fossil fuels imports. Thus, the introduction of ethanol blends can reduce the consumption of fossil fuels, while creating jobs and developing local industry. Notwithstanding, first generation ethanol can still contribute to developed countries, especially US and European countries, to commit with biofuels use mandates. The aim of this chapter is to present the state of the art of ethanol production in Latin America and African countries, identifying the main obstacles to the development and discussing policies that could be implemented to overcome such barriers.

Keywords Biofuels • First generation • Developing countries • Barriers

P. Guardabassi
Sustainability Science Program, Kennedy School of Government, Harvard University, 79 John F. Kennedy Street, Cambridge, MA 02138, USA
e-mail: Patricia_Guardabassi@hks.hrvard.edu

J. Goldemberg (✉)
Institute of Eletrotechnics and Energy, University of São Paulo, Av. Prof. Luciano Gualberto, 1289, 05508-010 São Paulo/SP, Brazil
e-mail: goldemb@iee.usp.br

M. C. McCann et al. (eds.), *Plants and BioEnergy*, Advances in Plant Biology 4,
DOI: 10.1007/978-1-4614-9329-7_1, © Springer Science+Business Media New York 2014

1.1 Introduction

The economic growth of developing countries and the maintenance of consumption patterns in developed countries continuously increase world's energy consumption. Concomitantly, the depletion of oil reserves has been observed as well as the impacts of climate change caused by human actions, especially due fossil fuels consumption. In the specific case of energy, this topic is of special importance since the projections indicate, in a scenario based on current policies for the energy sector, a growth of the world's energy consumption by 47 % between the years 2008 and 2035 (OECD/IEA 2010), based on increased use of coal and natural gas. In this context it is necessary to develop alternative sources of energy and modern technologies that can replace fossil fuels and mitigate greenhouse gas (GHG) emissions.

The most promising alternatives are those based on renewable sources of energy such as solar panels and wind turbines. However for the transportation sector, which accounted for 13 % of greenhouse gases global emissions of 2004 (IPCC 2007), biofuels are the only worldwide commercially available option, since electric vehicles or hydrogen still demand technological development. Thus, biofuels are gaining an increasingly importance to reduce GHG emissions as well as the dependence of fossil fuels. In view of the impacts from the extensive use of fossil fuels, and the opportunity to reduce their emissions through the adoption of renewable energy in their energy matrixes, developed countries have established targets for use of clean fuels (Goldemberg 2007).

The most ambitious one was established by the European Union, whose goals include a 20 % share of renewable energies in overall Community consumption and, in the transportation sector, at least 10 % biofuels by 2020, equivalent to 14 billion liters. However, alleging aiming to reduce the impact of first generation biofuels on food prices, the EU cut by a half its biofuels target. Within this bioenergy program the European Parliament and the European Union Council established the European Union Directive 28/2009 on the promotion of the use of energy from renewable sources. This Directive sets sustainability criteria that must be accomplished by countries willing to provide biofuels to the European Union country members. In the case of European Union, biofuels are expected to provide a great contribution in the achievement of renewable energy use target and GHG emissions in transport sector (European Expert Group on Future Transport Fuels 2011).

Another initiative that will also be central for the global biofuels market due to its dimensions was taken by the U.S. government in the Energy Independence and Security Act (EISA). Established in 2007, it set a minimum consumption in the country of 45 billion liters of biofuels by 2010, reaching 136 billion liters in 2022. The legislation identifies three types of biofuels, which should account for about 60 % of this volume (equivalent to 88 billion liters), they are: cellulosic ethanol, biomass diesel and "other advanced". To be classified as "advanced" biofuels must reduce by 50 % GHG emissions on a life cycle basis, compared to gasoline. The levels of GHG emissions reduction for biofuels were adopted by the Environmental Protection Agency (EPA), and according to the Agency's calculations, Brazilian ethanol reduces GHG emissions by 61 %. In summary, European and American initiatives together are responsible for the demand of 150 billion liters of ethanol in the year 2022.

However, due to edaphoclimatic conditions and restrictions on land availability, many countries do not produce volumes of biofuels sufficient to supply the domestic market. Consequently, the international trade of biofuels has been growing and more producing countries are likely to be part of it. Ethanol can be produced from a series of feedstocks; however its access to American and European markets will depend on its GHG emissions balance (Worldwatch Institute 2007). Among different raw materials to produce ethanol, sugarcane is the most effective in GHG emissions reduction, because small amounts of fossil fuels are used in its production chain. According to (Smeets et al. 2007), for the year 2050, Sub-Saharan Africa and Latin America and the Caribbean regions have the greatest potential for production of agricultural residues that could be used as bioenergy feedstock.

The "Global Agro-Economic Zones (GAEZ)" system developed by FAO, in conjunction with the International Institute for Applied Systems Analysis (IIASA), identified and quantified areas with potential to produce raw materials for biofuels based on climate and soil conditions. In the case of rain-fed sugar cane, GAEZ identified 135 million suitable or very suitable hectares and 130 million hectares moderately suitable hectares (of which only 22 million are currently planted with sugar cane). For unprotected areas the overall potential is estimated at 87 million hectares suitable or very suitable, of which 26 million hectares are located Africa and 54 million hectares South America (Fischer et al. 2009). The regions with greater aptitude for the cultivation of rain-fed sugar cane are located in South and Central America, the Caribbean, Central Africa and some countries in West and South Africa, South Indian and Southeast Asia. Although suitable for the production of biofuels these regions produce small amounts of ethanol.

From the world's total production of ethanol, in 2010, of 86.8 billion liters; United States and Brazil were responsible for 76.2 billion; and despite its potential the African continent produced only 0.16 billion liters and Latin America and Caribbean 2.6 billion liters (RFA 2011). In fact, there are barriers that must be overcome in order to allow the development of a biofuels production sector. In order to surpass such barriers, it is necessary to settle a legal framework that ensures an appropriate institutional condition, economically attractive to investors and to promote the production minimizing environmental and social impacts.

1.2 The State of the Art of Ethanol Production

Ethanol production has been based in the so-called "first generation technologies" either by direct fermentation (in the case of sugar cane) or saccharification of starch (in the case of corn and wheat) followed by fermentation. New technologies, or second generation technologies, to produce biofuels include the enzymatic conversion of lignocellulosic material. Despite huge amounts invested and the demand created by US Renewable Fuel Standard, there are no commercial plants operating yet. According to the Advanced Ethanol Council "Advanced Ethanol Council (AEC) (2012)", industry is reaching commercial deployment phase, however the "high capital risk from OPEC-induced price distortions, constrained blending markets and policy uncertainty continues to slow down the rate of deployment".

Currently, the US is the largest ethanol producer with an annual amount of 49 billion liters, using corn as feedstock, followed by Brazil, which uses sugarcane as feedstock, producing 28 billion liters (REN21 2012). So far, sugarcane ethanol has proved to be the most competitive raw material in use, due to its positive energy balance, and consequently positive reduction of greenhouse gas emissions on a life cycle basis, the lower costs of production and higher production yields when compared to corn (Goldemberg, The Brazilian biofuels industry 2008). Therefore, it is plausible to extend a successful experience, such as the Brazilian ethanol program, to other developing countries.

1.3 Existing Mandates to Ethanol Production and Use

1.3.1 Africa

Ethanol is produced in Africa to replace gasoline since de 1970s, in countries highly dependent on imports, such as Zimbabwe, Malawi and Kenya. Malawi produces ethanol since 1982 and uses E10 blends. The country is running tests with ethanol dedicated engines. The local production reaches 18 million liters, half of it domestically consumed and the rest exported to African countries. There are two distilleries operating in the country with a total producing capacity of 32 million litres. Government aims to estimulate the production of sugarcane in order to use this idle capacity (Janssen and Rutz 2012).

Other nations are introducing policies aiming to leverage the production of biofuels aiming to increase energy security due to energy matrix diversification and reduction of fossil fuels imports. Hence, such policies are also instruments to promote the development of rural areas, through job creation, income generation, local industry expansion and investments in infrastructure. South Africa is the largest market in the region. The country established "The Biofuels Industrial Strategy of the Republic of South Africa", in December 2007, which aims to develop a biofuels industry that could supply the domestic market of 2 % ethanol blends (equivalent to 400 million liters per year) within 5 years. Sugarcane and sugar beet are edible crops to ethanol production, and sunflower, rapeseed and soybeans to biodiesel. The program will demand about 1.4 % of country's agriculture land. The country will only adopt mandatory blends when domestic biofuels production is ensured (Department of Minerals and Energy 2007).

In 2009, Mozambique Council of Ministers approved the National Biofuels Policy and Strategy. According to the Minister of Energy, the use of E10 and B3 would started in 2012, and the country's productive capacity to meet domestic demand, but the interest of the government is to continue to promote ethanol production aimed at export in the coming years (Gil 2011). Kenyan biofuel policy aims to reduce oil imports by 25 % by 2030 and to increase access to energy through sustainable biofuels production. The 2009 draft of Biofuels Strategy stress a general capacity to produce ethanol from molasses to supply E10 blends. In 2010, the Kenyan biofuel policy strategy defined an E10 blending mandate (equivalent to 93 million liters). The national installed capacity of 125,000 liters per day is producing only 60,000 litres per day due to the limited current supplies of molasses (GTZ and Ministry of Agriculture Kenya 2008).

In Tanzania, a country that presents suitable climate conditions and available arable land and water, the Biofuels Guidelines of December 2009, addresses key issues related to: institutional framework; application procedure for investors; land acquisition and use; contract farming; sustainability of bioenergy development; avoidance of food versus fuel conflicts and sufficient value creation for the local rural population. In Zambia, energy security and matrix diversification are the main drivers of biofuels introduction. The Sixth National Development Plan defines biofuel blending ratios for bioethanol and biodiesel, for the current period up to 2015: up to E10 and B5. Ethiopia has three state owned sugar factories which have been operational for long time. The country introduced E5 in 2009, then increasing to E10 early 2012. Government says that the country's ethanol blending policy has saved the country $20.5 million in fuel imports since the policy began in 2008 (Biofuels Digest 2012).

1.3.2 Latin America and Caribbean

In the Latin America and Caribbean there are a growing number of nations adopting biofuels. In Argentina, national legislation defines the blend of 5 % ethanol to gasoline and the same percentage of biodiesel to diesel oil. However, regarding ethanol, the mandate must be introduced progressively due the lack of domestic production capacity to attend the demand (Fundación Bariloche 2011). Uruguay has approved Law 18,195/2007 to introduce gradually introduce ethanol blends up to a mixture of 5 % in 2015. Domestic production is not sufficient nowadays to supply the internal market, though a project being developed by the state-owned oil company, is promoting the development of sugarcane crops, especially in least developed regions of the country (Fundación Bariloche 2011). Paraguay defined by Law 2,748/2005, from the year of 2006, the blend of ethanol to gasoline ranging from 20 % up to 24 %. In May 2008, Decree 12,240/2008 determined the reduction of taxes to biodiesel and ethanol e cut out importing taxes on flexible fuels vehicles and E85 new and used vehicles (USDA 2009). In Colombia, the promotion of biofuels initiated in 2001, due Law 693/2001 that stipulates rules to ethanol use and determines incentives to production, trade and consumption of this fuel (CENBIO and CENTROCLIMA 2011).

In Mexico, besides the Bioenergy Promotion and Development Law, which aims to promote the diversification of energy matrix, the use of biofuels is not mandatory (BIOTOP 2009). In Costa Rica, the use of ethanol was initiated in the 1970s aiming at reduce the country's oil dependence, however it had not succeed at that time (Nogueira 2004). In January 2008, the "Biofuels National Program" was launched aiming the increase of country's energy security and greenhouses emissions reduction through the blend of 7.5 % ethanol to the gasoline and 5 % of biodiesel to the diesel in that year, progressively increasing to E10 and B20 blends in 2010. The lack of infrastructure obligated government to postpone the goals (Aguero 2011), however ethanol is available in few regions in the north of the country yet (Villegas e Campos 2011) and apparently government is not interested in expanding ethanol supply. Other countries in the region, such as El Salvador, Honduras, Nicaragua, Guatemala, Peru, Ecuador and the Dominican Republic have legislation stimulating the production and use of biofuels, however there are no mandatory blends and local production is low, absent or devoted to foreign markets (especially European Union and United States under advantageous trade agreements).

1.4 Main Obstacles to the Development of Ethanol Industry

To analyze the production of biofuels two main aspects must be considered, the feedstock production (agricultural side) and its processing into biofuel (industrial side). The current yield of agriculture production in Africa is low due to the lack of adequate agricultural management, derived from the lack of access to fertilizers, seeds, water and training. One of the causes is the lack of access to credit for small holders, that prevents then to buy agriculture supplies such as fertilizers, seeds and equipment hence reducing production costs (Mitchell 2011).

Other aspects contributing to slow down development are related to the risks to investors and include absentee or weak land tenure policies (Cotula et al. 2004) precarious infrastructure for distribution of the final product (Cornland et al. 2001); lack of policies to guarantee the existence of a consumer market (e.g. mandatory blending). Regarding the industrial production there are many technologies to produce ethanol from various feedstocks. The so-called first generation biofuels technologies based on sucrose fermentation are well known and largely used worldwide. The implementation of such technologies requires few adjustments to local operation condition and the main barrier in this case would be the lack of trained personnel.

An additional benefit from the production of biofuels is the possibility of using crops residues, e.g. sugarcane bagasse, as fuel in cogeneration systems. Those units can supply the energy needs of the biofuel facility and even produce surplus electricity to feed the surrounding areas, with the advantage that the feedstock is readily available at low (or no) cost. It can be advantageous in African countries where access to energy services is limited, especially in rural areas. Cogeneration technology is commercially available in a wide range of power capacities. Low technical skills and the lack of a supportive policy and regulatory framework prevent investments in more efficient production. The absence of attractive and pre-determined tariffs are the major barriers to the development of a to sell the exceeding production (Cogen for Africa n.d.).

1.5 Policy Proposals

The development of a stable institutional and political environmental is required in order to attract companies aiming to invest in biofuels production in such markets. To establish a captive market for biofuels through the adoption of mandatory blends is essential, however, in many countries the domestic market can be so small, due to economic conditions and a reduce vehicles fleet, that looking at exporting could present an interesting opportunity. In this case, fuel quality standards have to be carefully considered. A policy of prices that enables ethanol to compete with both gasoline and sugar prices is essential, especially in the initial stages of its introduction into the market. Also, an extensive network for distribution and retail has to be designed in order to easily offer the product to consumers.

Considering the utilization of third parties, such as small farmers, investments in training and equipment tend to be necessary; nevertheless many of these farmers do not have access to credit to invest in its production. This problem could be partially amended through the development of funding policies and tools, such as microcredit and rural credit. There are countries where land tenure rights are unclear, thus the introduction and improvement of land use and ownership legal framework is required in order to protect and guarantee small farmers rights.

Regarding environmental aspects, the development of a zoning that defines edible areas to grow sugarcane crops and those areas that must be protected. In the case of other environmental issues related to the use of fertilizers, pesticides and atmospheric emissions, the use of commercial available technologies can be adopted as best practices. Water use tends to be a more delicate topic, especially in African countries. The reuse of the vinasse (an ethanol production by-product) for fertirrigation can reduce the need of water withdraw.

A fundamental aspect is related to labor conditions. It is well know that usually sugarcane cutters, that represent the largest amount of workforce in this sector, have low education levels, thus continuous training programs are essential to increase productivity and avoid accidents. Also, the endorsement of international working treaties and the correct enforcement must avoid child labor and inappropriate and degrading working conditions.

The promotion of electricity surplus production through cogeneration systems based on sugarcane bagasse still needs the creation of institutional and regulatory policies aiming to minimize market risks for investors. Mechanisms such the definition of standard power purchase agreements and feed-in tariffs can create confidence in the market, and stimulate investments in modern and efficient biomass cogeneration projects.

Acknowledgments This work was partially conducted while the author was a Giorgio Ruffolo Fellow in the Sustainability Science Program at Harvard University. Support from Italy's Ministry for Environment, Land and Sea is gratefully acknowledged.

References

Advanced Ethanol Council (AEC) (2012) Cellulosic biofuels industry progress report 2012/2013. http://ethanolrfa.3cdn.net/d9d44cd750f32071c6_h2m6vaik3.pdf, 2012

Aguero M (2011) Recope descarta plan de etanol a corto plazo. La Nación. 21 de abril de 2011. www.nacion.com/2010-07-14/ElPais/NotasSecundarias/ElPais2446754.aspx (acesso em 21 de abril de 2011)

Biofuels Digest (2012) Ethiopia's ethanol blending saves $20.5 million in three years. 9 de 7 de 2012. http://www.biofuelsdigest.com/bdigest/2012/07/09/ethiopias-ethanol-blending-saves-20-5-million-in-three-years/ (acesso em 5 de 10 de 2012)

BIOTOP (2009) Biofuel policies and legislation in Latin America. Relatório de projeto, 42

CENBIO and CENTROCLIMA (2011) Bioenergy, rural development and poverty alleviation in Brazil and Colombia. São Paulo, Brazil

Cogen for Africa. Key drivers and barriers for biomass cogeneration development. s.d

Cornland DW et al (2001) Sugarcane resources for sustainabledevelopment: a case study in Luena, Zambia. Stockholm Environment Institute, Stockholm

Cotula L, Toulmin C, Hesse C (2004) Land tenure and administration in Africa: lessons of experience and emerging issues. IIED, Londres

Department of Minerals and Energy (2007) Biofuels industrial strategy of the Republic of South Africa

European Expert Group on Future Transport Fuels (2011) Future transport fuels. European expert group on future transport fuels

Fischer G, Hizsnyik E, Prieler S, Shah M, van Velthuizen H (2009) Biofuels and food security. The OPEC International, Vienna

Fundación Bariloche (2011) Bioenergy for rural development and poverty alleviation—GNESD. San Carlos de Bariloche

Gil F (2011) Biocombustíveis e petróleo: Mistura obrigatória a partir de 2012. Moçambique para todos. 14 de agosto de 2011. http://macua.blogs.com/moambique_para_todos/2011/08/biocombust%C3%ADveis-e-petroleo-mistura-obrigatoria-a-partir-de-2012.html (acesso em 15 de novembro de 2011)

Goldemberg J (2007) Ethanol for a sustainable energy future. Science 315:808–810

Goldemberg J (2008) The Brazilian biofuels industry. Biotechnol Biofuels, 1 de May de 2008: 1–6

GTZ, Ministry of Agriculture Kenya (2008) A roadmap for biofuels in Kenya

IPCC (2007) Fourth assessment report: climate change 2007—synthesis report. Contribution of Working Groups I, II and III to the Fourth Assessment Report of the Intergovernmental Panel on Climate Change. IPCC, Geneva

Janssen R, Rutz D (2012) Bioenergy for sustainable development in Africa. Springer, Berlin

Mitchell D (2011) Biofuels in Africa: opportunities, prospects, and challenges. The World Bank, Washington, DC

Nogueira LAH (2004) Perspectivas de un programa de biocombustibles en América. Cidade do México: CEPAL

OECD/IEA (2010) World energy outlook 2010. International Energy Agency, Paris

REN21 (2012) Renewables 2012 global status report. REN21 Secretariat, Paris
RFA (2011) World fuel ethanol production 2011. http://ethanolrfa.org/pages/World-Fuel-Ethanol-Production (acesso em 15 de novembro de 2011)
Smeets EMW, André PC, Faaij APC, Lewandowski IM, Turkenburg WC (2007) A bottom-up assessment and review of global bio-energy potentials to 2050. Prog Energy Combust Sci 33:56–106
USDA (2009) Paraguay biofuels annual. Buenos Aires
Villegas JS, Campos C (2011) Gasolina con alcohol seduce a choferes de Guanacaste. La Nación. 21 de 04 de 2011. http://www.nacion.com/2011-04-19/ElPais/Relacionados/ElPais2753442.aspx (acesso em 25 de maio de 2011)
Worldwatch Institute (2007) Biofuels for transport. Springer, London

Chapter 2
Can Energy Policy Drive Sustainable Land Use? Lessons from Biofuels Policy Development Over the Last Decade

Jeremy Woods and Nicole Kalas

2.1 Introduction

The mandated increase in bioenergy as a means to decarbonise our energy supply, enhance energy security, and promote rural development has raised concerns regarding the impacts biomass feedstock production may have on food security. These national mandates appear to have placed bioenergy feedstock production in competition for resources required to feed a growing global population. In turn, concerns over the direct and indirect impacts of bioenergy, particularly conventional biofuels,[1] have pushed policy makers to try to direct biomass crop production for energy onto marginal, degraded and 'unused' land. Moving bioenergy onto marginal lands will inevitably raise the costs of feedstock production, but it may also be contradictory to food security where sustainable intensification and reduced losses require increased energy inputs into agriculture. This marginalisation ignores the beneficial role that perennial energy crops could play in managing the sustainable intensification of overall agricultural production required to feed over 9 billion people by 2050. Chapter 2, therefore, explores the role and drivers of bioenergy in future world energy production, land use change and wider sustainability issues, and proposes an alternative, integrated approach toward a resource efficient and sustainable provision of agricultural products, including food, feed, biobased chemicals, materials and energy.

[1] Conventional biofuels are produced through the fermentation of sugars or starches to bioethanol from commodity crops such as sugarcane, maize, wheat, and beet, or through the methyl esterification of vegetable oils to biodiesel from palm, soy or oilseed rape.

J. Woods (✉) · N. Kalas
Centre for Environmental Policy, Centre for Energy Policy and Technology, Imperial College London, London, UK
e-mail: jeremy.woods@imperial.ac.uk

N. Kalas
e-mail: n.kalas@imperial.ac.uk

M. C. McCann et al. (eds.), *Plants and BioEnergy*, Advances in Plant Biology 4,
DOI: 10.1007/978-1-4614-9329-7_2, © Springer Science+Business Media New York 2014

2.1.1 Changing Patterns in Global Energy Supplies

Two concurrent developments in energy use are changing the pattern of global energy supplies (see Fig. 2.1). On one side, climate policy and energy security driven increases in efficiency and uptake of renewable energy technologies in the USA, European Union and Japan are slowly reversing the upward trend of oil imports observed to date. In the USA, increased domestic production of oil and the recent intensification in shale gas exploration (hydraulic fracturing) have placed the country on a path to energy self-sufficiency. On the other side, rapid economic growth in China and India, driven primarily by fossil fuels, has led to increased consumption of cheap coal and imported oil. These opposing trends in oil consumption are raising the competition for energy security and shifting the global balance of oil imports from OECD to non-OECD countries, where China is expected to become the world's largest oil importer by 2020.

As Fig. 2.2 illustrates, new coal has provided nearly 50 % of incremental energy supply since 2000 and in increasingly inefficient power plants (to lower capital costs). In the USA, shale gas has started to drive coal out of the electricity generation mix and is also degrading the role for dedicated biomass and other sources of renewable energy. In the UK, power generators are moving rapidly towards large scale biomass co-firing in existing electricity plants and reducing demand for dedicated biomass.

Cheap coal and the 'shale gas revolution' are the biggest challenges to climate change mitigation and the meaningful deployment of renewables. The world is not on track to meet the internationally agreed target to limit the long-term rise in the global average temperature to 2 °C. Over 80 % of global energy consumption is based on fossil fuels, and the energy sector accounts for approximately 2/3 of Greenhouse Gas (GHG) emissions (IEA 2013).

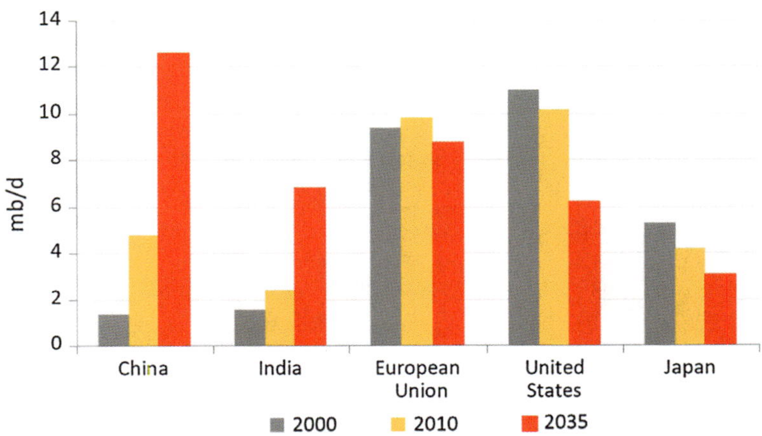

Fig. 2.1 Net imports of oil (2000–2035). *Source* IEA (2011)

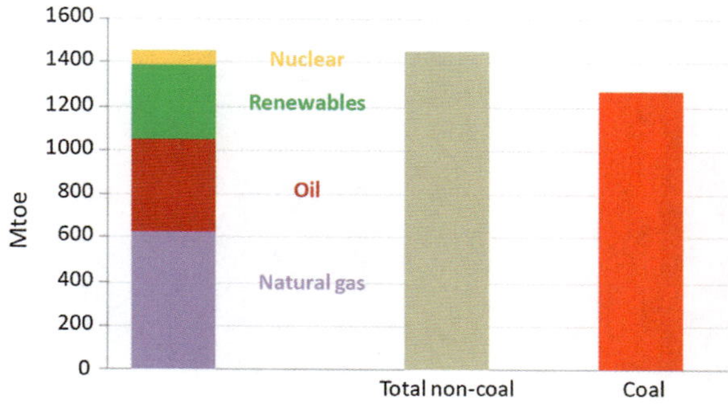

Fig. 2.2 Growth in global energy demand (2000–2010). *Source* IEA (2011)

2.1.2 Future World Energy Production and Price Trends: Is Bioenergy Policy Swimming Against the Tide?

2.1.2.1 World Energy Production and Price Trends

The IEA's Current Policies Scenario[2] shows an increase in world oil production by 26 % from 82.3 mb/d[3] (2007) to 103.8 mb/d in 2030 (see Fig. 2.3) (IEA 2008). To meet demand growth and offset decline, an additional 64 mb/d would be needed, which corresponds to six times Saudi Arabia's current capacity.

Figure 2.4 shows the global trends in prices for fossil fuels ($/GJ) indicating a continued increase in oil prices, and recent decrease in both gas and coal (BP 2012). Gas and coal are expected to resume their upward trend, but stay below oil prices.

2.1.2.2 Global Bioenergy Policy and Consumption

Bioenergy policies are motivated by climate change mitigation targets, energy access and security, and rural development. While global demand for biomass feedstocks is predominantly driven by policies in the EU and USA, at least 33 countries have now implemented mandates for biofuels (blending requirements) (Biofuel Digest 2012).

In the EU's *Renewable Energy Directive* (RED), Member States have committed to reduce their CO_2 emissions by 20 % and to target a 20 % share of renewable energies in the EU energy mix (including 10 % of transport fuels by 2020 as part of the 2007 *The EU climate and energy package* (EC 2009a)). Biofuel demand is projected to be 7,307 ktoe[4] (14,450 million litres) of ethanol, and 21,650 ktoe

---

[2] The IEA's Current Policies Scenario (previously called the Reference Scenario) assumes no changes in energy and GHG emission reduction policies (IEA 2010).

[3] mb/d = million barrels/day (1 barrel = 159 litres).

[4] Ktoe = kilo tonnes of oil equivalent (1 ktoe EtOH = 1.978 million litres, 1 ktoe biodiesel = 1.32 million litres).

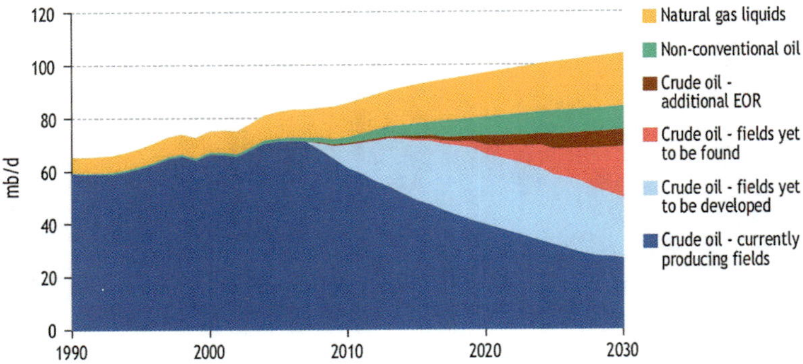

Fig. 2.3 World oil production by source in the Current Policies Scenario (1990–2030). *Source* IEA (2008)

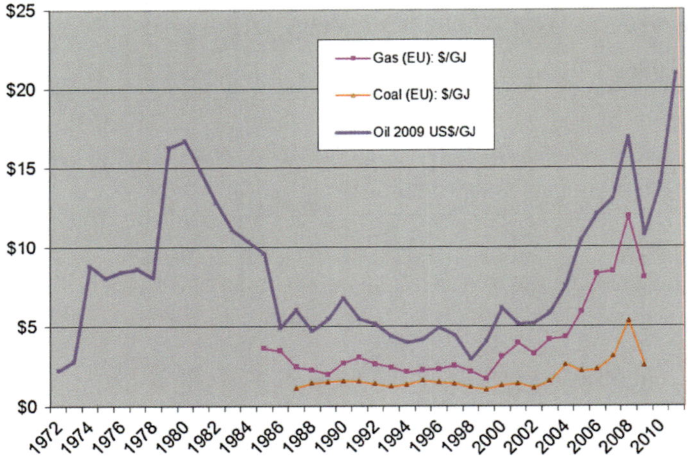

Fig. 2.4 Global trends in fossil fuel prices (1970–2011). *Source* BP (2012)

(28,600 million litres) of biodiesel (Beurskens et al. 2011). The dominant driver for the RED is GHG mitigation, but energy security is also a serious concern. In conjunction with the *Fuel Quality Directive* (FQD), sustainability criteria for GHG emission reductions and biodiversity conversion are applied to each supply chain through assurance and certification schemes (EC 2009a, b, 2012).

In the USA, biofuel blending is mandated by the *Renewable Fuels Standard* (RFS2) to achieve the targets established in *The Energy Independence and Security Act of 2007* (EISA). The RFS2 has laid the foundation for achieving significant GHG emissions in the transport sector and for promoting the development of the US renewable fuels sector. It provides volumetric standards for renewable fuels, including advanced biofuels, and includes GHG emission thresholds producers are

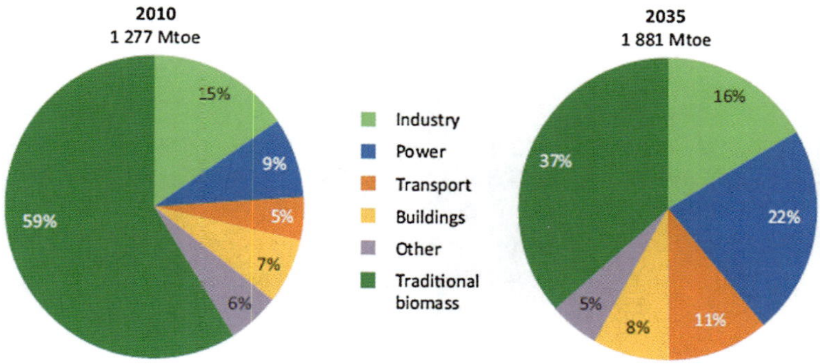

Fig. 2.5 World bioenergy use by sector and use of traditional biomass in the IEA new policies scenario (2010 and 2035). *Source* IEA (2012)

required to meet. Under the RFS2, annual biofuel production, which in the USA is predominantly ethanol, is to increase from currently 13.2 billion gallons (60 billion litres) (2012) to 36 billion gallons (164 billion litres) by 2022 (EPA 2010).

In 2010, total global bioenergy consumption amounted to 1,277 Mtoe,[5] or excluding traditional biomass,[6] 526 Mtoe (see Fig. 2.5). The IEA New Policies Scenario[7] estimates that in 2035 total world bioenergy use will increase to 1,881 Mtoe, or 1,200 Mtoe excluding traditional biomass (at an average annual growth rate of 3.3 %). Currently, the industrial sector is the largest consumer of bioenergy (196 Mtoe), but the power sector will dominate bioenergy consumption in 2035 (414 Mtoe). Together, the power and industrial sector will demand approximately 2/3 of global bioenergy in 2035. The use of traditional biomass will continue to decline, as access to modern and more efficient energy technologies, including modern bioenergy,[8] increases in developing countries. Excluding the use of traditional biomass, the EU will be the single largest consumer of bioenergy, increasing its consumption from 130 Mtoe (2010) to 230 Mtoe (2035), whereas the US will follow closely with 210 Mtoe by 2035 (IEA 2012). Global biofuel (or liquid bioenergy) consumption, dominated by ethanol, is estimated to increase by 250 % to 210 Mtoe during that period, driven primarily by blending mandates (IEA 2012).

[5] Mtoe = million tonnes of oil equivalent (1 Mtoe = 41.9 PJ).

[6] Traditional biomass includes wood, charcoal, crop residues and animal dung and is mainly used for heating and cooking (IPCC 2011).

[7] The New Policies scenario is IEA's central scenario and takes into account the cautious implementation of broad policy commitments and plans to address energy and GHG emission reduction challenges (IEA 2012).

[8] Modern bioenergy is utilised at higher efficiencies than traditional biomass and includes liquids and gases as secondary energy carriers to generate heat, electricity, combined heat and power (CHP), and transport fuels (IPCC 2011).

2.2 What is Sustainable Bioenergy and What to Measure

The main areas of concern for policymakers regarding the sustainability of bioenergy production (in particular that of biofuels) are its impacts on food security and global commodity prices, life cycle GHG emissions reductions, resources depletion, 'land grabbing', ecosystem services and biodiversity. Figure 2.6 shows that sustainability of bioenergy needs to be considered systematically and holistically across the three pillars of sustainability (environmental, social and economic). It also points out the importance of scale and geographic context in the sustainability assessment of bioenergy value chains.

The EU, which depends more on imported feedstocks than the USA, both in terms of amounts and variety, to meet its bioenergy demands has been on the forefront of formulating broad environmental sustainability safeguards into its regulations (FAO 2013). However, the implementation of these criteria is complicated by the fact that many feedstocks have multiple, substitutable end-uses, e.g., wheat is used for food, feed, and fuel production, whereas the criteria apply to a single end-use thus creating the potential for leakage (Frank et al. 2012). Furthermore, at present, social sustainability safeguards are only realised as part of voluntary schemes adopted by selected biofuels producers.

The sustainability of bioenergy in terms of their efficacy to reduce GHG emissions by substituting fossil fuels hinges on two main factors: land use and biomass production practices. Land use change has direct (positive or negative) implications on terrestrial carbon stocks, and management practices encompassing zoning, crop selection and cultivation, energy and fertiliser inputs impact the GHG balance of the end-product. The core of the debate about the efficacy of bioenergy (again, with a particular focus on biofuels) continues to centre on the issue of indirect land use change (ILUC). While some modelling results indicate no ILUC impacts (e.g., Kim and Dale 2011), other studies show significantly lower impacts than previously estimated (e.g., INRA 2013) or very high GHG

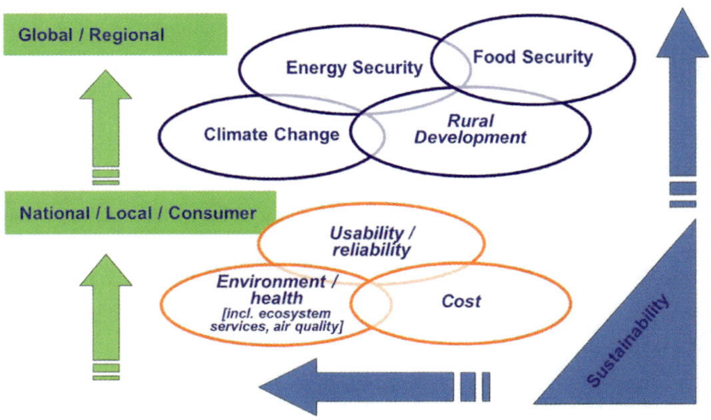

Fig. 2.6 Measures of sustainable bioenergy

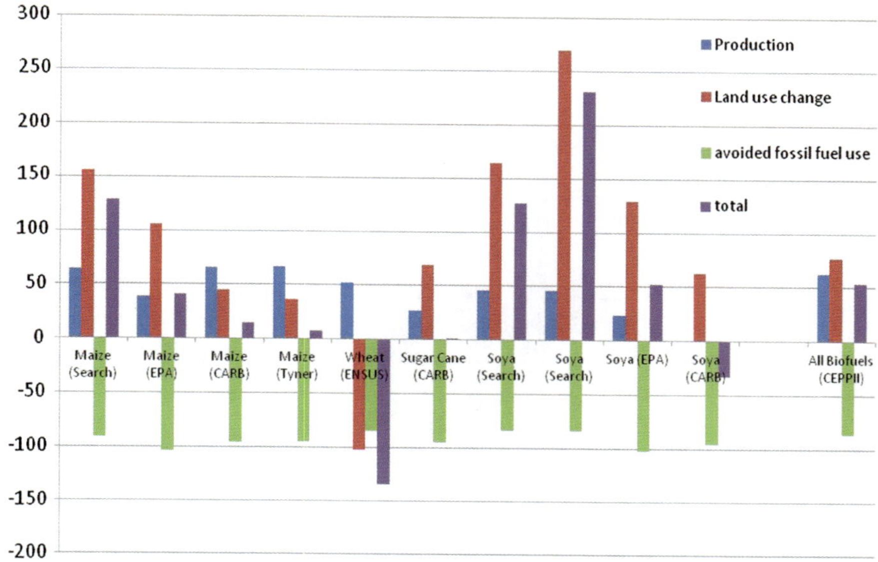

Fig. 2.7 Uncertainties of estimated indirect land use change GHG emission for selected biofuels (g CO_2 eq/MJ). *Source* adapted from EC DG-Tren (2010)

emissions (e.g., Searchinger 2010). Figure 2.7 further illustrates the divergence in the results of different ILUC modelling studies. The debate around the significant scientific uncertainties as to the magnitude and effect of ILUC has slowed down the development of bioenergy supply chains and diverted attention from wider issues of the sustainability of bioenergy and agricultural production more broadly.

'Land grabbing', defined as "the transfer of the right to own or use the land from local communities to foreign investors through large-scale land acquisitions" (Rulli et al. 2012) has also been attributed to the increase in demand for bioenergy feedstocks (GRAIN 2013). While numerous cases of illegal appropriations and human rights violations with disastrous impacts on smallholders and local communities have been reported and must be prevented in the future, a recent analysis by the Land Matrix (2013) suggests that the scale of the problem may have been largely exaggerated. Land Matrix reviewed 950 large-scale land acquisitions (LSLA) of 200 ha or more since 2000. Of the 750 concluded deals, covering a total area of 32.6 Mha, their research concludes that only approximately 5 % (or 1.63 Mha) have gone into agricultural production. Figure 2.8 shows that that while biofuel production has had an impact, food crops accounted for a larger share of deals and area. Forestry and tourism were also important sources of demand for land. A study by IIED on the socio-economic impacts of such land acquisitions concludes that the impact of these investments depends on the way they are structured, and "can either create new opportunities to improve local living standards, or further marginalise the poor (IIED 2009)".

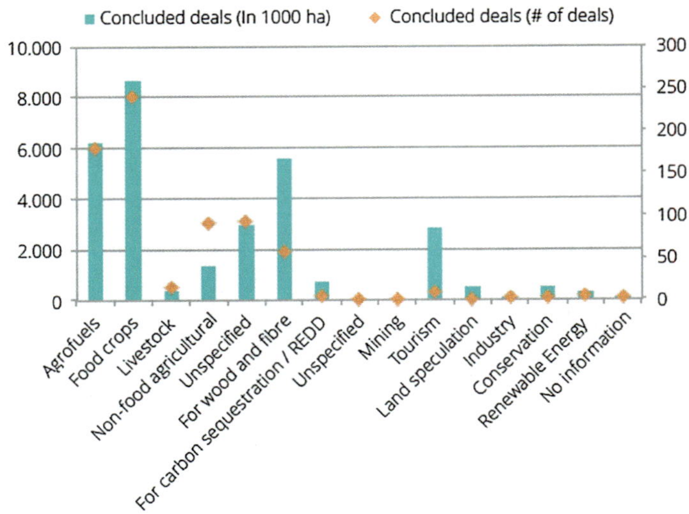

Fig. 2.8 Main drivers of large-scale land acquisitions. *Source* Land Matrix (2013)

Biofuels have also been blamed for the 2008-2009 spikes in food prices (e.g., Pimentel et al. 2009; ActionAid 2010). However, recent studies indicate that the causal relationships are more complex and that the increase in commodity prices can be primarily attributed to high crude oil prices (affecting energy and fertiliser costs), exchange rate movements, stock-to-use ratios, unusually frequent adverse weather events, and only in small part to EU and US demands for conventional biofuel feedstocks (Baffes and Dennis 2013; Oladosu and Msangi 2013). Nevertheless, concerns over global food security have dichotomised the issue and effectively placed the production of food and fuel in opposition (Rosillo-Calle 2012).

Sustainable bioenergy production must also adequately consider the protection of biodiversity. According to the Millennium Ecosystem Assessment (2005), see Fig. 2.9, "current rates of species extinction are at least two orders of magnitude above background rates and are expected to rise to at least three orders above background rates". In the UK, 60 % of monitored species have declined over the past 50 years and 10 % of species are threatened by extinction (UK 2013). Drivers of this unprecedented rate of biodiversity loss are habitat conversion and fragmentation, primarily due to agricultural expansion and urban development; increasingly, climate change, which contributes to habitat change, is becoming the dominant driver of extinction.

To address the aforementioned concerns regarding the sustainability of bioenergy and to provide policymakers and producers with a comprehensive framework to promote and monitor the development of bioenergy supply chains, the Global Bioenergy Patnership (GBEP) Proposed 24 indicators for the sustainable production and use of modern bioenergy (GBEP 2011). Table 2.1 summarises these indicators by pillars and themes. These indicators are thus far the only comprehensive framework for the sustainable development of bioenergy.

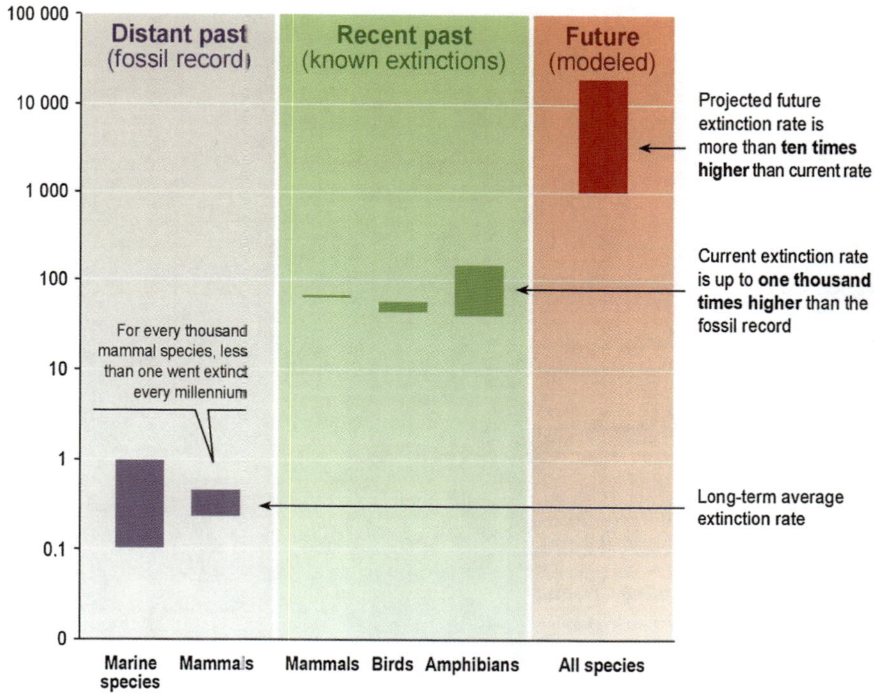

Fig. 2.9 Species extinction (per thousand species per millennium). *Source* MEA (2005)

2.2.1 Sustainable Agricultural Intensification: The Future of Food and Farming: Five Challenges for Global Sustainability

The production of bioenergy sits within a larger system of agricultural production. Bioenergy policies, given their narrow scope and mandate, cannot address the inefficiencies of global agricultural production overall. However, the controversies surrounding the large scale deployment of bioenergy, such land use change, food versus fuel, 'land grabbing', biodiversity loss, etc. may have assisted in recognising the necessity for a profound shift from conventional agricultural practices to a more sustainable, resource efficient and climate-smart, multi-product agricultural production system.

The need to provide food, shelter, energy and other resources for 9.2 billion people in 2050 against the backdrop of climate change requires concerted efforts today to avoid future shocks to global food production (Foresight 2011; Garnett et al. 2013). The Future of Food and Farming report highlights five key challenges for global food system (Foresight 2011):

A. Balancing future demand and supply sustainably—to ensure that food supplies are affordable.

Table 2.1 GBEP sustainability indicators

Pillars		
GBEP's work on sustainability indicators was developed under the following three pillars, noting interlinkages between them:		
Environmental	*Social*	*Economic*
Themes		
GBEP considers the following themes relevant, and these guided the development of indicators under these pillars:		
Greenhouse gas emissions, productive capacity of the land and ecosystems, air quality, water availability, use efficiency and quality, biological diversity, land-use change, including indirect effects	Price and supply of a national food basket, access to lard, water and other natural resources, labour conditions, rural and social development, access to energy, human health and safety	Resource availability and use efficiencies in bioenergy production, conversion, distribution and end use, economic development, economic viability and competitiveness of bioenergy, access to technology and technological capabilities, energy security/diversification of sources and supply, energy security/Infrastructure and logistics for distribution and use
Indicators		
1. Lifecycle GHG emissions	9. Allocation and tenure of land for new bioenergy production	17. Productivity
2. Soil quality	10. Price and supply of a rational food basket	18. Net energy balance
3. Harvest levels of wood resources	11. Change in income	19. Gross value added
4. Emissions of non-GHG air pollutants, including air toxics	12. Jobs in the bioenergy sector	20. Change in consumption of fossil fuels ard traditional use of biomass
5. Water use and efficiency	13. Change in unpaid time spent by women and children collecting biomass	21. Training ard regualification of the workforce
6. Water quality	14. Bioenergy used to expand access to modern energy services	22. Energy diversity
7. Biological diversity in the landscape	15. Change in mortality and burden of disease attributable to indoor smoke	23. Infrastructure and logistics for distribution of bioenergy
8. Land use and land-use change related to bioenergy feedstock production	16. Incidence of occupational injury, illness and fatalities	24. Capacity and flexibility of use of bioenergy

Source GBEP (2011)

B. Ensuring that there is adequate stability in food supplies—and protecting the most vulnerable from the volatility that does occur.
C. Achieving global access to food and ending hunger—food security for all.
D. Managing the contribution of the food system to the mitigation of climate change.
E. Maintaining biodiversity and ecosystem services while feeding the world.

Because of the size of future threats and increasing demands that go beyond food production, such as climate change mitigation and ecosystem service provision, a radical redesign is needed. No action is not an option—if the food system fails to deliver against these future challenges the implications will be profound, and, aside from the human tragedy, threaten political stability and security.

2.3 Bioenergy and Land Use

2.3.1 Technical Bioenergy Potentials and Global Land Availability

Figure 2.10 is based on IEA (2009) data and shows the relative contribution of global bioenergy (including electricity, district and onsite heat, and biofuels) to global energy demand, assuming a 1.5 % CAGR[9] for conventional energy, a 5 % CAGR for modern bioenergy, and 1.2 % for traditional biomass. In this scenario, bioenergy provision would equal approximately 250 EJ, the equivalent to roughly 25 % of global primary energy demand, if business as usual trends continue. Based on these results, it was calculated that in 2050 (beyond the 2035 IEA scenarios), approximately 100–650 Mha (1–6.5 million km^2) of land would be required for the production of the necessary biomass (Murphy et al. 2011).

The range corresponds to similar, lower range estimates found in the literature (see Table 2.2), and can be explained in part by the inherent uncertainty about land availability and productive potentials in the future. However, a reasonable share of the range in published potentials can also be explained by the individual assessments' differing focus on fundamental theoretical potentials versus realistic exportation potentials. When evaluating the bioenergy resource and potential for exploiting biomass for bioenergy at national to global scales, Sathaye (in Brown et al. 1996) outlines the following five step progression from theoretical to practical/realisable potentials:

1. Biological/theoretical potential
2. Technological potential
3. Economic potential
4. Ecological potential

[9] CAGR: Compounded annual growth rate.

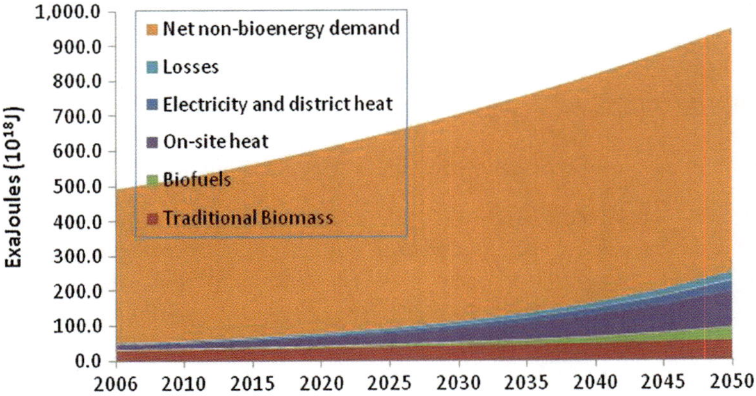

Fig. 2.10 Global bioenergy versus primary energy demand (2006–2050). *Source* calculations based on IEA (2009)

5. Realistic potential/implementation.

Furthermore, as Fig. 2.11 shows, land availability is unevenly distributed among different regions. Latin America and the Caribbean and Sub-Saharan Africa are the only two regions where substantial amounts of suitable land may still be 'available' and where agricultural yields could be significantly increased through improved inputs and management practices.

2.3.2 *Energy Crop Yield Estimates*

The productivity of agricultural crops depends on climate, soil conditions, and agricultural management practices, including types of cultivars, the quality of seeds, availability of water, agrochemical inputs, and pest and control. Given the competing demands for land to meet of human needs for (1) supply of resources, (2) provision of ecosystem services, and (3) space for human infrastructure (Dunlap and Catton 2002), further expansion of cropland is limited in the longer term. Instead, future yield increases will have to be achieved through higher per hectare productivity. In part, this can be attained by closing the yield gap between attainable and observed yields across regions (Mueller et al. 2012). Haberl et al. (2007) estimate that global productivity of cropland is currently 35 % below its potential productivity. Additional gains will come from a more efficient utilisation of all harvested plant parts in integrated production systems to meet human demand for food, feed, fibre, chemicals, and energy.

Figure 2.12 maps the results of multiple studies on energy crop potentials against available land areas. Generally, the data points indicating yields below 5 odt[10]/ha assume production on marginal and degraded lands, while those above

[10] odt = oven dry tonne.

Table 2.2 Overview of recent studies on technical potentials of biomass from energy crops

Reference	Type of potential	Regions	Time frame	Sustainability constraints	Land use types	Land area used [mio. Km^2]	Productivity [tonnes dry matter/ha/yr]	Potential of energy crops [EJ/yr]
van Vuuren et al. (2009)	Technical	Global	2050	Biodiversity, food security, soil degradation, water scarcity	Abandoned agricultural land (75 %) Grassland (25 %)	13	Depending on land suitability and climate factors 1.0-3.2 kg dry matter/m^3/yr	120-300 EJ/yr (unconstrained) 65-115 EJ/yr (constrained)
WBGU (2008)	Technical	Global	2050	Biodiversity, C balance, deforestation, degraded land, food security, water scarcity	Land suitable for bioenergy cultivation according to the crop functional types in the model, considering sustainability	2.4–5.0	7.5-12.6 t/ha/yr	34-120 EJ/yr
Campbell et al. (2008)	Technical	Global	2000 (not clearly mentioned)	Agricultural lands, ecosystems, food security, releasing carbon stored in forests, water scarcity	Abandoned agricultural land (100 %)	3.9–4.7	4.3 t/ha/y (AGB)	32–41 EJ/yr (AGB)
Field et al. (2008)	Technical	Global	2050	Biodiversity, food security. ecosystems, deforestation	Abandoned agricultural land (100 %)	3.9	3.2 tC/ha/yr	27 EJ/yr (AGB)
Dornburg et al. (2010)	Technical	Global	2050	Land for food excluded various assumptions on (non-) exclusion of degraded and protected land	Not explicitly specified	Not specified	Not specified	Energy crops: 120 EJ/yr

(continued)

Table 2.2 continued

Reference	Type of potential	Regions	Time frame	Sustainability constraints	Land use types	Land area used [mio. Km^2]	Productivity [tonnes dry matter/ha/yr]	Potential of energy crops [EJ/yr]
Smeets et al. (2007)	Technical	11 world regions	2050	Biodiversity, deforestation, food security	Surplus agricultural land (100 %)	7.3–35.9	16–21 odt(oven dry tonnes)/ha/yr	215–1,272 EJ/yr
Hoogwijk et al. (2005)	Technical	11 world regions	2050-2100	Biodiversity, food security	Abandoned agricultural land (100 %) remaining land not for food or material procution (10–50 %) extensive grassland	Abandoned: 0.6–1.5 rest land 0.3–1.4	Depending on land suitability and climate factors	Abandoned: 130–400 EJ/yr rest land 235–240 EJ/yr total: 300–650 EJ/yr
Erb et al. (2009)	Technical	11 world regions	2050	Excluded: land for food and feed, forestry and unproductive land	Cropland not needed for food and fiber supply intensification of grazing land	2.3–9.9 depending on food and feed demand (44 scenarios)	Equal to potential (cropland) or actual (grazing land) NPP	Bioenergy crops: 28–128 EJ/yr residues: 21–36 EJ/yr

Source IIASA (2012)

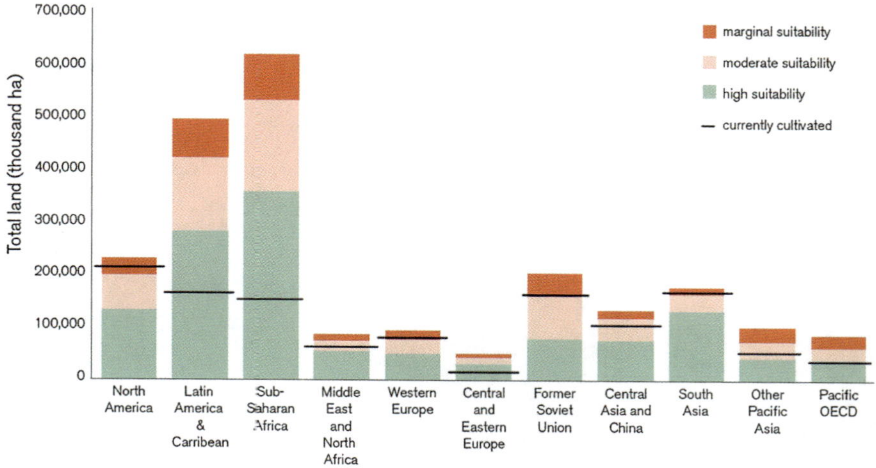

Fig. 2.11 Global land use and availability. *Source* PCFISU (2011—based on IIASA GAEZ study)

15 odt/ha assume production on good quality land and high yielding crop varieties (Slade et al. 2011). In some regions, e.g., Brazil, bioenergy crop yields (sugarcane) already exceed 18 odt/ha.

2.3.3 *Boxing in Bioenergy*

Concerns over the direct and indirect impacts of bioenergy, particularly conventional biofuels, have pushed policy makers to try to direct biomass crop production to marginal, degraded and 'unused' land. This will not only raise the costs of feedstock production and transport, but may also be contradictory to food security where sustainable intensification and reduced losses require increased energy inputs. Furthermore, this marginalisation ignores the beneficial role that perennial energy crops could play in integrated production systems by managing intensification through nutrient capture and water quality benefits, watershed, soil and erosion protection, and increased biodiversity.

2.3.4 *Integrating Biomass Supply Chains and Sustainable Biorenewables Innovation*

To transcend the inefficiencies and inadequacies of current bioenergy supply chains, bioenergy production needs to be embedded within a broader, integrated system of biomass productions systems which are optimised for the sustainable

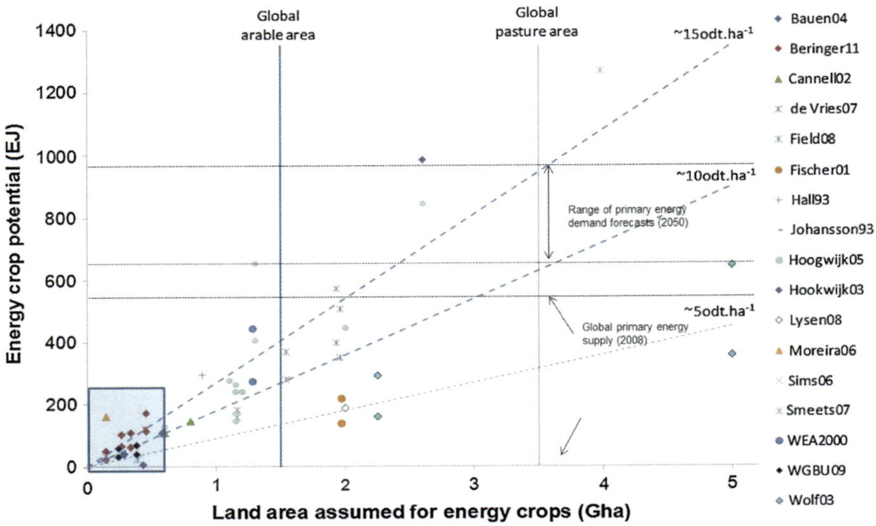

Fig. 2.12 Energy crop yield estimates. *Source* adapted from Slade et al. (2011)

provision of food, feed, fuel, fibres, chemicals, energy and ecosystem services. The integration of biomass supply chains aims to optimise the use of resources in agricultural supply chains and exploit its maximum value along each step, as illustrated in Fig. 2.13. Initial studies indicate that integrated food and energy systems—or multi-functional production systems—can increase overall yields per unit of land area, provide important socio-economic benefits through the diversification on marketable products for farmers, and have positive impacts on the environment, including enhanced carbon sequestration (FAO 2010; Bogdanski 2012; Dale et al. 2010). Research is underway on how the inclusion of perennial buffer strips can trap nutrient run-off along arable land adjacent to rivers and streams while providing valuable feedstocks for biobased products and advanced biofuels, or using perennials for the restoration of degraded land (Gopalakrishnan et al. 2009, 2012).

Novel approaches to integrated feedstock production are only the first step in sustainable biomass supply chains and will only be able make a difference at scale if integrated with efficient downstream processing and access to markets. Figure 2.14 shows the main technological conversion pathways and inter-linkages for biomass feedstocks which enable an optimised, or 'cascading' (Haberl and Geissler 2000), use of all plant components, including the utilisation co-products (Black et al. 2011).

2.4 How Can Bioenergy Policy Drive Sustainable Land Use?

Bioenergy policies were designed to promote the use of bioenergy as a means to reduce GHG emissions, increase energy access and security and stimulate rural development. Overall, sustainability considerations, with the exception of GHG emission

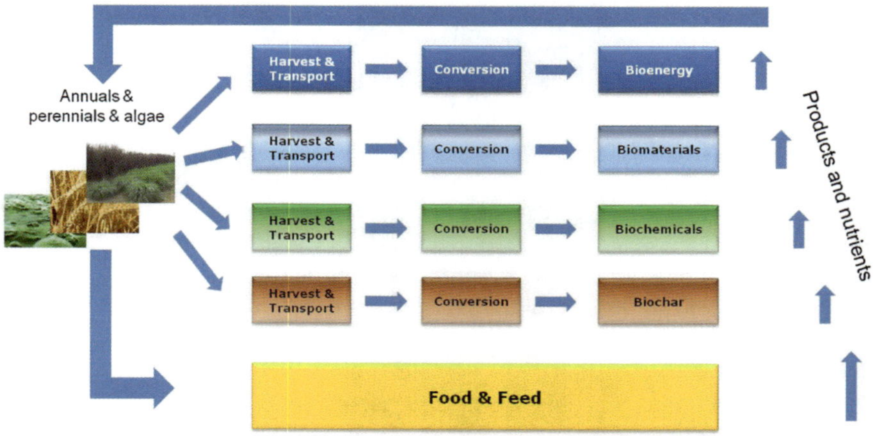

Fig. 2.13 Integrating biomass supply chains

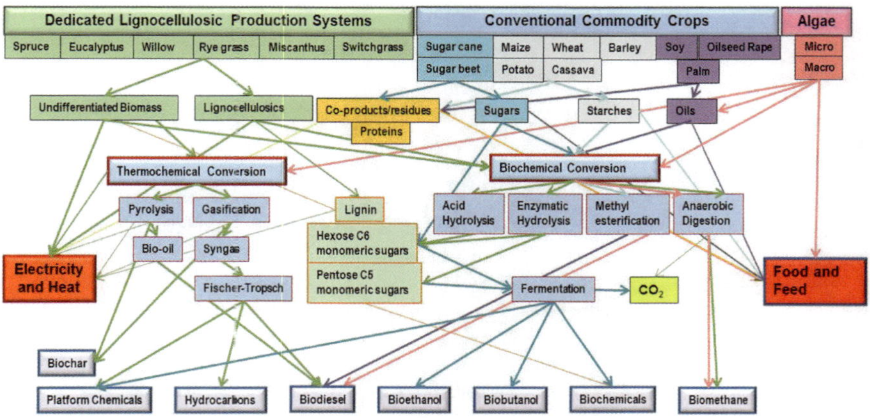

Fig. 2.14 Feedstock and technology pathways for biorenewables—many options. *Source* Adapted from Black et al. (2011)

reduction requirements and some provisions for the protection of high biodiversity areas, have not been adequately addressed to date. Managing sustainability effectively is challenging, and needs to balance all three of its pillars—and possible trade-offs between them—across different spatial and temporal scales and management systems.

In contrast to other renewable energy technologies, bioenergy is inextricably linked to land use, agricultural production and forestry. The attempt to regulate bioenergy feedstock production without addressing the sustainability of agricultural production overall, will continue to result in leakage and do little to resolve the deadlocked food versus fuel versus biodiversity debate.

Current agricultural land use practices are not sustainable and pushing future energy crop production onto marginal lands, as some scenarios envision, may in the long run increase the economic, social, and environmental costs of bioenergy production. Furthermore, it would miss the opportunity to use bioenergy markets as a vehicle to promote the necessary sustainable intensification of agriculture in areas that lag behind in terms of agricultural productivity, are least food secure and suffer disproportionately from the impacts of climate change.

Instead of continuing with business as usual, this chapter offers an alternative path forward, and shows that by integrating the production of food, feed, fuel, fibres, chemicals, and energy the environmental, social and environmental sustainability of agricultural production systems could be greatly enhanced.

Bioenergy policy alone cannot effectively drive sustainable land use. Given the many demands on land, land use planning is inherently complex and intensive, and past experience has shown that complex policy is often ineffective. One way for regulators to overcome this dilemma in the area of bioenergy and land use, is to incentivise (through a mix of regulation and market-based incentives) a shift to more resource use efficient, integrated production systems as an integral component in the transition to a low carbon economy, with potential linkages to payments for ecosystem services and carbon sequestration schemes.

References

ActionAid (2010) Meals per gallon—the impact of industrial biofuels on people and global hunger. London. http://www.actionaid.org.uk/sites/default/files/doc_lib/meals_per_gallon_final.pdf

Baffes J, Dennis A (2013) Long-term drivers of food prices. Policy research working paper 6455. World Bank, Washington DC. http://papers.ssrn.com/sol3/papers.cfm?abstract_id=2268316##

BP (2012) BP statistical review of world energy. June 2012. London. http://www.bp.com/content/dam/bp/pdf/Statistical-Review-2012/statistical_review_of_world_energy_2012.pdf

Beurskens LWM, Hekkenberg M, Vethman P (2011) Renewable energy projections as published in the national renewable energy action plans of the European member states—summary report. ECN, EEA, Copenhagen. http://www.ecn.nl/docs/library/report/2010/e10069.pdf

Biofuel Digest (2012) Biofuels mandates around the world: 2012. http://www.biofuelsdigest.com/bdigest/2012/11/22/biofuels-mandates-around-the-world-2012/ (accessed on 4 July 2013)

Black MJ, Whittaker C, Hosseini SA, Diaz-Chavez R, Woods J, Murphy RJ (2011) Life cycle assessment and sustainability methodologies for assessing industrial crops, processes and end products. Ind Crops Prod 34(2):1332–1339. doi:10.1016/j.indcrop.2010.12.002

Bogdanski A (2012) Integrated food–energy systems for climate-smart agriculture. Agri Food Secur 1:9 (http://www.agricultureandfoodsecurity.com/content/pdf/2048-7010-1-9.pdf)

Brown S, Sathaye J, Cannell M, Kauppi P, Burschel P, Grainger A, Heuveldop J, Leemans R, Moura Costa P, Pinard, Nilsson S, Schopfhauser W, Sedjo R, Singh N, Trexler M, van Minnen J, Weyers S (1996) Management of forests for mitigation of greenhouse gas emissions. In: Watson RT, Zinyowera MC, Moss RH, Dokken DJ (eds) IPCC climate change 1995—impacts, adaptations and mitigation of climate change: scientific-technical analyses, contribution of working group II to the 2nd assessment report of the intergovernmental panel on climate change. Cambridge University Press, Cambridge, pp 773–797. https://docs.google.com/uc?export=download&confirm=no_antivirus&id=0B1gFp6Ioo3akQS1GX3owSk0tLU0

Campbell JE, Lobell DB, Genova RC, Field CB (2008) The Global Potential of Bioenergy on Abandoned Agriculture Lands. Environ Sci Technol 42(15):5791–5794

Dale B, Bals BD, Kim S, Eranki P (2010) Biofuels done right: land efficient animal feeds enable large environmental and energy benefits. Environ Sci Technol 44:8385–8389. doi:10.1021/es101864b

Dornburg V, van Vuuren D, van de Ven G, Langeveld H, Meeusen M, Banse M, van Oorschot M, Ros J, van den Born GJ, Aiking H, Londo M, Mozaffarian H, Verweij P, Lyseng E, Faaij A (2010) Bioenergy revisited: Key factors in global potentials of bioenergy. Energy Environ Sci 3:258–267

Dunlap RE, Catton WR (2002) Which function(s) of the environment do we study? A comparison of environmental and natural resource sociology. Soc Natural Resour 15(3):239–249. doi:10.1080/089419202753445070

EC (2009a) Directive 2009/28/EC of the European Parliament and of the council of 23 April 2009 on the promotion of the use of energy from renewable sources and amending and subsequently repealing Directives 2001/77/EC and 2003/30/EC. http://eur-lex.europa.eu/LexUriServ/LexUriServ.do?uri=CELEX:32009L0028:en:NOT

EC (2009b) Directive 2009/30/EC of the European Parliament and of the council of 23 April 2009 amending Directive 98/70/EC as regards the specification of petrol, diesel and gas-oil and introducing a mechanism to monitor and reduce greenhouse gas emissions and amending Council Directive 1999/32/EC as regards the specification of fuel used by inland waterway vessels and repealing Directive 93/12/EEC. http://eur-lex.europa.eu/LexUriServ/LexUriServ.do?uri=CELEX:32009L0030:EN:NOT

EC (2012) Biofuels—sustainability schemes. Recognised voluntary schemes. http://ec.europa.eu/energy/renewables/biofuels/sustainability_schemes_en.htm (accessed on 4 July 2013)

EC DG-Tren (2010) Assessment of the impact of land use change on greenhouse gas emissions from biofuels and bioliquids. Incomplete draft; version 4 1 10

EPA (2010) 40 CFR Part 80. Regulation of fuels and fuel additives: changes to renewable fuel standard program, final rule, 3 Feb 2010 (RFS2). http://www.gpo.gov/fdsys/pkg/FR-2010-03-26/pdf/2010-3851.pdf (accessed on 4 July 2013)

Erb K, Haberl H, Krausmann F, Lauk C, Plutzar C, Steinberger JK, Müller C, Bondeau A, Waha K, Pollack G (2009) Eating the Planet: Feeding and fuelling the world sustainably, fairly and humanely–a scoping study. Institute of Social Ecology, Vienna

FAO (2010) making integrated food-energy systems work for people and climate. Rome. http://www.fao.org/docrep/013/i2044e/i2044e.pdf

FAO (2013) Biofuels and the sustainability challenge: a global assessment of sustainability issues, trends and policies for biofuels and related feedstocks. Rome. http://www.fao.org/docrep/017/i3126e/i3126e.pdf

Field CB, Campbell JE, Lobell DB (2008) Biomass energy: the scale of the potential resource. Trends Ecol Evol 23(2):65–72

Foresight (2011) The future of food and farming. The Government Office for Science, London. http://www.bis.gov.uk/assets/foresight/docs/food-and-farming/11-546-future-of-food-and-farming-report.pdf

Frank S, Böttcher H, Havlík P, Valin H, Mosnier A, Obersteiner M, Schmid E, and Elbersen B (2012) How effective are the sustainability criteria accompanying the European Union 2020 biofuel targets? GCB Bioenergy

Garnett T, Appleby M, Balmford A, Bateman I, Benton T, Bloomer P, Burlingame B, Dawkins M, Dolan L, and Fraser D (2013) Sustainable intensification in agriculture: premises and policies. Sci 341(6141):33–34

GBEP (2011) The global bioenergy partnership sustainability indicators for bioenergy. Rome. http://www.globalbioenergy.org/fileadmin/user_upload/gbep/docs/Indicators/The_GBEP_Sustainability_Indicators_for_Bioenergy_FINAL.pdf

Gopalakrishnan G, Negri MC, Wang M, Wu M, Snyder SW, LaFreniere L (2009) Biofuels, land, and water: a systems approach to sustainability. Environ Sci Technol 43(15):6094–6100. doi:10.1021/es900801u

Gopalakrishnan G, Negri MC, Salas W (2012) Modeling biogeochemical impacts of bioenergy buffers with perennial grasses for a row-crop field in Illinois. Glob Change Bioenergy 4(6):739–750. doi:10.1111/j.1757-1707.2011.01145.x

GRAIN (2013) Land grabbing for biofuels must stop—EU biofuel policies are displacing communities and starving the planet. Barcelona. http://www.grain.org/article/entries/4653-land-grabbing-for-biofuels-must-stop
Haberl H, Geissler S (2000) Cascade utilization of biomass: strategies for a more efficient use of a scarce resource. Ecol Eng 16:111–121. doi:10.1016/S0925-8574(00)00059-8
Haberl H, Erb KH, Krausmann F, Gaube V, Bondeau A, Plutzar C, Gingrich S, Lucht W, Fischer-Kowalski M (2007) Quantifying and mapping the human appropriation of net primary production in earth's terrestrial ecosystems. Proc Nat Acad Sci 104(31):12942–12947 (http://www.pnas.org/content/104/31/12942)
Hoogwijk M, Faaij A, Eickhout B, de Vries B, Turkenburg W (2005) Potential of biomass energy out to 2100, for four IPCC SRES land-use scenarios. Biomass Bioenergy 29(4):225–257
IEA (2008) World energy outlook 2008. Paris. http://www.worldenergyoutlook.org/media/weowebsite/2008-1994/weo2008.pdf
IEA (2009) World energy outlook 2009. Paris. http://www.worldenergyoutlook.org/media/weowebsite/2009/WEO2009.pdf
IEA (2010) World energy outlook 2010. Paris. http://www.worldenergyoutlook.org/media/weo2010.pdf
IEA (2011) World energy outlook 2011. Paris. http://www.worldenergyoutlook.org/publications/weo-2011/
IEA (2012) World energy outlook 2012. Paris. http://www.worldenergyoutlook.org/publications/weo-2012/
IEA (2013) World energy outlook special report. Redrawing the energy–climate map. Paris. http://www.worldenergyoutlook.org/energyclimatemap/#d.en.36900
IIASA (2012) Global energy assessment (GEA). Global energy assessment—toward a sustainable future. Cambridge University Press, Cambridge, UK and New York, NY, USA and the International Institute for Applied Systems Analysis, Laxenburg, Austria. http://www.globalenergyassessment.org/
IIED (2009) 'Land grabs' in Africa: can the deals work for development? IIED Briefing, Sept 2009. London. http://pubs.iied.org/pdfs/17069IIED.pdf
INRA (2013) Le changement d'affectation des sols induit par la consommation européenne de biodiesel: une analyse de sensibilité aux évolutions des rendements agricoles. Rennes. http://prodinra.inra.fr/ft?id=%7B7858309A-82DB-4D52-9B5C-966887EBDEC3%7D
IPCC (2011) Renewable energy sources and climate change mitigation. special report of the intergovernmental panel on climate change. In: Edenhofer O, Pichs-Madruga R, Sokona Y, Seyboth K, Matschoss P, Kadner S, Zwickel T, Eickemeier P, Hansen G, Schloemer S, von Stechow C (eds). Cambridge University Press, Cambridge and New York. http://www.ipcc.ch/pdf/special-reports/srren/SRREN_Full_Report.pdf
Kim S, Dale BE (2011) Indirect land use change for biofuels: testing predictions and improving analytical methodologies. Biomass Bioenergy 35(7):3235–3240. doi:10.1016/j.biombioe.2011.04.039
Land Matrix (2013) Land matrix newsletter—June 2013. http://www.landmatrix.org/media/filer_public/2013/06/10/lm_newsletter_june_2013.pdf
Millennium Ecosystem Assessment. (2005) Ecosystems and Human Well-being: Synthesis. Washington, DC, Island Press
Mueller ND, Gerber JS, Johnston M, Ray DK, Ramankutty N, Foley JA (2012) Closing yield gaps through nutrient and water management. Nature 490:254. doi:10.1038/nature11420
Murphy R, Woods J, Black M, McManus M (2011) Global developments in the competition for land from biofuels. Food Policy 36 (Suppl 1):S52–S61. doi: 10.1016/j.foodpol.2010.11.014
Oladosu G, Msangi S (2013) Biofuel-food market interactions: a review of modeling approaches and findings. Agriculture 3:53–71. doi:10.3390/agriculture3010053
PCFISU (2011) What price resilience? Towards sustainable and secure food systems. The Prince's Charities' International Sustainability Unit, London. http://pcfisu.org/wp-content/uploads/pdfs/TPC0632_Resilience_report_WEB11_07_SMALLER.pdf

Pimentel D, Marklein A, Toth MA, Karpoff MN, Paul GS, McCormack R, Kyriazis J, Krueger T (2009) Food versus biofuels: environmental and economic costs. Human Ecol 37:1–12. doi:10.1007/s10745-009-9215-8

Rosillo-Calle F (2012) Food versus fuel: toward a new paradigm—the need for a holistic approach. ISRN Renew Energy 2012:Article ID 954180. doi:10.5402/2012/954180

Rulli MC, Saviori A, and D'Odorico P (2013) Global land and water grabbing. Proc Nat Acad Sci 110(3):892–897. http://www.pnas.org/content/110/3/892.short

Searchinger T (2010) Biofuels and the need for additional carbon. Environ Res Lett 5:024007. doi:10.1088/1748-9326/5/2/024007

Slade R, Saunders R, Gross R, Bauen A (2011) Energy from biomass: the size of the global resource. Imperial College Centre for Energy Policy and Technology & UK Energy Research Centre, London. http://www.ukerc.ac.uk/support/tiki-download_file.php?fileId=2095

Smeets EM, Faaij AP, Lewandowski IM, Turkenburg WC (2007) A bottom-up assessment and review of global bio-energy potentials to 2050. Prog Energy Combust Sci 33(1):56–106

UK (2013) State of nature. http://www.rspb.org.uk/ourwork/science/stateofnature/index.aspx

van Vuuren DP, van Vliet J, Stehfest E (2009) Future bio-energy potential under various natural constraints. Energy Policy 37(11):4220–4230

WBGU (2008) Future Bioenergy and Sustainable Land Use. London and Sterling, VA, Earthscan

Chapter 3
Advanced Biofuels: Economic Uncertainties, Policy Options, and Land Use Impacts

Wallace E. Tyner and Farzad Taheripour

Abstract Second generation or cellulosic biofuels have potential to become a reliable source of renewable fuel. In this chapter we describe five major uncertainties that currently inhibit the development of these biofuels: (1) future oil prices, (2) feedstock availability and cost, (3) conversion technology efficiency and cost, (4) environmental impacts, and (5) government policy. In each of these areas, there are significant issues that impede development and commercialization of the second generation biofuels industry. However, all of these uncertainties could be managed if society were willing to pay the higher cost of cellulosic biofuels.

3.1 Introduction

During the past decade production of first generation biofuels, mainly corn and sugarcane ethanol and biodiesel from oilseeds, have increased rapidly in response to government policies and market incentives around the world and in particular in U.S., Brazil, and European Union. In 2000, the global production of biofuels was limited to about 17.5 billion liters of ethanol and 0.8 million liters of oilseed biodiesel. The global ethanol and biodiesel production in 2012 are about 86 and 21.5 billion liters, respectively. Development of these biofuels has raised concerns and debates on their environmental and economic consequences. Much of the debate has focused on induced land use change emissions due to cropland expansion for biofuel production, implications of using food crops for biofuel

W. E. Tyner (✉) · F. Taheripour
Department of Agricultural Economics, Purdue University, 403 West State St.,
West Lafayette, IN 47907-2056, USA
e-mail: wtyner@purdue.edu

F. Taheripour
e-mail: tfarzad@purdue.edu

M. C. McCann et al. (eds.), *Plants and BioEnergy*, Advances in Plant Biology 4,
DOI: 10.1007/978-1-4614-9329-7_3, © Springer Science+Business Media New York 2014

production, and welfare impacts of biofuel production and policy. With respect to these issues, second generation biofuels, which can convert cellulosic materials to liquid fuel, have gained favor among politicians. Some cellulosic biofuel pathways have the potential to generate more environmental gains compare to the first generation. In addition, they may have fewer negative impacts on food production and prices. However, there is very little commercial production of these biofuels today because they are not economic, and policies have not been deemed adequate to stimulate investment. This chapter discusses these issues following a brief review of evolution in global biofuel production and policy during the past decade.

3.2 Background

Production of biofuels was limited until a decade ago. In 2000, biofuel production was limited to 10.5 billion liters of Brazilian sugarcane ethanol, 6.2 billion liters of corn ethanol produced in U.S., and 0.8 million liters of oilseed biodiesel mainly produced in the European Union. During the past decade these regions have significantly increased their biofuel production, and several other countries have launched biofuel production. The global ethanol and biodiesel production will reach 86 and 21.5 billion liters in 2012. Figure 3.1 shows distributions of these biofuels among leading biofuel countries. The U.S., Brazil, and European Union are the top three ethanol producers, and European Union, U.S., and Argentina are the main biodiesel producers.

In general, the biofuel industries were launched under a mix of support policies including subsidies, mandate, and trade barriers across the world. However, biofuel mandates became the dominant support policy over time. Biofuels have

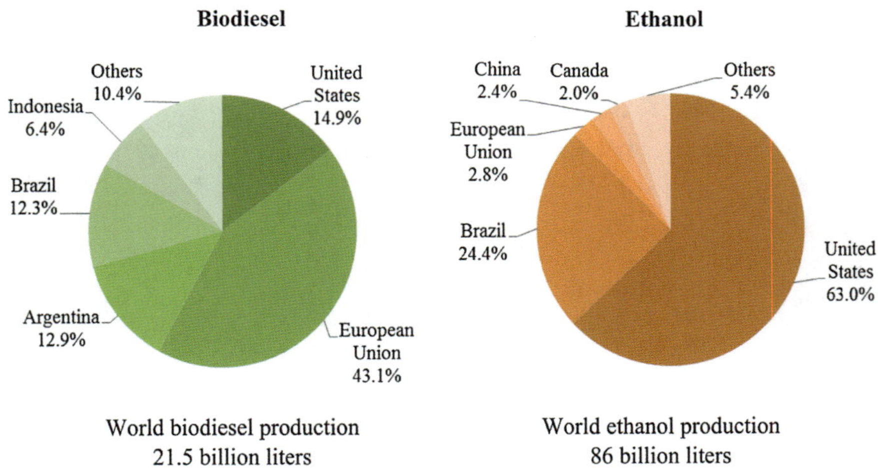

Fig. 3.1 Global biofuel production in 2012

been a part of U.S. energy policy since 1978 (Tyner 2008a). Brazil also has a long history of promoting biofuels (Tyner 2008b), and Europe also has promoted biofuels. Biofuels were developed for three main reasons: (1) reduction in greenhouse gas emissions (GHG); (2) upgrading national security and reduction in oil dependency; and (3) improvement in income for farmers and rural areas. U.S. development of these biofuels has not been without controversy, as debates have raged concerning the extent to which biofuels actually reduce GHG emissions and on the extent to which biofuels have contributed to increased food prices—the food-fuel debate (Abbott et al. 2008, 2011; Tyner 2012). Most of these debates have pertained to first generation biofuels, and much less so to second generation.

Second generation biofuels, also called cellulosic biofuels, can be produced from a wide range of cellulosic feedstocks such as corn stover, switchgrass, miscanthus, tree crops and residues, and municipal solid waste. While all of these feedstocks are possible sources of second generation biofuels, there is little or no commercial production of second generation biofuels today. As was the case with first generation biofuels, second generation biofuels cannot grow without government support. The biofuels industry in the U.S. and elsewhere can be viewed as a government created industry. In the U.S., corn ethanol production began in the early 1980s after the passage of the Energy Tax Policy act of 1978, which included a subsidy for corn ethanol (Tyner 2008a, b). The subsidy continued at different levels through the end of 2011. Today the major driver of biofuels development is the Renewable Fuel Standard (RFS) (U.S. Congress 2007), which mandates certain levels of biofuels of different types each year as shown in Fig. 3.2. Conventional biofuel (mainly corn ethanol) should reach 56.8 billion liters in 2015 and remain at this level until 2022. U.S production capacity is now close to the 56.8 billion liters mandate for 2015. EPA has recently increased the biodiesel

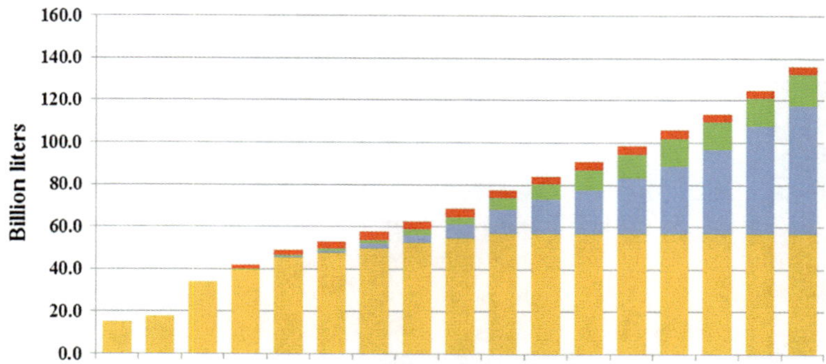

	2006	2007	2008	2009	2010	2011	2012	2013	2014	2015	2016	2017	2018	2019	2020	2021	2022
Coventional biofuel	15.1	17.8	34.1	39.7	45.4	47.7	50.0	52.2	54.5	56.8	56.8	56.8	56.8	56.8	56.8	56.8	56.8
Cellulosic advanced	0.0	0.0	0.0	0.0	0.4	0.9	1.9	3.8	6.6	11.4	16.1	20.8	26.5	32.2	39.7	51.1	60.6
Non cellulosic advanced	0.0	0.0	0.0	0.4	0.8	1.1	1.9	2.8	3.8	5.7	7.6	9.5	11.4	13.2	13.2	13.2	15.1
Biomass-based diesel	0.0	0.0	0.0	1.9	2.5	3.0	3.8	3.8	3.8	3.8	3.8	3.8	3.8	3.8	3.8	3.8	3.8

Fig. 3.2 U.S. renewable fuel standard

mandate level to 5.2 billion liters, and capacity exists for that level as well. The government subsidy for corn ethanol ended at the end of 2011, and the biodiesel subsidy also has been in limbo. For corn ethanol, the RFS has not been binding (i.e. production has been more than RFS) so far, but it has for biodiesel. The reason is that the wholesale price of corn ethanol minus ethanol tax credit was less than the wholesale price of gasoline until 2011, and now the wholesale price of ethanol is less than the wholesale price of gasoline. That is not the case for biodiesel, so the mandate continues to bind for this biofuel. As long as crude oil price remains at about $85 per barrel or higher, corn ethanol can survive and produce according to the RFS, even without government subsidies. However, second generation biofuels are more expensive to produce with current prices and production costs. According to the RFS schedule the U.S. is supposed to produce about 1.9 billion liters of cellulosic ethanol in 2012 and expand it to 60.6 billion liters in 2022. As mentioned earlier, the US currently produces very little cellulosic biofuel, and its future expansion path is very uncertain.

So what are the prospects for the development and deployment of second generation biofuels? That is the subject of this chapter, and it depends critically on five areas of uncertainty (Tyner 2010a). The uncertainties are: (1) future crude oil prices, (2) feedstock availability and cost, (3) conversion technology efficiency and cost, (4) environmental and GHG impacts, and (5) government policy. We will review each of these uncertainty areas in turn and then provide some overall conclusions on the prospects for second-generation biofuels.

3.3 Future Oil Price

The price of crude oil is a key variable which directly affects the prices of biofuels and their profitability. Other factors being constant, the higher the price of crude oil the higher the prices for cellulosic biofuels. To be economic without subsidies or mandates, given the existing production costs, second generation biofuels likely will need crude oil to be $130–$150 or higher. Figure 3.3 provides the current U.S. Department of Energy crude oil price forecast out to 2040 (U.S. Department of Energy 2013). In the reference case, crude oil price grows from about $100/bbl. in 2012 to $163/bbl. in 2040. In the high oil price case, it grows to $235/bbl., and in the low price case it becomes $75/bbl. All of these prices are in real terms; that is, excluding the effect of general inflation. The forecast range between $75/bbl. and $235/bbl. is very high, but even the reference case forecast would be problematic for investors in second generation biofuels, because at least for the next decade the price of crude oil will remain under $125/bbl. With crude oil prices below $130/bbl., production of cellulosic biofuel likely will not be profitable without government support.

What is driving the wide price forecast range? The high case assumes more rapid economic growth globally and more limited new oil and other liquid fuel discoveries. The low case assumes slower economic growth and more abundant crude oil and other liquid fuel supplies. Recent developments in oil supply

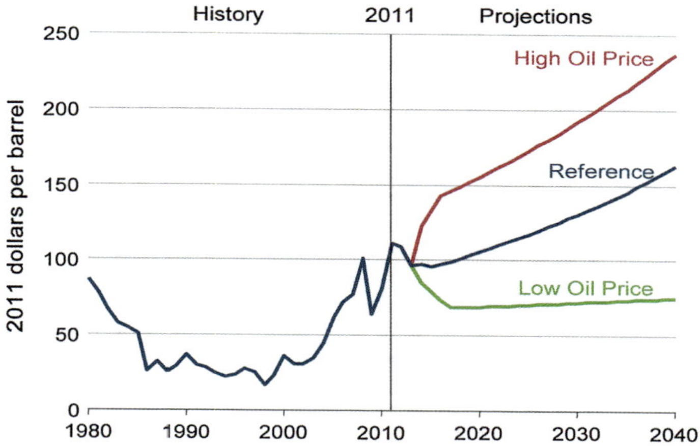

Fig. 3.3 Average annual Brent spot crude oil prices in three projection cases, 1980–2040

increases might push us closer to the lower end. The International Energy Agency predicts that the U.S will become the world's largest oil producer around 2020 and that North America will become a net petroleum exporter by around 2035 (International Energy Agency 2012a). These trends would eventually tend to push crude oil prices lower unless there are corresponding reductions in production elsewhere. Also, the natural gas revolution in the U.S. will create additional energy supplies that will, to some extent compete with crude oil (International Energy Agency 2012b). The bottom line is that crude oil price is highly uncertain, but even the reference case is problematic for second generation biofuels. It is likely that some form of government intervention will be necessary to stimulate investment in these biofuels.

3.4 Feedstock Supply and Cost

The cost of feedstock is another key variable which affects profitability of producing cellulosic biofuel. Current estimates of biomass feedstock costs are much higher than earlier estimates (Congressional Research Service 2010; National Research Council 2011; Thompson and Tyner 2013). For years, the standard figure used by DOE was $33 per dry metric ton delivered to the plant. Today that cost is estimated at $83–$147 per dry metric ton, more than three times earlier estimates. Since feedstock cost is the major component of variable cost, this difference has a major impact on cellulosic biofuel cost. Table 3.1 provides a range of feedstock costs for various biofuel cellulosic feedstocks (National Research Council 2011). Column 2 of this table shows the feedstock cost estimate, and column 3 is the estimate of what the biofuel producer could afford to pay for the feedstock on a break-even basis. Revenue was calculated from an assumed 2022 crude oil price of $111

Table 3.1 Willingness to pay and willingness to accept for alternative cellulosic feedstocks

Feedstock	Willingness to accept $/dry metric ton	Willingness to pay $/dry metric ton	Price gap $/dry metric ton	Price gap $/liter
Corn stover	101	28	74	0.25
Alfalfa	130	29	101	0.35
Switchgrass in the Midwest	147	29	118	0.40
Switchgrass in Appalachia	110	29	82	0.28
Miscanthus in the Midwest	127	29	98	0.34
Miscanthus in Appalachia	116	30	86	0.29
Wheat straw	83	30	53	0.18
Short rotation woody crops	98	26	72	0.25
Forest residues	86	26	60	0.20

Source National Research Council (2011)

per barrel. Conversion yield was assumed to be 292 l of ethanol per dry metric ton of feedstock. All estimated capital and operating costs except feedstock were backed out leaving what the biofuel producer could afford to pay for feedstock. The fourth column is the price gap or the difference between the cost to supply the feedstock and what processors could afford to pay, expressed in dollars per metric ton. The last column is that same price gap expressed in dollars per liter of ethanol. These figures include no government subsidy. The bottom line is that with these feedstock prices and technologies available today, cellulosic biofuels would not be economic and will not be launched without government supports. Indeed, the last column of Table 3.1 shows required subsidy per liter of ethanol for alternative feedstocks when the crude oil price is $111/bbl. The required subsidies are estimated to range between $0.18 and $0.40 per liter of ethanol.

While the cellulosic feedstock prices vary significantly, the quantity of available feedstock is not an issue (National Research Council 2011; U.S. Department of Energy 2011). All the major estimates indicate that there would be plenty of biomass available to meet and exceed the RFS requirement for cellulosic biofuels.

Another issue for feedstocks is variability in feedstock supply. This issue applies to all agricultural feedstocks, but may be more problematic for corn stover. It is likely that corn stover will be harvested after the corn harvest. Experts suggest that there is about a three week window after corn harvest for removing the stover and putting it in storage. If we have wet weather during that period such that harvest is limited and/or the stover must be baled wet, then the stover yield will be less than expected. Supply variability raises issues for contracting for biomass supply (Alexander et al. 2010). Contracts must work both for the farmer and for the processor. Contracts will need to be designed to share risk between farmers and processors such that both parties have incentives to enter into long term contracts and abide by the contract terms. Contracts may be based on acreage, tonnage, quality, and other factors. The bottom line is that while availability of biomass is not an issue for biofuels production, feedstock cost and feedstock supply variability could be significant impediments.

3.5 Conversion Technology

There are two major conversion pathways for biofuel production: biochemical conversion and thermochemical conversion. In addition, there are hybrid processes that are partly biochemical and partly thermochemical. The biochemical pathway normally results in ethanol as the biofuel product. The thermochemical pathway normally produces green diesel, bio-gasoline, or jet fuel. These products are sometimes called drop-in fuels as they are closer in composition to the existing fossil based fuels and unlike ethanol can be more easily integrated into the existing fuel supply chain. Currently, the U.S. ethanol industry is faced with a restriction, known as the blend wall. Since the early days of the U.S. ethanol industry, the regulated maximum ethanol content for standard vehicles has been 10 %. Standard vehicles were designed to handle blends only up to that level. Currently, U.S. annual gasoline consumption is about 500 billion liters. This means the total consumption of ethanol, with current mix of vehicles, will be limited to 50 billion liters, which is about the current U.S. ethanol production. Consuming ethanol beyond this level will be constrained by the blend wall. Unlike ethanol, drop in fuels do not have blending limits. They can be blended in much higher percentages than is the case for most ethanol use in the U.S. Also, they can be shipped in pipelines, which is not possible for ethanol.

Earlier in the biofuels era, most of the research and development interest was focused on the biochemical pathway. However, in recent years, much more interest has emerged on the thermochemical pathway or hybrid pathways (Tyner 2010b). Drop in fuels are seen as more attractive for many of the reasons described above.

For the conversion technologies, it is known that biofuels can be produced via either the biochemical or thermochemical pathways. The question is at what cost? Since there are no commercial plants, any cost estimate is uncertain. Research continues on both biochemical and thermochemical process as well as several hybrid processes. Major breakthroughs are possible at any point in the future. The estimates for conversion costs vary among alternative conversion pathways and feed stock. The capital cost for the thermochemical pathway is higher than the biochemical, \$0.30 per liter versus \$0.15 per liter (Taheripour et al. 2011). On the other hand, the variable costs (excluding feedstock) are higher for biochemical pathways. For example, the variable cost of converting corn stover to biofuel using a biochemical process is estimated to be about \$0.37 per liter. The corresponding cost for the thermochemical is about \$0.13 per liter (Taheripour et al. 2011). If feedstock is \$110 per metric, and ethanol yield is 292 l per metric ton, then feedstock cost would be \$0.38 per liter. This means that the total conversion cost (including capital, operating, and feedstock costs) for converting cellulosic materials to biofuel is around \$0.88 per liter plus or minus a few cents variation among alternative pathways and feedstock. This is significantly higher than the total conversion cost of corn ethanol which is about \$0.65 per liter at current prices (Iowa State University 2012).

3.6 Environmental Issues

Biofuels were developed to mitigate GHG emissions. For example, prior to 2007, the general consensus was that corn ethanol can reduce GHG emissions a bit more than 20 % after considering all emissions generated through the production process of ethanol and its consumption (Wang et al. 1999; Farrell et al. 2006). However, early analyses ignored the consequences of biofuel induced land changes for GHG emissions. Once early estimates of GHG emissions due to land use change were included, it appeared that corn ethanol was more GHG intensive than gasoline (Searchinger et al. 2008). These authors estimated that producing corn ethanol generates more than 100 g CO_2 e/MJ emissions due to induced cropland expansion for required corn production. However, the more recent studies find that the early estimates have overstated the induced land use emissions due to biofuels (Hertel et al. 2010; Tyner and Taheripour 2013). For example, Taheripour and Tyner (2013) estimated that producing corn ethanol generates about 13–23 g CO_2 e/MJ emissions due to cropland expansion for required corn production, depending on the implemented land sue emission factors and assumptions on land use carbon fluxes. While the more recent estimates for induced land use change for the first generation biofuels are usually lower than their earlier estimates (Wicke et al. 2012), they are significantly different from zero. This indicates that first generation biofuels do not contribute significantly to reducing GHG emissions, if we take into account their related induced land use change emissions. As technologies improve, it is possible that the GHG emission profiles for first generation biofuels will improve as well.

The first generation biofuels could cause other environmental concerns as well. Corn production is intensive in fertilizer and chemicals, and increased corn acreage likely would lead to higher fertilizer and chemical runoff and soil erosion (National Research Council 2011). These effects would lead to a reduction in downstream water quality. There also has been some concern regarding local air quality issues associated with ethanol production.

While many studies evaluated environmental impacts of the first generation biofuels, only a few attempts have been made to assess these impacts for second generation biofuels. In general, second generation biofuels are believed to have more positive environmental impacts. Perennial grasses like miscanthus and switchgrass need fewer nutrients added and do a better job of preventing soil erosion. In addition, producing dedicated crops on marginal lands can increase their carbon sequestration capabilities (Anderson-Teixeira et al. 2009). As mentioned earlier, cellulosic biofuels also can be produced from crop or forest residues. If biofuels are produced from these materials, then land use implications will be around zero as shown in Taheripour and Tyner (2013).

However, if cellulosic biofuels are produced from dedicated crops, their land use change emissions will not be zero. Taheripour and Tyner (2013) have estimated induced land use emissions for several biofuel pathways under alternative assumptions of soil carbon emissions factors and assumptions on land use change carbon fluxes. The core of their results is shown in Table 3.2. This table shows induced land use change emissions for seven biofuel pathways, corn ethanol

Table 3.2 Estimated induced land use emissions for alternative biofuels (g CO_2 e/MJ)

Feedstock	Biofuel	Without CP-EF			With CP-EF		
		WH	CARB	TEM	WH	CARB	TEM
Corn	Ethanol	12.9	15.1	17.0	15.5	18	22.6
Miscanthus	Bio-gasoline	6.1	7.1	7.3	18.1	19.4	25.6
Switchgrass	Bio-gasoline	21.4	24.9	23.4	43.7	47.6	57.0
Miscanthus	Ethanol	5.8	10.1	10.1	15.7	25.4	32.3
Switchgrass	Ethanol	20.3	35.5	33.1	38.2	63	74.0

Source Taheripour and Tyner (2013)

plus 6 cellulosic biofuel pathways. The induced land use changes due to producing required feedstock for each pathway are converted to GHG emissions using three different published land use change emission factors (EFs). The implemented EFs are: Woods Hole Emission Factors (WH), emission factors developed by California Air Resources Board experts (CARB), and a set of emission factors developed using a Terrestrial Ecosystem Model (TEM). Finally, for each emission factor calculations are done under two cases. The first case assigns zero EFs to changes in converted marginal land (cropland pasture) to dedicated crops. In Table 3.1 this case is shown under the title of "Without CP-EF". The second case assumes emission EF for cropland pasture in each agro-ecological zone is equal to the half of the emissions factor associated with the pasture land in than zone. In Table 3.1 this case is shown under the title of "With CP-EF".

The Table 3.1 shows that unlike the common belief, producing biofuels from dedicated crops generates induced land use change emissions. The results also indicate that the land use change emissions vary significantly among alternative sets of EFs and biofuel pathways. When we assume converting cropland pasture converted to dedicated crops does not generate carbon emissions, then producing ethanol and or bio-gasoline from miscanthus generate the lowest induced land use emissions. However, in the counterpart case, when non-zero emission factors are assigned to converted cropland pasture, then corn ethanol generates the lowest emissions. This means that second generation biofuels produced from dedicated crops could actually generate more induced land use change emissions than first generation. Finally, this table indicates that switchgrass generates the higher rate of land use emissions (even larger than corn), regardless of the type of implemented EF in both cases of with and without CP-EF.

These results show another set of uncertainties associated with the second generation biofuels. These results show that at the end of the day, the limiting resource is land, and second generation feedstocks (except residues) do require land. For residues such as corn stover, there is an unresolved question of the extent to which residue removal reduces soil carbon stock, and thus has adverse GHG impacts. For land using perennial crops like switchgrass, the land required often would come from the livestock sector. If more land is used for biofuel feedstocks, less would be available for cattle grazing, hay production, etc. Thus, it is not entirely true that second generation biofuels do not generate land use change emissions. In addition, the second generation biofuels, if produced from dedicated crops, will compete with food crops in the market for land. Hence, the issue of food versus fuel will not disappear, if the second generation biofuels are produced from dedicated crops. Most of the corn produced is actually used for animal feed, not for direct human consumption. So corn use for ethanol competes with food use primarily via the livestock sector and meat consumption and prices. The same is true, perhaps to a lesser degree, for second-generation dedicated energy crops.

3.7 Government Policy

The U.S. corn ethanol industry, Brazilian sugarcane ethanol industry, and European Union biofuels industry (mainly biodiesel) were all industries created via the support from governments. Support for these industries generally began in the 1980s. In all three regions, support initially came in the form of government subsidies. However, over time governments have evolved towards implementing mandates in lieu of subsidies (Tyner 2008a, b). As biofuels production increased, the government cost of the subsidies became a significant burden. The cost of mandates is off-line; that is, the cost is imposed on consumers via the mandate, and does not show up in government budgets.

In the early years, there was strong political support for government support of biofuels. Agricultural groups had a disproportionate amount of political power, and the biofuel support was not questioned. In recent years, however, the political support in the U.S. has waned. Sectors affected by higher food prices–food manufacturers and retailers, restaurants, etc. have come out against biofuels support. Even the agricultural livestock producers have broken rank with the rest of the agricultural sector in opposing biofuels. So the reality today is that biofuel support is less certain than it was in the early years.

The major biofuel support policy today in the U.S., Brazil, and the European Union is some sort of mandate or target for biofuels use. In the U.S., the mandate is the RFS. As shown in Fig. 3.1, it requires 136.3 billion liters ethanol equivalent of biofuels to be blended by 2022. The RFS that can be filled by corn ethanol in 2015 is 56.8 billion liters, and U.S. production capacity is about to that level today. For cellulosic biofuels, the RFS is 60.6 billion liters ethanol equivalent. There is another category called other advanced biofuels (with 15.1 billion liters target for 2022) that can be met by sugarcane ethanol among other possibilities. The final mandate is for 3.8 billion liters of biodiesel.

To enforce the mandates the Environmental Protection Agency (EPA) uses a procedure which tracks biofuel and consumption. In this process, the EPA issues a unique Renewable Identification Number (RIN) when a batch of biofuel is produced or imported. Each obligated party (blender) has a quota for blending each year that is based upon their projected fuel market share. At the end of the year, obligated parties (blenders) must present RINs equivalent to their quota to the EPA to prove that they meet their annual biofuel obligations. At the end of each year, if a blender does not have enough RINs to meet its obligation, it will be punished a fine. RINs can be traded in market place. Hence, if a blender needs RINs to meet its obligation, it can buy from the market and if it has extra RINs it can save it for future use or sell it to the market.

While the RFS has defined annual targets for each biofuel, the law allows the EPA to waive them for cellulosic biofuels, under some certain circumstances. EPA has waived the cellulosic biofuel mandates every year so far, mainly because no cellulosic biofuel is produced commercially. The cellulosic portion of the RFS has a special provision in which for any year in which any part of the cellulosic RFS is waived, it is possible to buy a credit from EPA instead of actually blending cellulosic biofuel. A company can buy a credit from EPA and buy an advanced biofuel blending RIN from another company and use the combination of the EPA credit plus the advanced biofuel (e.g. sugarcane) RIN to satisfy the cellulosic blending mandate (U.S. Congress 2007). The cost of the credit from EPA is determined by the difference between the previous year wholesale gasoline price and a base gasoline price (about \$0.80 per liter) adjusted for inflation since 2007. In other words if the 2012 value is \$0.85, and the 2012 average wholesale gasoline price was \$0.70, then an obligated blender could purchase from EPA a credit for \$0.15 per liter. The RIN would have to be purchased on the open market, and might have a value of around \$0.10 per liter. Thus, an obligated blender could blend cellulosic biofuel or purchase the credit plus RIN for \$0.25 per liter and meet the obligation in that way. If the cost of the cellulosic biofuel is about \$1 per liter, and the wholesale price of gasoline is \$0.70, the difference is \$0.30 per liter. The obligated blender would have the choice of blending at a cost of \$0.30 or purchasing the credit and RIN for \$0.25. Clearly, under these assumptions, it would be more attractive to purchase the credit and RIN instead of purchasing cellulosic biofuel. This means that the mandate is not really a mandate in reality. This out-clause creates huge uncertainty for potential investors in cellulosic biofuels. Cellulosic biofuels are not competitive on the open market. The hope and expectation was that the government mandate would create the market, but with this out-clause, it clearly does not, at least under the assumptions used here. Thus, it will be difficult to attract private sector investment into the industry.

The above analysis applies anytime EPA waives any part of the cellulosic RFS. In reality, EPA has waived most of the cellulosic mandate in every year it has existed. It will be necessary for EPA to waive part of the mandate all the way out to 2022 because the industry, even if it became attractive, could not possibly grow as fast as the mandate.

There have also been requests to waive the corn ethanol mandate with the most recent being due to the 2012 U.S. drought (Tyner et al. 2012). The criterion EPA is required to use in this case is economic harm caused by the RFS. In November 2012 EPA ruled that the 2013 corn ethanol mandate would not be waived. However, the possibility of a waiver at any time also creates uncertainty for the industry.

Another major impediment to growth of the ethanol industry in the U.S. is what is called the blend wall. Most gasoline in the U.S. is blended at 10 % with ethanol. The U.S. currently consumes about 500 billion liters of gasoline type fuel.

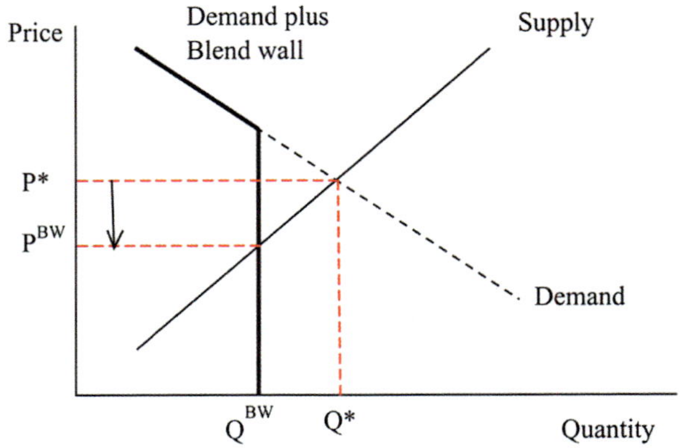

Fig. 3.4 Ethanol market in the presence of blend wall

Blending at 10 % means that the maximum amount of ethanol that can be blended is about 50 billion liters. A very small amount of E85 (85 % ethanol) is sold for use only in flex fuel vehicles, but these vehicles represent a small portion of the U.S. vehicle fleet. While EPA has approved moving to 15 % blends, very little progress has been made in implementing the decision. The EPA approval was for vehicles built since 2001, and that amounts to about 2/3 of the vehicle fleet (Tyner et al. 2010). If a service station were to change to E15 (15 % ethanol), they would lose 1/3 of their customers. E15 also cannot be used in any small engines such as marine engines, lawn mowers, chain saws, etc. This means that the new rule has not been able to remove the exiting blend wall so far.

The economics of the blend wall are illustrated in Fig. 3.4. The kinked bold line represents the demand curve in the presence of blend wall. When we reach the aggregate blending limit, there is no more room for additional ethanol in the market place. We are then in a position of too much ethanol (production capacity) chasing too little market (the blend wall). At that point, the ethanol price becomes its breakeven price with the corn price. In fact, we see that relationship existing today, and the recent historical link between gasoline and ethanol is largely broken. Because of the blend wall, ethanol is largely priced today on corn.

The blend wall has very important implications for cellulosic ethanol. Even if the blending limit of 15 % can be applied more broadly, that still limits the total ethanol blending to about 72 billion liters. There is no place to put a significant amount of cellulosic ethanol. That is one reason for the increased attention to thermochemical conversion and drop-in biofuels.

Finally, there is one government policy option that merits serious attention and reduces uncertainties in markets for cellulosic biofuels significantly. It is called a reverse auction. The U.S. Air Force and Navy have expressed a strong interest in biofuels (U.S. Air Force 2010). The military could use a reverse auction to procure biofuels. In so doing, it would specify the fuel properties, delivery location, and quantity to be delivered each year for about 15 years. A long term contract would be necessary to gain private sector participation in the bidding. Once the call for bids was out, private companies could secure provisional feedstock contracts with farmers and submit a bit. The lowest bid from a qualified bidder wins the contract. This approach reduces the adverse impact of the uncertainties described above in several ways. First, it sets the fuel price, so it does not really matter what happens to oil price in the future. Second, feedstock uncertainties are handled because bidders would have provisional contracts with farmers that would be executed if they won the competition. Third, presumably companies would not be bidding unless they were confident of the technical and economic dimensions of their conversion technology. Fourth, government policy and support would not matter because this becomes a private contract between the military and the company. Thus, the reverse auction could be a mechanism to get the cellulosic biofuel industry moving. In a sense, the difference between the winning bid price and the projected fossil fuel price for the same commodity becomes the implicit subsidy. This implicit subsidy is a least cost option since it is market determined. At this point, the military does not have Congressional approval to use this policy mechanism, so any movement in this direction is blocked.

3.8 Conclusions

Second-generation (cellulosic) biofuels have potential to become a reliable source of renewable fuel. In this chapter we have described five major uncertainties that currently inhibit the development of these biofuels: (1) future oil prices, (2) feedstock availability and cost, (3) conversion technology efficiency and cost, (4) environmental impacts, and (5) government policy. In each of these areas, there are significant issues that impede development and commercialization of the second generation biofuels industry. However, all of these uncertainties could be managed if society were willing to pay the higher cost of cellulosic biofuels.

References

Abbott P, Hurt C, Tyner WE (2008) What's driving food prices? Farm foundation report. https://www.farmfoundation.org/webcontent/Farm-Foundation-Issue-Report-Whats-Driving-Food-Prices-404.aspx. Accessed 6 May 2013

Abbott P, Hurt C, Tyner WE (2011) What's driving food prices in 2011? Farm foundation report. http://www.farmfoundation.org/webcontent/Whats-Driving-Food-Prices-in-2011-1742.aspx. Accessed 6 May 2013

Alexander C, Ivanic R, Rosch S, Tyner WE, Wu S, Yoder J (2010) Contract theory and implications for perennial energy crop contracting. Energy Econ 34:970–979

Anderson-Teixeira K, Davis S, Mater M, Delucia E (2009) Changes in soil organic carbon under biofuel crops. GCB Bioenerg 1:75–96

Congressional Research Service (2010) Cellulosic ethanol: feedstocks, conversion technologies, economics, and policy options. Washington, D.C., Contract No: R41460. http://www.nationalaglawcenter.org/assets/crs/R41460.pdf. Accessed 6 May 2013

Farrell AE, Plevin RJ, Turner BT, Jones AD, O'Hare M, Kammen DM (2006) Ethanol can contribute to energy and environmental goals. Science 311:506–508

Hertel T, Golub A, Jones A, O'Hare M, Plevin R, Kammen D (2010) Effects of US maize ethanol on global land use and greenhouse gas emissions: estimating market-mediated responses. Bioscience 60:223–231

International Energy Agency (2012a) World energy outlook executive summary. Paris, France. http://www.iea.org/publications/freepublications/publication/English.pdf. Accessed 6 May 2013

International Energy Agency (2012b) Golden rules for a golden age of gas. Paris, France. http://www.worldenergyoutlook.org/media/weowebsite/2012/goldenrules/weo2012_goldenrulesreport.pdf

Iowa State University (2012) Ethanol profitability 2012. Available from: http://www.agmrc.org/renewable_energy/

National Research Council (2011) Renewable fuel standard: potential economic effects of U.S. biofuel policy. National Academy of Sciences. http://www.nap.edu/openbook.php?record_id=13105&page=1. Accessed 6 May 2013

Searchinger T, Heimlich R, Houghton RA, Dong F, Elobeid A, Fabiosa J, Tokgoz S, Hayes D, Yu T (2008) Use of U.S. croplands for biofuels increases greenhouse gases through emissions from land use change. Science 319:1238–1240

Taheripour F, Tyner WE (2013) Induced land use emissions due to first and second generation biofuels and uncertainty in land use emission factors. Econ Res Int, vol 2013, http://www.hindawi.com/journals/econ/2013/315787/

Taheripour F, Tyner WE, Wang MQ (2011) Global land use changes due to the U.S. cellulosic biofuel program simulated with the GTAP model. Argonne National Laboratory and Purdue University. http://greet.es.anl.gov/files/luc_ethanol. Accessed 6 May 2013

Thompson J, Tyner WE (2013) Corn stover for bioenergy production: Cost estimates and farmer supply response. Biomass Bioenergy (in press)

Tyner WE (2008a) The US ethanol and biofuels boom: Its origins, current status, and future prospects. Bioscience 58:646–653

Tyner WE (2008b) The global impacts of US and EU biofuels policies. In: Auurbier P, van de Vooren J (eds) Sugarcane ethanol: contributions to climate change mitigation and the environment. Wageningen Academic Publishers, Wageningen, pp 181–197

Tyner WE (2010a) Why the push for drop-in biofuels? Biofuels 1:813–814

Tyner WE (2010b) Cellulosic biofuels market uncertainties and government policy. Biofuels 1:389–391

Tyner WE (2012) Biofuels and agriculture: a past perspective and uncertain future. Int J Sustain Devel World Ecol 19:389–394

Tyner WE, Dooley F, Viteri D (2010) Alternative pathways for fulfilling the RFS mandate. Amer J Agric Econ 92:1–8

Tyner WE, Taheripour F, Hurt C (2012) Potential impacts of a partial waiver of the ethanol blending rules. Farm foundation report. http://www.farmfoundation.org/news/articlefiles/1841-Purdue%20paper%20final.pdf. Accessed 6 May 2013

Wang M, Saricks C, Santini D (1999) Effects of fuel ethanol use on fuel-cycle energy and greenhouse gas emissions. Center for Transportation Research, Energy Systems Division, Argonne National Laboratory. http://www.transportation.anl.gov/pdfs/TA/58.pdf. Accessed 6 May 2013

Wicke B, Verweij P, van Meijl H, van Vuuren DP, Faaij APC (2012) Indirect land use change: review of existing models and strategies for mitigation. Biofuels 3:87–100

U.S. Air Force (2010) Air force energy plan 2010. http://www.safie.hq.af.mil/shared/media/document/AFD-091208-027.pdf. Accessed 6 May 2013

U.S. Congress (2007) Energy Independence and Security Act of 2007, in H.R. 6, 110 congress, 1st session

U.S. Department of Energy (2011) U.S. billion ton update—biomass supply for a bioenergy and bioproducts industry. http://www.usbiomassboard.gov/pdfs/eaton_tac_q3_2011.pdf. Accessed 6 May 2013

U.S. Department of Energy (2013) Annual energy outlook 2013: early release. Washington, D.C. 2012. http://www.eia.gov/forecasts/aeo/er/index.cfm. Accessed 6 May 2013

Chapter 4
Algae Farming and Its Bio-Products

Gal Hochman and David Zilberman

Abstract Many expect algae to contribute to food, feed, health, and fuel, as well as to remove or transform pollutants in water or air. But what did we really achieve after several decades of research and development? What ended up being commercialized and consumed in large volumes? We try and shed light on these questions by surveying and assessing the current state of algae uses.

4.1 Introduction

Increased concern about climate change, energy security and non-renewability associated with the dependence on fossil fuel has led to investigate alternative sources of renewable fuels. Algae have been seen as an attractive feedstock for biodiesel because of their capacity to produce oil under stresses, frequently using waste products as nutrients (Sheehan et al. 1998). Wiley and Campbell (2011) and Li et al. (2008) presented some of the processes that are considered for utilization of algae as a fast growing feedstock for biofuel. But Gallagher (2011) argues that productivity, capital costs, technical constraints, and uncertainty, constrain the economic viability of algae as biofuel. While future research will identify new cost reducing strategies to utilize algae for fuel production, algae has a large economic potential in other applications some of which can complement the use of algae as

G. Hochman (✉)
Department of Agriculture, Food, and Resource Economics, School of Environmental and Biological Sciences, Rutgers University, 55 Dudley Road, Cook College, New Brunswick, NJ 08901, USA
e-mail: gal.hochman@rutgers.edu

D. Zilberman
Department of Agricultural and Resource Economics, University of California, 207 Giannini Hall, Berkeley, CA 94720, USA
e-mail: zilber11@berkeley.edu

M. C. McCann et al. (eds.), *Plants and BioEnergy*, Advances in Plant Biology 4,
DOI: 10.1007/978-1-4614-9329-7_4, © Springer Science+Business Media New York 2014

feedstock for fuel. This chapter overviews a wide range of applications of algae that has or likely to have significant economic potential. The development of algae as feedstock for biofuel will benefits from technological breakthroughs in other applications, and understanding these other application is important in developing sustainable and diversified algae products and technologies.

Algae are large and diverse group of organisms, typically autotrophic organisms that produce complex compounds such as lipids, carbohydrates, and proteins, using simple substances located in their surroundings. Although most algae are photosynthetic "plant like" that lack the distinct cell and organ types present in land plants, some produce energy from uptake of organic carbon.

Algae have raised much hope and scholars have argued that there is much potential of using algae for wastewater treatment, as well as feed food health and fuel production. Many barriers, however, exist for these processes to become economically viable and environmentally friendly. The development of molecular genetics (biotechnology) raised much hope in improving algae strains and making them commercially viable but also brought much fear, because large portions of the population are suspicious and oppose this technology. Biotechnology is a field of applied science and technology, which employs living organisms and their subcellular components for industrial applications and environmental management. Biotechnology makes use of viruses, bacteria, yeasts, fungi, algae, plant and animal cells, and enzymes as components of industrial processes.

This chapter survey current developments of algal bio-products. The chapter surveys the economic performance of the algae industry thus far and the economic potential of alternative product lines. The analysis, especially the economic analysis and the cost estimates, are based on interviews with scientists and businessmen. But because of confidentiality, the findings are not attributed directly to their sources. The list of sources, however, is provided in Appendix A.

4.2 Algae Farming

Existing patterns of use of algae are insightful and useful for future projections. Still, present success stories are only the tip of the iceberg. Much more research and knowledge are needed for full societal gains from the economic potential of algal products.

When surveying algae, one has to distinguish between two major signs–microalgae and microalgae. While macroalgae are basic seaweeds, microalgae are microscopic organisms, which make up the world's phytoplankton and form rapidly growing populations in water when supplied with the necessary nutrients for their culture.

4.2.1 Macroalgae

Macro-algal grow on rocky substrates, forming multilayer perennial vegetation capturing almost all available photons. The macro-algal maximum productivity is 10 times higher that of plankton population (Carlsson and Bowles 2007). The maximum chlorophyll content corresponds to an algal biomass of about 10 kg per m^2 (Luning and Pang 2003).

About 200 species of macro-algae are used around the globe, where 10 of which are intensively cultivated (Luning and Pang 2003),

- Brown algae: *Laminaria japonica* and *Undaria pinnatifida*;
- Red algae: *Porphyra* spp., *Eucheuma* spp., *Kappaphycus* spp., and *Gracilaria* spp., and;
- Green algae: *Monostroma* spp., and *Enteromorpha* spp.

Suitable macroalgal species used for large-scale cultivation include species of *Alaria*, *Corallina*, *Cytoseria*, *Ecklonia*, *Egregia*, *Eucheumia*, *Gracillaria*, *Laminaria*, *Macrocystis*, *Pterygophora*, and *Sargassum* (Carlsson and Bowles 2007).

The world production of macroalgae for commerce amounts to \$5.5–6 billion a year (McHugh 2003; Pulz and Gross 2004). Many types of dried seaweed are being used as food products, mainly in salads and for seasonings, with currently species of *Rhodophhyta* and *Phaeophyta* used industrially to produce 7.5–8 million tons of wet seaweed annually. While the food industry is estimated to generate \$5 billion a year, a further \$600 million is estimated to have been generated from hydrocolloids extracted from the cell wall of the macro-algae at an average value of about \$10,900 a ton. Sales of one of the dried kelp, called Nori, are estimated to be \$1 billion–a high value product worth \$16,000 a ton. While Nori is consumed mostly in Japan and the Far East, its consumption is spreading to the West and it is now being produced and consumed in California. Much of the Nori in the market is harvested from the sea, but there are substantial efforts in cultivating it through mariculture.

But the economic value of the macroalgae derivatives goes much beyond their usefulness as food products. *Agar* is the most diversely used macroalgae derivative with substantial worldwide sales. *Agar* is a class of vegetable gums that is derived from the two varieties of seaweed–*Gelidium* and *Gracilaria*. It is a very strong gelling agent with unique properties. *Agar* is not poisonous to humans, has no nutritional content, does not rot, and can absorb liquids and swell. The gel it generates may survive a wide range of temperatures; is indigestible by most bacteria; and is very elastic, resilient, and clear (Chapman 1970; Renn 1984). *Agar* and its derivatives have a wide variety of uses. The value of *agar* products varies substantially according to their quality.

The relatively low-value *agar* is used in the food industry (emulsifier, gelling agent, and preserving agent) as a component of many laxatives and as an impression material (especially in dentistry). The medium-value use of *agar* is in bacteriological and microbiological applications. The highest-value derivative of *agar* is called *agarose* and is used in a microbiological genetic-engineering application.

The demand for *agar* and *agarose* is continuing to grow, and there is much interest in alternative sources of *agar* and some experimentation with domestication of it. *Agar* served as an example to the larger market for algae derivatives and versatility of products of different quality that can be produced from individual algae. It also serves to emphasize that this is a time where domestication of seaweed production attracts much interest.

Carrageenan and *alginates* are two other macroalgae derivatives much used as gums, emulsifiers, and gels. Annual sales of these products are in the hundreds of millions of dollars. According to McHough (2003), the market value of carrageenan is $240 million, that of alginate is $213 million, and that of *agar* is $137 million. Producing methane from macro-algae has also been discussed in the literature; however, it has been argued that algal biomass production, as a stand-alone product, is not economically viable (Chynoweth 2002).

4.2.2 Microalgae

There are more than 8,000 microalgae species which are divided into four types: *cynaobacteria* (blue-green algae), *rhodophytes* (red algae), *chlorophytes* (green algae), and *chromophytes* (all other algae). Each of these types contains hundreds of species. Each species may be thousands of genetically distinct strains. Only a small fraction of these varieties have been studied for possible beneficial use, and there is much ignorance and uncertainty regarding the behavior and properties of most micro-algae species. The most frequent used micro-algae include *Cyanophyceae* (blue-green algae), *Chlorophyceae* (green algae), *Bacillariophyceae* (including diatoms), and *Chrysophyceae* (including golden algae).

The world market value of micro-algae has been estimated at $5–6.5 billion, out of which about 2.5 billion dollars have been generated by the health food sector, 1.5 billion dollars from the production of *docosahexanoic acid* (DHA) and 700 million dollars from aquaculture (Pulz and Gross 2004).

4.3 Commercial Uses of Algae

Much interest is currently expressed in production of algal biomass. Present calculations of production costs of algal biomass suggest that with current technology it not economically viable as a stand-alone product, although several studies have argued that co-production of algal biomass may become viable under certain scenarios (Carlsson and Bowles 2007; Lundquist et al. 2010; Hochman et al. 2013). However, co-producing algal biomass limits the scale of energy production to the profitable application (Reith 2004). Lundquist et al. (2010) argued that co-producing algal biomass with wastewater treatment is less limiting, but that co-producing algal biomass with value added products such as astaxanthin and β-carotene does significantly limit algal biomass production.

Algae can be used to produce raw material for co-firing to produce electricity, liquid fuel production via pyrolysis and thermochemical liquefaction (bio-oil), or biomethane generation through fermentation. While these processes cannot, yet, compete with fossil fuels and their heating value is low with 29 MJ/kg compared with 42 MJ/kg (Miao et al. 2004), algae energy content and heating value is higher than that of other biomass feedstock. Major limitations of commercialization of algae biofuels include (1) algae mass culture oil content; (2) harvesting methods; (3) separation techniques; and (4) supply of CO_2 and other nutrients (Miao et al. 2004; Lundquist et al. 2010).

However, small-scale cultivation and industrial scale production of microalgae has evolved in the last few decades. Several substantial applications have been established. These applications include:

4.3.1 Waste Water Treatment

Micro- and macro-algae can be used to sequester, remove, or transform pollutants including excess heavy metal, nutrients, and xenobiotics from wastewater, or CO_2 from exhausts. The yield derived from this process can be algal biomass used to produce chemicals, biofuels, bio-oil, and biogas as co-products (Munoz and Guieysse 2006).

Many studies and several commercial facilities have demonstrated the viability of microalgae in sewage treatment (for example, Oswald 1987a, b). Oxygen production by microalgae for waste oxidation by bacteria in ponds is generally recognized. There are also promising results demonstrating algae contribution in enhancing sedimentation, disinfection, nutrients, and in removing heavy metal and organic toxins. Oswald estimates the savings associated with the use of algae (in place of electricity) for oxygen production in sewage ponds to be between $3,300 and $14,000 (1985) per hectare (based on an energy price of 10 cents per kilowatt-hour).

The algae biomass produced in sewage pools can be used to produce energy by fermentation. There have been several large-scale experimentations in the combined use of algae for waste management and energy production, and a combined system seems especially appropriate and economical to locations with expensive and scarce energy resource.

The use of alga for sewage oxidation (including use of the resulting biomass for energy production or other economic purposes) is likely to increase substantially as energy prices increase and more knowledge about the technology becomes available through experience. Increased productivity is an important factor in determining the future of the technology and will influence its fate.

There is substantial demand for technologies capable of removing chemicals from bodies of water. For example, there has been extensive search in California for technologies capable of ridding water of selenium and other minerals and toxins. The volume of the problem is immense, and hundred of millions of dollars are allotted annually to waste treatment. Similar problems occur elsewhere and suggest a good area for future applications for algal use (given that through research algal technologies can provide effective and economical solutions).

There are many links and dependencies between microalgae for waste research. There is much potential for economic gain combining the use of algae for waste management and other activities (Shelef 1982). Many of the technologies developed in the use of microalgae for waste management systems are appropriate for algal utilization in other production activities. Insights regarding the use of microalgae for waste management have commercial value and can be sources of income. There are markets for expertise in water and waste management.

4.3.2 Fine Chemicals

Like *agar*, algal products vary substantially in price and value, according to their use and refinement.

β-*Carotene* is a metabolite with a wide range of commercial applications. It is used as a food coloring (with a major application in providing the yellow color to margarine), as a good additive to enhance the color of the flesh of fish and the yolk of eggs, and to improve the health and fertility of grain-fed cattle (see survey by Borowitzka and Borowitzka 1987).

Until the early 1980s, commercial production of β-carotene was synthetic, and Hoffman Laroche had a virtual monopoly on the production and marketing of β-carotene. During the 1970s, researchers (Borowitzka and Brown 1974; BenAmotz and Avron 1980) realized that, under nutrient stressed, high salt, and highlight conditions, the microalgae, *Dunaliella salina*, accumulates up to 14 % of dry weight as β-carotene. This discovery led to commercial derivation of natural β-carotene from this organism.

In the 1980s, the price of extracted and purified natural β-carotene was much higher than that of synthetic β-carotene ($1,000 to $2,000 per kg for natural versus $400 to $800 per kg for synthetic), reflecting the preference consumers and buyers have for natural products. Even though the price difference between natural and synthetic declined in the future as the supply of natural β-carotene increased, the differences continued and the natural product always fetched the higher price. Moreover, natural β-carotene has physical properties that make it superior to synthetic. In particular, natural β-carotene is fat-soluble. β-carotene, as well as some other carotenoids, are touted to be anti-carcinogenic and are effective in controlling cholesterol and in reducing risks of heart disease (Nishino et al. 2002). If substantiated, these desirable medical properties increase even more the demand and desirability of natural β-carotene.

Calculations for a fifty 1,000 m^2 *Dunalliela salina* farm suggest that the fixed setup cost (site preparation, pond construction, production, and harvesting and processing equipment) is between $1 and $1.5 million. Using a five-year return-of-investment period as a criterion to distribute the fixed cost, a conservative estimate of the annual fixed cost per 1,000 m^2 would be $6,000 ($0.2 \times 1{,}500{,}000/50$). Annual variable cost (labor, CO_2, nutrients, and electricity) is estimated to be between $15,000 and $24,000 per 1,000 m^2. Using a very conservative approach, total costs are estimated to be about $30,000 per 1,000 m^2. The most conservative estimate of yield we have seen is 50 kg of β-carotene per 1,000 m^2. Thus, under these conservative estimates, the break-even point is reached when a kilogram of natural β-carotene fetches $600 per kg.

The estimates used thus far are quite conservative. Ben-Amotz and Avron (1980) estimated that, with more experience and fine-tuning in production, annual yields could rise to 120 kg per 1,000 m^2 (based on harvesting 400 t of *Dunaliella salina* on 50,000 m^2). It is assumed that algae have 30 % dry matter, and 5 % of it is β-carotene. Thus, β-carotene per 1,000 m^2 is 0.05(400/500)(1,000) = 120 kg per 1,000 m^2. Thus, using this yield estimate, assuming an annual variable cost of $15,000 per 1,000 m^2 and a fixed cost of $21,000 per 1,000 m^2, the investment in β-carotene production can be recaptured in 1 year assuming a natural β-carotene price of $300 per kg.

Phycobiliproteins are algal derivatives that have utility in diagnostic tools. Specifically, *biliproteins* from microalgae (*phycobiliproteins*) are used as fluorescent markers for genetic screening in cell analysis and *imionchemialasi* (Pulz and Gross 2004). This application is based on a discovery made by Professor Alexander Glazer of the University of California at Berkley and published in 1982. According to Professor Glazer, the first application started in 1983 when two laboratories started producing phycobiliproteins. The case of phycobiliprotein, like the case of *agar*, demonstrates the large range of commercial opportunities algal products are starting to have, with sophistication and growth in biological and genetic research, experimentation, and commercial application.

4.3.3 Food and Feed Products

Many microalgae have a high nutritional value. They contain proteins, vitamins and minerals, and non-saturated fats. Moreover, they can yield higher outputs for the same levels of water and land. These characteristics led to the success of the production of *Spirolina*, a microalgae that grown isolated in a monocultural setting (another type of micro-algae grown isolated in a monoculture is *Dunaliella salina*). It has commercial success as a "health food" and is a component of many health food products. Worldwide, *Spirulina* is grown in many countries for animal feeds and food nutrition supplement (FAO 2010). Other important species include *Chlorella*, *Dunaliella*, *Nostoc*, and *Aphanizomenon*.

Culture of the freshwater algal *Haematococcus pluvialis* was developed in a few countries and is used for the extraction of *astaxanthin* (FAO 2010), a natural antioxidant–see also Del Campo et al. (2007). A price tag of $8–15 per kg has been quoted in the literature (Benemann and Oswald 1996; Vonshak 1997; Lee 2001; Carlsson and Bowles 2007). Carlsson and Bowles (2007) suggest that currently the delivery price to the US from China for 20-ton containers is $5 per kg for *Spirulina* and twice as high for *Chlorella*. Production costs have been estimated at $2–5 US per kg.

Algae produced as a co-product of waste management plants may be used for animal/livestock as well as other value added products (Lundquist et al. 2010; Hochman et al. 2010). Much research is needed to develop large-scale food and feed production from algae—one needed to identify species, production, procedures, etc. The food surplus problems in the United States and strong political influence of grain farmers in America prevent production of feed from microalgae. Note that, while there is no public support for research on obtaining food from algae, there is much support for fuel production from algae. Growing lipid-rich species of freshwater algae for biofuel production is the latest development in freshwater algae. Compared with seaweed farming, the culture of freshwater algae is generally poorly reported worldwide.

4.3.4 Fatty Acids

Certain unsaturated fatty acids in triglycerides have desirable therapeutic and health-promoting properties. Research has shown that omega-3 fatty acids reduce cholesterol and fat levels in the blood and "cleanse" the lining of blood vessels (Simopoulos 1991). The medical use of omega-3 fatty acids for prevention and treatment of heart disease is increasing via prescribing fish oil to heart patients.

Usually, this treatment continues throughout the lifetime of the patients. Moreover, some doctors prescribe similar dosages to individuals with high-risk profiles with respect to coronary diseases. As evidence of the effectiveness of this treatment spreads, its adoption is likely to grow. Studies (for example, Yetir 1988) have shown that omega-3 fatty acids have the effective therapeutic properties dealing with rheumatoid arthritis and immunodeficiency diseases, and doctors are considering prescribing pills derived from fish oils to combat these diseases. Cod and other fish are not the direct producers of microalgae and extraction of omega-3 fatty acids from the microalgae. The product extraction directly from the algae is likely to be superior to the cod liver oil as (1) it will not have the off flavor of cod liver and (2) it will be more "pure" product and thus more effective. The use of microalgae should not be restricted to direct extraction of omega-3 fatty acids. They can also be used as feed for chickens and dairy cows with will then introduce omega-3 fatty acids to eggs and milk. The application to eggs may be especially useful since it will tend to reverse (and combat) the contribution of eggs to cholesterol buildup.

The medical discoveries about the therapeutic properties of omega-3 fatty acids suggest a very large market to algal-derived fatty acids. The superiority of the microalgae derivative and the continued growth in demand for omega-3 fatty acids suggest much higher sales potential for fatty acids derived from algae. There is a substantial market for omega-3 fatty acids, and they can be marketed through distributional channels of drugs and health products. Currently, species used to produce fatty acids include *Odontella aurita/Bacillariophyta*, as well as *Isochrysis galbana/Chlorophyta* and *Phaedactylum Tricornutum/Bacillariophyta* (Molina Grima et al. 1994; Pulz and Gross 2004).

To assess the profitability of their production from microalgae, one has estimates based on information available from other products. For the profitability analysis, let P denote price per kilo of fatty acid, OC denote operational cost per 1,000 m^2, I denote investment per 1,000 m^2, and Y denote output of fatty acid per 1,000 m^2. Output of fatty acid is the product of a share of fatty acid in dried weight of algae (denoted by S) and production of dry matter of algae per 1,000 m^2 (denoted by Q), i.e., $Y = Q * S$.

Based on the price of cod liver oil pills, the retail price of omega-3 fatty acid is estimated to be around $600 per kg. Next, it is assumed that producer price is only 20 % of retail price (the other 80 % covers transportation, processing, storage, and marketing costs). Based on these conservative assumptions, the producer price we use for profitability analysis is $P = \$120$ per kg.

Based on several sources, 3 t of dry eight of algae per acre is a low-end estimate of annual production of most existing systems, and average output should be about 6 t per 1,000 m^2. (For brevity, the time dimension of yield and cost figures are omitted, but all yield and cost figures given here are annual.) In the longer run, with better knowledge and experience, dry weight algal production per 1,000 m^2 would reach 10 t per 1,000 m^2. We use three levels of dry weight algal production:

Low output: $Q = Q_L = 3$ t per 1,000 m^2
Medium output: $Q = Q_M = 6$ t per 1,000 m^2
High output: $Q = Q_H = 10$ t per 1,000 m^2.

Experiments done (Koren et al. 1988) obtained results consistence with our assumed levels. That team grew *Nonochloropsis salina* which is intended as a source for fatty acid in a 100, a 2.5 m^2 pond, and a photoreactor. Production levels are equivalent to annual yields per acre per year of between 3 t of dry weight (large pond) to 8 t of dry weight (photoreactor).

Based on several sources, the share of fatty acids in dry weight of algae moved from a low of 0.03 to high of 0.05. The experiments in Koren et al. (1988) resulted in shares that are closer to 0.03. That makes us somewhat more cautious in the values we use for S in our assessment. The values we use are:

Low share: $S = S_L = 0.03$
High share: $S = S_H = 0.04$.

Combining the dry weight and share estimates, we obtain six values of yield per acre, denoted for Y_1 (lowest) to Y_6 (highest). These values are:

Y_1 = low output–low share = 90 kg/1,000 m^2/year
Y_2 = low output–high share = 120 kg/1,000 m^2/year
Y_3 = medium output–low share = 180 kg/1,000 m^2/year
Y_4 = medium output–high share = 240 kg/1,000 m^2/year
Y_5 = high output–low share = 300 kg/1,000 m^2/year
Y_6 = high output–high share = 400 kg/1,000 m^2/year.

As argued previously (in the case of β-carotene), variable costs (operation cost) per 1,000 m^2 are guesstimated to range from a low of $OC = OC_L = \$15{,}000$ per 1,000 m^2 to a high of $OC = OC_H = \$24{,}000$ per 1,000 m^2. Investment per 1,000 m^2 (which has no time dimension) is estimated to be $I = \$30{,}000$ per 1,000 m^2. Then guesstimates have several implications.

Under the lowest yield assumption ($Y = Y_1$) revenues cannot cover variable costs even when assuming low variable cost level. The deficit in this case is

\$4,200 annually (15,000 − 90 × 120). The low variable costs are almost covered even with the lowest algae production if the fatty acid in the dry algae is higher. In the case of $Y = Y_2$ and $OC = OC_2$, the deficit is only \$600 dollars annually (15,000 − 17,400).

When algae output is at a medium level the low level of variable cost is covered and there is substantial surplus even assuming low fatty acid share. Specifically, the surplus above variables cost when $Y = Y_3$ and $OC = OC_3$ is \$6,600 annually (180 × 120 − 15,000), and variable costs are at the high level, the annual deficit is only \$2,400 annually. Average algae output and higher fatty acid ratio ($Y = Y_4$) generated a surplus of \$4,800 annually even when variable costs are at their high level (4,800 − 240 × 1,2000 − 2,400).

When Algae output is in the high level, it leaves substantial surplus above variable cost, even when share of fatty acids in dry weight is low. When $Y = Y_5$ and $OC = OC_H$, the annual surplus is \$6,000 and fixed cost can be recovered in 5 years. When algae output is high and share us high ($Y = Y_6$), surplus above the high level of variable costs can be recovered in less than a year and a half. Moreover, even when fatty price will decline by 50 % (down to \$60 per kg), revenues will cover the variable costs.

4.3.5 Polysaccharides

Polysaccharides are chemicals that are used as *viscosifiers* (thickening agents), fluctuating agents, and lubricants. The value of polysaccharides varies according to their use, availability, and purity. They include macroalgal derivatives such as carrageenan and *agar*. Polysaccharides are derived from bacteria, fungi, and algae. The bacteria and fungi are much more productive than algae, and genetic manipulation and engineering of bacteria is in a much more advanced stage than genetic engineering with algae. Still, algae generate complex and unique polysaccharides and many algal derivatives are irreplaceable. Microalgae are the source of important and commercially used polysaccharides, and the market for these algal derivatives are in the hundred of millions of dollars.

Microalgae (such a *Porphyridium cruentum/Rhodophyta*—Fuentes et al. 1999) are commercially used to produce polysaccharides. Under the right conditions, 15–55 % of the weight of the microalgae can be extracted as polysaccharides. Taking a very conservative approach—assuming 15 % polysaccharides share in dry weight, medium yields (5 t of dry weight algae per 1,000 m^2), and high cost (\$30,000 per 1,000 m^2 annual total cost)—the break-even price for polysaccharides production is \$40 per kg, which is within the medium range of market value for polysaccharides. Taking a slightly more optimistic view—30 % polysaccharides share in weight, medium yield (5 t per 1,000 m^2), and low cost (\$20,000 per 1,000 m^2)—the break-even price is less than \$15 per kg, quite a modest price for many polysaccharides. Thus, polysaccharides from microalgae have good economic potential.

4.3.6 Food Coloring

There is a growing demand worldwide for organic food coloring. Regulating agencies constantly limit the range of permissible chemical food coloring, and the regulatory process will be even stricter if and when organic substitutes are available. The volume of the market for food color is immense—in the billions of dollars annually.

Microalgae can be used as a source of many organic food coloring. As Borowitzka and Borowitzka (1987) show, some microalgae contain substantial amounts of other types of carotenes in addition to β-carotene. Other types of coloring appear in microalgae as well. In pure form it can fetch up to $1,000 per kg. It has been argued that the potential of microalgae, as a source of food coloring, is limited because algal-derived food coloring is not photostable. Namely, they tend to bleach with cooking. Nevertheless, in spite of this limitation, the potential market for microalgae-derived food coloring is vast.

4.3.7 Osmoregulators

These are carbohydrates that can affect osmotic processes. Glycerol is the most notable member in this compound category, which included other commercially viable products as well. Substantial weight of the dry weight of several algae, e.g., up to 50 % (*Dunaliella salina*), can be transformed to osmoregulators under the appropriate conditions. Microalgae compete with bacteria and animal fat as sources of osmoregulators. Research should and is likely to discover valuable osmoregulators that can be produced from microalgae.

4.3.8 Energy

The idea of using micro-algae to produce biodiesel is not a new idea and much research has been allotted to the topic (Gallaghar 2011). A project at the National Renewable Energy Laboratory has collected roughly 3,000 strains of algae from northwest and southeast regions of the continent of the U.S. and Hawaii (Sheehan et al. 1998). Carlsson and Bowles (2007) notes that the *vlc-PUFAs* may be less appropriate for the production of biodiesel, since the polyunsaturation leads to oxidation concerns in the fuel.

Gallaghar (2011) computed the Net-Present Value (NPV) using production and cost figures reported in the literature. He concludes that high yield and high oil prices, together with moderate government support and carbon prices, make biodiesel production viable economically. Assuming high yield (134 mt/ha) with lipid concentration of 40 % leads to 6,430 gal/ac. Then, is we assume $1.00 subsidy per gallon and a price of carbon of $44 per ton, a NPV of 17.4 million dollars is achieved with a payback period of 16.7 and an IRR of 12.4. Gallaghar (2011) analysis suggests the once the environment and the social cost of pollution is introduced into calculations, biodiesel production from algae can become economically viable with moderate subsidies. See also work by Demirbas and Demirbas (2011).

4.3.9 Other Applications

Microalgae contain many useful chemical compounds, and its derivatives can be used in the future for many other applications in addition to the ones mentioned above. They include cosmetic and skin products, food and feed supplements, vitamins, and fertilizers, as well as fuel. Microalgae may be less productive than bacteria, and our ability to manipulate it is much smaller. But microalgae contains unique and complex products not available otherwise. Therefore, science potentially can offer much research regarding the use and manipulation of microalgae. Today, in addition to the products discussed above, *Spirulina* is used to produce phycocyanin and biomass (Lee 2001; Costa et al. 2003) while *Chlorella vulgaris/Chlorophyta* is used to produce biomass (Lee 2001).

4.4 Production Systems

4.4.1 Open Ponds

We constructed some tentative estimates. First we present the following two estimates of annual operational costs per 1,000 m^2,

	High cost	Low cost
Labor	$10,000	$4,500
Feed (CO_2 and nitrogen)	10,000	8,000
Energy, oil, and water	15,000	1,000
Machinery cost (short term) repair and maintenance	2,500	1,500
	$24,000	$15,000

These operational costs include production, harvesting, and drying. The big difference between the high and low cost is in the labor cost estimates. Over time, as experience is gained, work procedures will become better established and labor costs will decline much further. Feed costs are the bug cost item in the long run, especially the CO_2 cost. The cost of repair and maintenance will decline with time as more efficient production technologies and machinery are developed. Energy efficiency is likely to increase over time, but energy consumption will rise as production becomes more automated and labor intensive. As mentioned before, these are very gross guesstimates; more knowledge on cost structures is required. Still, all experts we discuss with agree that operational costs can be reduced further $10,000 per 1,000 m^2 and that reducing CO_2 and feed costs is major challenge.

When it comes to capital cost, it may be up to $30,000 per 1,000 m^2. Based on 50–1,000 m^2 ponds, up to $15,000 per 1,000 m^2 will be required for land, land preparation, and pond construction. Equipment (piping, paddles, drying equipment etc.) will require another $10,000 per acre, design management and coordination will require the rest. Capital cost may be reduced over time to one third as experience is added. Obviously, rise of yield and capital-intensive technologies (plastic testing) may increase capital cost but, in this case, with substantial change in yields.

4.4.2 Photobioreactors

These systems are different types of tanks or closed systems, in which algae is grown (Richmond 2004). In these systems, water, nutrients and CO_2 are supplied in a controlled way, while oxygen is removed. Janssen et al. (2003) and Choi et al. (2003) review developments in work that optimize photobioreactors systems for algae cultivation.

4.4.3 Heterotrophic

Others work suggests that algae can be grown in conventional fermentors instead of photobioreactors to produce high value products (Wen and Chen 2003). Instead of use of light and photosynthesis, heterotrophic utilize carbon sources in the medium for the carbon and energy generation (Ward and Singh 2005).

4.5 Summary and Conclusions

This chapter suggests that commercial utilization of algae, beyond biofuels, is economically viable, and that there is a worldwide market for algal derivatives that is estimated to be in the billion of US dollars. While application of algae as biofuel has gained much attention, the literature suggests that some algal derivatives, which researchers worked on during the last several decades, matured and proved quiet lucrative. Others are still at the research and development stage, or are just been thought of.

Energy production from algae has gained much attention. Many algal species are rich in oil content and algal is more productive at producing oil than any of the existing terrestrial plants. While aquatic biomass may be used as raw material for co-firing and producing electricity power and heat, much research and public and private funds are channeled to the commercial development of the production of bio-oil, biomethane, biodiesel or biogas. However, currently these technologies cannot compete with fossil fuels. Policy can facilitate the adoption of these technologies of algae technologies as part of a green economy (Bangalore et al. 2012). Such policy emphasis may misfire if the algae sector cannot stand on it's own feet and compete after a relatively short period of transition and learning. Hybrid applications using algae to produce fuels and other product will increase the productivity of biofuel sector and made bio-algae more efficient.

This chapter briefly surveyed the economic performance of the algae industry thus far. It offers a snapshot of the algae industry and its potential. Further work, however, is needed to better assess the economic viability of various algae derivatives and to understand the true potential of algae and its impact on the energy sector. We leave this for future work.

Appendix A: Interviews

Abuoav J (Dr.,). Chief of Surgery, Mount Zion University, San Francisco, California.
Amit U. Ein Yahav, Arava, Israel.
Arad S (Dr.). Ben Gurion University, Box 1025, Beer Sheva, Israel, 84110.
Borowitzka M. School of Environmental and Life Sciences, Murdoch University, Perth, Australia.
Ben-Amotz A (Professor). Israel Oceanographic Institute, Tel Shikmona, P.O. B. 8030, 31080 Haifa, Israel.
Foget RD. Manager Marketing Services, Bio Products, FMC Corporation, Marine Colloids Division, 2000 Market Street, Philadelphia, PA, 19103.
Glazer AN (Professor). Department of Microbiology and Immunology, University of California, Berkley, California, 94720.
Guron Y. B.A.R.D. Fund, P.O. Box Bet Dagan 50250, Israel.
Martinez W. USDA-ARS, Room 226, Building 005, BARC-West, Beltsville, Maryland, 20705.
Neushul M (Professor). Marine Science Institute, University of California, William J Department of Civil Engineering, University of California, Berkley, California, 94720.
Oswald WJ Department of Civil Engineering, University of California, Berkley, California, 94720.
Ramus J. Duke University Marine Lab., Beaufort, North Carolina, 28516.
Renn DW. (Dr.). FMC Corporation, 5 Maple Street, Rockland, Maine, 04841.
Sfat MR. Bio-Technical Resources Inc., 1035 South Seventh Street, Mainitowoc, Wisconsin, 54220.
Vreeland V (Dr.). Department of Biology, University of California, Berkley, California, 94720.

References

Bangalore M, Hochman G, Zilberman D (2012) Differences in the adoption of agricultural anaerobic digestion in Europe and the United States. Working paper

Ben-Amotz A, Avron M (1980) Glycerol, β-carotene, and dry algae meal production by commercial cultivation of *Dunaliella*. In: Shelef G, Soeded SJ (eds) Algae biomass. Elsevier North-Holland Biomedical Press, Oxford, pp 603–610

Benemann JR, Oswald WJ (1996) Systems and economic analysis of microalgae ponds for conversion of CO2 to biomass. Final report (No. DOE/PC/93204--T5). Department of Civil Engineering, Pitburgh Energy Technology Centre, US. Available at http://www.osti.gov/energycitations/product.biblio.jsp?osti_id=493389

Borowitzka MA, Borowitzka LJ (1987) Vitamins and fine chemicals from micro-algae. In: Borowitzka MA, Borowitzka LJ (eds) Micro-algal biotechnology. Cambridge University Press, New York

Borowitzka LJ, Brown AD (1974) The salt relation of marine and halophilic species of *Dunalialla*: the role of glycerol as a compatible solute. Arch Icrobiol 96:37–52

Carlsson AS, Bowles DJ (2007) Micro-and macro-algae: utility for Industrial applications: outputs from the EPOBIO project, Sep 2007. CPL Press: Science Publisher, UK. (Report)

Chapman VJ (1970) Seaweed and the uses. Methuen and Company, London

Choi SL, Suh IS, Lee CG (2003) Lumostatic operation of bubble column photobioreactors for *Haematococcus pluvialis* cultures using a specific light uptake rate as a control parameter. Enzym Micro Technol 33:403–409

Chynoweth DP (2002) Review of biomethane from marine biomass. In: Reith JH, Hal JW, Lenstra WJ (eds) History, results and conclusions of the "US Marine Biomass Energy Program", (1968 to 1990), 194 pp

Costa JAV, Colla LM, Duarte P (2003) Spirulina platensis growth in open raceway ponds using fresh water supplemented with carbon, nitrogen and metal ions. Zeits Naturforsch, C-A J Biosci 58:76–80

Del Campo JA, Garcia-Gonzales M, Guerrero MG (2007) Outdoor cultivation of microalgae for carotenoid production: current state and perspectives. Appl Microbiol Biotechnol 74:1163–1174

Demirbas A, Demirbas MF (2011) Importance of algae oil as a source of biodiesel. Energy Conserv Mgmt 52:163–170

FAO Fisheries and Aquaculture Secretariat (2010) The state of world fisheries and aquaculture 2010. Food and Agriculture Organization of the United Nation, Rome, Italy. Available at http://41.215.122.106/dspace/handle/0/210

Fuentes MMR, Sanchez JLG, Sevilla JMF, Fernandez FGA, Perez JAS, Grima EM (1999) Outdoor continuous culture of Porphyridium cruentum in a tubular photobioreactor: quantitative analysis of the daily cyclic variation of culture parameters. J Biotechnol 70:271–288

Gallagher Brian J (2011) The economics of producing biodiesel from algae. Renewable Energy 36(1):158–162

Hochman G, Trachtenberg MC, Zilberman D (Forthcoming) Algae crops: co-production of algae biofuels. In: Dierig D, Cruz VM (eds) Industrial crops: breeding for bioenergy and bioproducts. Springer Science, New York

Janssen M, Tramper J, Mur LR, Wijffels RH (2003) Enclosed outdoor photobioreactors: light regime, photosynthetic efficiency, scale-up and future prospects. Biotechnol Bioeng 81:193–210

Koren A, Amit U, Ilani T, Black B, Kahvvan A (1988) Research activities in the microalgae growth in the Arava R&D center for aquaculture. Ein Yahav, Israel. (Report)

Lee YK (2001) Microalgal mass culture systems and methods: their limitation and potential. J Appl Phycol 13:307–315

Li Yanqun, Horsman Mark, Nan Wu, Lan Christopher Q, Dubois-Calero Nathalie (2008) Biofuels from microalgae. Biotechnol Prog 24(4):815–820

Lundquist TJ, Woertz IC, Quinn NWT, Benemann JR (2010) A realistic technology and engineering assessment of algae biofuel production. Robert E. Kennedy Library, Cal Poly, San Luis Obispo, CA. Available at http://digitalcommons.calpoly.edu/cenv_fac/188/

Luning K, Pang S (2003) Mass cultivation of seaweeds: current aspects and approaches. J Appl Phycol 15:115–119

McHugh DJ (2003) A guide to the seaweed industry. FAO fisheries technical paper no. 441, Rome, FAO

Miao X, Wu Q, Yang CY (2004) Fast pyrolysis of microalgae to produce renewable fuels. J Anal Appl Pyrol 71:855–863

Molina Grima EM, Perez JAS, Camacho FG, Sevilla JMF, Fernandez FGA (1994) Effect of growth-rate on the eicosapentaenoic acid and docosahexaenoic acid content of *Isochrysis galbana* in chemostat culture. Appl Microbiol Biotechnol 41:23–27
Munoz R, Guieysse B (2006) Algal-bacterial processes for the treatment of hazardous contaminants: a review. Water Res 40:2799–2815
Nishino H, Murakoshi M, Ii T, Takemura M, Kuchide M, Kanazawa M, Mou XY, Wada S, Masuda M, Ohsaka Y, Yogosawa S, Satomi Y, Jinno K (2002) Carotenoids in cancer chemoprevention. Cancer Metastasis Rev 21(3–4):257–264
Oswald WJ (1987a) Micro-algae and waste-water treatment. In: Borowitzka MA, Borowitzka LJ (eds) Micro algal biotechnology. Cambridge University Press, New York, pp 305–328
Oswald WJ (1987b) Large scale algal culture systems. In: Borowitzka MA, Borowitzka LJ (eds) Micro algal biotechnology. Cambridge University Press, New York, pp 357–394
Pulz O, Gross W (2004) Valuable products from biotechnology of microalgae. Appl Microbiol Biotechnol 65:635–648
Renn EW (1984) Agar and agarose: the indispensible partners in biotechnology. I&EC Res Devel 23:17–21
Richmond A (2004) Principles for attaining maximal microalgal productivity in photobioreactors: an overview. Hydrobiologia 512:33–37
Sheehan John, Dunahay Terri, Benemann John, Roessler Paul (1998) A look back at the US department of energy's aquatic species program: biodiesel from algae, vol 328. National Renewable Energy Laboratory, Golden
Shelef G (1982) High-rate algae ponds for waste water treatment and protein production. Water Sci Technol 14:439–452
Simopoulos AP (1991) Omega-3 fatty acids in health and disease and in growth and development. Am J Clin Nutr 54(3):438–463
Vonshak A (1997) Outdoor mass production of Spirulina: the basic concept. In: Vonshak A (ed) Spirulina platensis (Arthrospira): physiology, cell-biology and biotechnology. Taylor & Francis, London, pp 79–99
Ward OP, Singh A (2005) Omega-3/6 fatty acids: alternative sources of production. Process Biochem 40:3627–3652
Wen ZY, Chen F (2003) Heterotrophic production of eicosapentaenoic acid by microalgae. Biotech Adv 21:273–294
Wiley PE, Campbell JE, McKuin B (2011) Production of biodiesel and biogas from algae: a review of process train options. Water Environ Res 83(4):326–338
Yetir JZ (1988) Clinical application of fish oils. J Amer Med Assoc 260:665–670

Part II
Biomass Biology

Chapter 5
Regional Gene Pools for Restoration, Conservation, and Genetic Improvement of Prairie Grasses

Andrew R. Jakubowski and Michael D. Casler

Abstract Switchgrass (*Panicum virgatum*), big bluestem (*Andropogon gerardii*), and Indiangrass (*Sorghastrum nutans*) are native warm-season grasses that have been identified as potential cellulosic bioenergy feedstock crops due to their potential for high yields, perennial life habit, and nutrient use efficiency. This chapter outlines the role that improved cultivars and unimproved locally collected ecotypes can play in meeting agronomic and conservation goals. Improved cultivars grown for use as a bioenergy feedstock will be established in areas where introgression will occur with native populations. The concerns regarding the introgression of transgenes or non-adaptive alleles are outlined along with several avenues for mitigating these concerns. The agronomic and breeding history of each species is reviewed, as well as their importance in the conservation and restoration efforts of the prairie ecosystems of North America. We argue that both improved and locally collected ecotypes can coexist on the landscape and help to jumpstart the shift to a bioenergy based economy that provides sufficient biomass to meet cellulosic bioenergy goals, restore native ecosystems, and provide an array of regulating, cultural, and supporting ecosystem services while increasing the sustainability of agriculture.

A. R. Jakubowski (✉)
Department of Agronomy, University of Wisconsin-Madison, 1575 Linden Dr., Madison, Wisconsin 53706, USA
e-mail: Jakubowski@wisc.edu

M. D. Casler
USDA-ARS, U.S. Dairy Forage Research Center, 1925 Linden Dr., Madison, Wisconsin 53706, USA

M. D. Casler
DOE-Great Lakes Bioenergy Research Center, University of Wisconsin-Madison, 445 Henry Mall, Madison, Wisconsin 53706, USA

M. C. McCann et al. (eds.), *Plants and BioEnergy*, Advances in Plant Biology 4,
DOI: 10.1007/978-1-4614-9329-7_5,

Keywords Big bluestem • Cellulosic bioenergy • Conservation biomass • Cultivars • Indiangrass • Introgression • Local ecotypes • Plant adaptation regions • Plant breeding • Switchgrass

5.1 Introduction

Perennial warm-season grasses have been identified as preferred cellulosic bioenergy feedstock crops in temperate environments due to their high potential annual yields and high water and nutrient use efficiency (Byrt et al. 2011). Their perennial life habit eliminates the need for annual tillage and establishment, reducing establishment costs and greatly increasing their carbon sequestration potential (Adler et al. 2007). In addition, their annual senescence cycle results in retranslocation of aboveground nutrients to belowground organs, resulting in reduced nutrient removal when plants are harvested following senescence (Clark 1977). While non-native grasses have tremendous potential as a feedstock (*Miscanthus x giganteus*, in particular), this chapter will focus on the role native grass species can play in achieving agronomic and conservation goals related to bioenergy.

This chapter will focus on three native North American taxa, switchgrass (*Panicum virgatum* L.), big bluestem (*Andropogon gerardii* Vitman), and Indiangrass (*Sorghastrum nutans* (L.) Nash). These three species were the dominant grasses of the North American tallgrass prairie, with native ranges extending from the Rocky Mountains to the Atlantic Ocean, north to central Canada and south to the Gulf of Mexico (Fig. 5.1). These three species were dominant grasses in central North America, but today more than 80 % of prairie has been converted to other land uses (primarily agriculture), with tallgrass prairie reduced by as much as 99 % in some states or provinces (Samson and Knopf 1994). These remnant prairie sites remain in unplowed areas, but have become increasingly

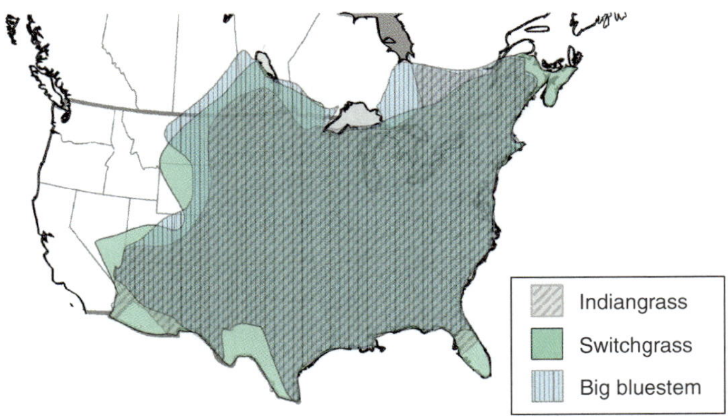

Fig. 5.1 Map of estimated native range of *switchgrass*, *big bluestem*, and *Indiangrass* in Canada and the USA. Data adapted from Barkworth et al. (2003)

isolated from one another (Simberloff and Gotelli 1984). To date, the vast majority of planting of these species has been for prairie restoration. While these efforts have grown exponentially over the past 20 years, they have reclaimed only a fraction of the land once in prairie (Anderson 2009). The use of native warm-season grasses for use as a bioenergy feedstock will greatly expand the presence of these species on the landscape. We argue that the planting and breeding of these three species, with an understanding of their geographic and genetic context, can meet agronomic and conservation goals while making these species common once again in North America.

5.2 Species Background

5.2.1 Switchgrass

Considerable interest has been shown in using switchgrass as cellulosic bioenergy feedstock crop. Switchgrass was identified in part because of its favorable traits and in part due to serendipity (Parrish et al. 2012). Of the three species discussed within this chapter, switchgrass has been the focus of the majority of research efforts. There are two primary taxonomic groups of switchgrass, upland and lowland ecotypes. Upland ecotypes are adapted to northern regions and consist of primarily two cytotypes. Upland octoploids ($2n = 8x = 72$) are the most common ploidy for wild and cultivated accessions, while tetraploids ($2n = 4x = 36$) are slightly less frequent (Zhang et al. 2011b). Plants ranging from 2x to 12x have been identified, but these cytotypes are rare (Costich et al. 2010). Lowland ecotypes are adapted to southern regions, are later flowering, and generally more productive than upland ecotypes. Lowland ecotypes are predominantly tetraploids, but several octoploid accessions have recently been confirmed (Zalapa et al. 2011; Zhang et al. 2011a). The center of diversity is in the Gulf Coast region and analysis of population structure suggests relatively large interbreeding populations occurring at the scale of hundreds of kilometers (Zhang et al. 2011b).

Yields of switchgrass vary widely depending upon the region and the ecotype grown. A meta-analysis of 39 field trials found an average yield of upland ecotypes of 8.7 ± 4.2 and 12.9 ± 5.9 Mg/ha in lowland ecotypes (Wullschleger et al. 2010).

5.2.2 Big Bluestem

Big bluestem has drawn interest as a potential biofuel crop recently (Anderson et al. 2008; Zhang et al. 2012). The species is primarily found as a hexaploid ($2n = 6x = 60$) with enneaploid ($2n = 9x = 90$) less common (Norrmann et al. 1997).

Hybridization between the two cytotypes is possible and aneuploids appear to be common (Keeler 1992). Established plants grown in agronomic settings have been identified as tall as 3 m, suggesting there is great potential for producing high yielding stands of bluestem (Stubbendieck et al. 1991). The species has also been shown to have higher fermentability potential when compared to other warm-season grass species (Jung and Vogel 1992; Weimer and Springer 2007). In agriculturally managed systems, big bluestem was found to produce between 9.9 and 15.7 Mg/ha of biomass depending on fertilization in Iowa, USA (Hall et al. 1982). In an evaluation of productivity in unmanaged systems in the Great Plains, big bluestem was found to produce more than twice as much biomass as either switchgrass or Indiangrass (Epstein et al. 1998). Big bluestem was also reported to have produced twice as much biomass per unit of applied nitrogen than switchgrass and Indiangrass (Perry and Baltensperger 1979). Recent work found three overlapping gene pools in the species from the Great Plains, Midwest, and Northeast USA, suggesting large interbreeding populations similar to those found in switchgrass (Price et al. 2012).

5.2.3 Indiangrass

Indiangrass has seen considerably less interest as a bioenergy feedstock. However, there has been research into its productivity and the heritability of traits for use as a forage crop. Three cytotypes have been identified ($2n = 2x = 20$, $2n = 4x = 40$, and $2n = 6x = 60$) with the tetraploid cytotype most common (Gould 1975; Riley and Vogel 1982). Biomass yields of three newly released Indiangrass varieties ('Chief', 'Scout', and 'Warrior') averaged 10.2 Mg/ha when grown in Nebraska, USA (Vogel et al. 2010). Heritability estimates for yield and in vitro dry matter digestibility (IVDMD) were calculated to be 0.43 and 0.42, respectively (Vogel et al. 1981). There have been no large-scale evaluations of the population structure of the species using modern genetic markers, but the similarity in geographic distribution and reproductive traits with switchgrass and big bluestem suggests there may be a similar population structure in Indiangrass. A comparison of switchgrass, Indiangrass, and big bluestem in Iowa, USA found Indiangrass and switchgrass sward plots to have similar productivities, while big bluestem produced significantly less (Wilsey 2010). Additional research is necessary to evaluate germplasm to determine the yield potential of Indiangrass outside of the Great Plains, USA.

5.3 Breeding

There are many potential benefits to using perennial warm-season grasses for biomass production, but all three species have similar characteristics in need of improvement. Establishment of all crops is slow and yields are a fraction of their potential yields during the establishment season (Parrish and Fike 2005). The seeds of all three species maintain dormancy and require up to a year of storage to achieve optimal germination (Beckman et al. 1993; Emal and Conard 1973). These species also emerge later in the spring than cool-season plants, creating potential weed pressure and an incomplete use of the growing season (Sanderson et al. 2004). All three species have seed shattering habits that will reduce the efficiency of breeding and large-scale seed production. However, sustained breeding efforts should be capable of improving all of these traits.

Previous breeding efforts in these species suggest that there is potential for improvements in yield and quality characteristics, with the majority of work occurring in switchgrass. Yields are expected to increase by 50–100 % through a combination of conventional breeding, genetic modification, and the development of hybrid switchgrass (crosses between upland and lowland ecotypes) in the next 10–20 years (Casler 2010; Martinez-Reyna and Vogel 2008; Vogel and Mitchell 2008). Evaluations of genetic variability for favorable bioenergy traits appear to be high in switchgrass (Das et al. 2004; Rose et al. 2008). The early recognition of these three species as potential forage crops has resulted in significant collection of germplasm for breeding that is publically available within the USDA Germplasm Resources Information Network (GRIN) system. As of November 2012, there are 332, 1,188, and 54 accessions in GRIN for switchgrass, big bluestem, and Indiangrass, respectively. Additional collections of the three species from remnant prairies are underway throughout the USA.

This tremendous increase in potential productivity can make the economics of growing perennial grasses for bioenergy much more favorable (Perrin et al. 2008). Previous breeding efforts in native warm-season grasses have proven successful, although programs have focused on improving forage quality and nutritional value for animal grazing until recent efforts in switchgrass for bioenergy (Anderson et al. 2008; Mitchell et al. 2005; Vogel et al. 2010). Cultivars of all three species have been released with improved forage quality using phenotypic selection of IVDMD (Hopkins et al. 1993; Mitchell et al. 2005; Vogel et al. 2006; Vogel et al. 2010). These cultivars also show minor improvements in yield. Only switchgrass has had improved varieties specifically for use as a bioenergy feedstock crop released by public breeders (Burns et al. 2010; Wu and Taliaferro 2009) and at least one private company (Ceres, Inc Thousand Oaks, CA).

The use of genetic modification to improve plants for use as a bioenergy feedstock has been called imperative for making bioenergy crops viable (Gressel 2008). Indeed, the down-regulation of genes involved in lignification improved ethanol yield in switchgrass by up to 38 % with significantly reduced pretreatment requirements and conversion costs (Fu et al. 2011). However, field testing of these varieties is required to confirm their improvement as conventional breeding for lower lignin in switchgrass and other perennial crops has been shown to dramatically reduce plant fitness (Casler 1997; Casler et al. 2002). Whether herbicide-resistance transgenes will be incorporated into publically available varieties in the future is unknown; however, herbicide resistant lines of switchgrass have been developed (Song et al. 2012).

5.4 Gene Flow Concerns and the Role of Unimproved and Local Varieties

While breeding efforts will be important for improving yield and quality in these species, there are significant concerns regarding the breeding and introduction of improved native perennial crops (Kwit and Stewart 2012; Lonsdale and FitzGibbon 2011). Because these species are native throughout North America, there is fear that improved or introduced populations will introgress with native populations and result in reduced fitness of wild populations or the development of novel invasive populations (Lesica and Allendorf 1999; Selbo and Snow 2005; Byrne and Stone 2011). Many of the traits identified as targets of breeding programs, such as rapid growth rates, plasticity across a range of environments, high yield, and cold and drought tolerance, have been identified as traits associated with increased invasion potential (Heaton et al. 2008; van Kleunen et al. 2010; Mclaughlin et al. 1999; Raghu et al. 2006; Sakai et al. 2001; Theoharides and Dukes 2007). These species also have weedy relatives with which hybridization may be possible (Ahrens et al. 2011). If genetically modified varieties of these species are released (particularly with herbicide resistance), this concern of introgression into weedy relatives is magnified further (Bagavathiannan et al. 2010; Stewart et al. 2003; Warwick et al. 2009).

Much work remains in evaluating the pollen flow dynamics of these species. The closest analog to pollen flow studies was undertaken with genetically modified creeping bentgrass (*Agrostis stolonifera*, Watrud et al. 2004; Zapiola et al. 2008). This species is a wind-pollinated perennial that is highly outcrossing and has a synchronous flowering period with weedy relatives with which hybridization is possible (Belanger et al. 2003). Systematic dispersal studies using sentinel plants recovered herbicide resistance seedlings at a distance of up to 21 km (Watrud et al. 2004). A follow up study collected plants from a 4.8 km area surrounding the initial planting area identified nine transgenic plants out of 20,400 tested (Zapiola et al. 2007). Switchgrass, Indiangrass, and big bluestem are likely to disperse pollen over longer distances than creeping bentgrass due to their taller morphology and higher pollen production (Kausch et al. 2010). Recent work has shown that switchgrass pollen can remain viable under ideal field conditions for up to 2.5 h (Ge et al. 2011).

There is little previous knowledge on which to assess the risks of creating novel invasive species through conventional breeding. One example used reed canarygrass (*Phalaris arundinacea* L.) as a model for risk assessment of breeding native perennial crops (Jakubowski et al. 2011). Reed canarygrass is native to North America and Eurasia, but is considered a noxious wetland invader throughout much of North America. The species is also planted as a forage crop and improved varieties have been released in the North American marketplace. Cultivation had been suggested as the origin of the development of invasiveness in the species (Merigliano and Lesica 1998). However, Jakubowski et al. (2011) showed that cultivars were more productive and fecund than wild

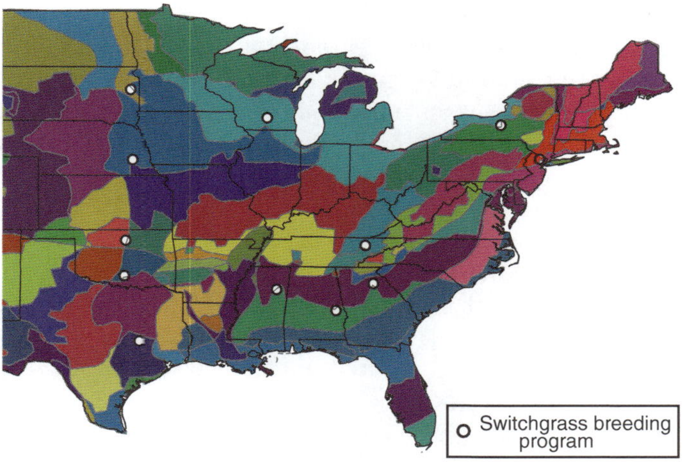

Fig. 5.2 Plant adaptation regions (PARs) to determine regional gene pools based on climatic and geographic data. The locations of the 12 public switchgrass breeding programs are identified by the *open circles*. Map developed using the methods of Vogel et al. (2005)

Eurasian populations in the environments for which they were selected (uplands with nitrogen addition), but no more productive in the environment in which they are considered invasive (wetlands). The authors conclude that breeding *per se* is not responsible for invasion by the species, but the widespread planting of reed canarygrass for forage and soil stabilization made the species common on the landscape.

These concerns, along with the growing chorus for the recognition of the additional ecosystem services beyond production provided by the use of perennial grasses in agriculture, encourage the use of locally collected ecotypes of native grasses (Boody et al. 2005; Gasparatos et al. 2011). Locally collected seed is defined as unimproved germplasm collected from remnant sites within a specified distance of the planting site. The geographic scale at which seed is considered local varies by species (McKay et al. 2005), but should be determined by genetic analysis as the scale at which interbreeding populations of a species exist. Vogel et al. (2005) developed general guidelines for determining the geographic scale of local germplasm using climatic and geographical variables by combining Bailey's Ecoregions with the USDA Plant Hardiness Zone maps. These regions were named Plant Adaptation Regions (PARs) and were developed to guide collection and selection of germplasm for natural area restoration when resources are unavailable for genetic analysis (Fig. 5.2). For the obligate out-crossing and wind-pollinated grasses discussed here, local seed would be collected within approximately 500 km of a site. As mentioned earlier, up to 99 % of land originally in tallgrass prairie has been converted to other land uses. The planting of locally collected varieties of these species would be an economically valuable way of maintaining genetic resources on the landscape. Most cultivars and local collections are widely adapted within one hardiness zone north and south of their origin (Casler 2012), which includes multiple plant adaptation regions. Thus, the concept of "local" can be quite broad, covering a broad geographic area for many cultivars. Public breeding programs can play an important role in conserving germplasm and releasing locally adapted varieties. The 12 programs involved in switchgrass breeding are scattered throughout the native range of the species and test populations in many of the PARs (Fig. 5.2).

Further, collection and use of locally collected varieties may prove to be more productive than improved varieties from distant locations. Numerous studies comparing production of local and non-local varieties in warm-season grasses have found that locally collected varieties are more productive than non-local ecotypes (Casler 2005; Casler et al. 2004; Gustafson et al. 2004; McMillan 1964, 1965). The breeding of perennial grasses is a slow process that requires many years of selection and evaluation before an improved variety is released (Casler 2010; Yamada et al. 2005). This process is expected to accelerate with the increasing availability of genomic tools, but is still limited by the generation time of perennial crops. This process is further complicated by genotype by environment interactions and the wide range of marginal lands upon which perennial grasses will be established. The development of improved varieties in which improvements are stable across a wide range of environments has proven difficult (Eberhart and Russell 1966; Simmonds 1991). A comparison of productivity and quality of locally collected wild accessions and cultivars (developed in Nebraska) of big bluestem found no consistent differences in production or grazing tolerance in pastures in Wisconsin, USA (Chamberlain et al. 2012). Emerging work in understanding interactions between plant and mychorrizal fungi provenance suggests that mismatched plant and fungal communities may have major impacts on plant production, particularly in nutrient limited environments (Collins-Johnson et al. 2010, Klironomos 2003). A 'home-field advantage' has been identified in a range of plant species when grown in the same soils from which they were collected (Smith et al. 2012; Whitham et al. 2012).

Increasingly, the use of biomass harvest in conservation and restoration has been identified as a sound management tool (Jakubowski et al. 2010; Gasparatos et al. 2011; Hull et al. 2011). Land managers often have limited resources to conduct intensive management, such as the use of burning and herbicide to control undesirable species. Partnerships between land managers and local farmers to harvest biomass from these lands historically considered closed to agricultural production are becoming more common. The farmer harvests biomass and is allowed to retain any benefits gained from the sale or processing of biomass, while the land manager is receiving weed control and nutrient management at no cost. The value of the biomass serves as an economic incentive to encourage restoration and conservation for protection of soil, water, and faunal resources. The development of best management practices with a concrete basis in conservation has been developed to assure appropriate use of these lands (Ventura et al. 2012). Because these lands are not managed for maximum agricultural production, the use of the highest yielding variety is not necessary, nor appropriate. In addition, these types of lands will be important in providing cellulosic biomass without establishment costs during the early phases of a shift to a bioenergy economy.

5.5 Coexistence of Improved Varieties and Conservation Priorities

There is room for the coexistence of improved varieties (including genetic modification) of native warm-season grasses and native populations. The development of improved cultivars with significant gains in yield and ease of establishment will be critical in realizing the economic and ecological benefits of harvesting perennial biomass crops for bioenergy. Yield gains achieved through breeding may prove to be the most important factor in making bioenergy crops economically feasible. However, concerns regarding development of novel invasive species and gene flow from improved or GM varieties into wild populations are well founded. Thoughtful policy and strategies are required to minimize these risks (Kausch et al. 2010).

Several agronomic and breeding strategies can also minimize these risks. Breeding often selects improvements for a very specific target population of environments, and stability across a wide range of environments has proved elusive for even the most highly bred crops (Simmonds 1991). Selecting cultivars for specific environments or requiring specific management (i.e. high nitrogen environments or cultivation) reduces the probability that a crop will escape cultivation. Breeders generally strive for cultivar stability across a range of environments, as these cultivars are more marketable and simplify selection for a producer. In contrast to typical cultivar development, a range of cultivars that are each designed for a specific environment may be preferable for bioenergy feedstock crops. In addition, the evaluation of crops in environments in which they have the potential to be invasive may be appropriate to select against the populations that are highly productive in these environments. This will require collaboration between breeders and weed scientists to determine which traits and environments are most important.

Another safeguard to reduce the chance of creating a novel invader is to select for reduced fecundity of perennial crops. Efforts to select for slowed maturity and reduced flowering to improve forage quality have been successful in several perennial pasture species (Buxton 1996; Casler et al. 2004). In addition, research to reduce or eliminate flowering by altering the genetic mechanisms involved in the flowering pathway has shown promise (Salehi et al. 2005; Jung and Müller 2009). This technique not only reduces the fecundity of the crop but can also increase biomass production due to a reallocation of resources. Selecting for reduced fecundity in biomass crops has the potential to reduce the risks of introgression of improved genes or traits into native populations and the escape of improved varieties into undesirable environments. However, this will reduce the ability to produce seed of new cultivars. Alternatively, the development of winter hardy lowland varieties of switchgrass is underway due to their significantly higher yields over upland varieties when grown at Northern latitudes (Casler 2012). Lowland varieties grown at Northern latitudes have a 4–6 week delay in flowering time compared to upland varieties. This *de facto* reproductive barrier should reduce the introgression of improved genes or traits into upland native populations in Northern latitudes, but not in Southern latitudes where native populations of lowland plants exist. The development of male-sterility systems would complicate breeding efforts, but would eliminate pollen dispersal and reduce the risk of transgene escape (Kausch et al. 2010).

5.6 Conclusion

The use of both improved and locally collected varieties can coexist on the landscape and help to jumpstart the shift to a bioenergy based economy that provides sufficient biomass to meet cellulosic bioenergy goals, restore native ecosystems, and provide an array of regulating, cultural, and supporting ecosystem services while increasing the sustainability of agriculture. Breeders should appreciate the concerns of the conservation community regarding the introgression of exotic germplasm into native populations and incorporate appropriate safeguards to reduce these risks into their breeding programs. Conservationists should appreciate the importance of breeding in enabling a shift to a bioenergy economy and the multitude of ecosystem services associated with such a shift. Most importantly, the communities should work together to develop strategies that appropriately utilize germplasm, protect natural ecosystems, and meet agronomic goals.

References

Adler PR, Del Grosso SJ, Parton WJ (2007) Life-cycle assessment of net greenhouse-gas flux for bioenergy cropping systems. Ecol Appl 17:675–691

Ahrens C, Ecker G, Auer C (2011) The intersection of ecological risk assessment and plant communities: an analysis of Agrostis and Panicum species in the northeastern U.S. Plant Ecol 212:1629–1642

Anderson R (2009) History and progress of ecological restoration in tallgrass prairie. In: Taylor C, Taft J, Warwick C (eds) INHS special publication 30: canaries in the catbird seat. University of Illinois, Champaign-Urbana, pp 217–228

Anderson W, Casler MD, Baldwin B (2008) Improvement of perennial forage species as feedstock for bioenergy. In: Vermerris W (ed) Genetic improvement of bioenergy crops. Springer, New York, pp 309–346

Bagavathiannan M, Spök A, Van Acker R (2010) Commercialization of perennial GE crops: looming challenges for regulatory frameworks. J Agric Environ Ethics 24:227–242

Barkworth M, Capels K, Long S, Piep M (2003) Flora of North America, vol. 25. Magnoliophyta: Commelinidae (in part). http://herbarium.usu.edu/webmanual

Beckman JJ, Moser LE, Kubik K, Waller SS (1993) Big bluestem and switchgrass establishment as influenced by seed priming. Agron J 85:199–202

Belanger FC, Meagher TR, Day PR et al (2003) Interspecific hybridization between *Agrostis stolonifera* and related *Agrostis* species under field conditions. Crop Sci 43:240–246

Boody G, Vondracek B, Andow DA et al (2005) Multifunctional agriculture in the United States. Bioscience 55:27–38

Burns JC, Godshalk EB, Timothy DH (2010) Registration of "colony" lowland switchgrass. J Plant Registr 4:189–194

Buxton DR (1996) Quality-related characteristics of forages as influenced by plant environment and agronomic factors. Anim Feed Sci Technol 59:37–49.

Byrne M, Stone L (2011) The need for "duty of care" when introducing new crops for sustainable agriculture. Curr Opin Environ Sustain 3:50–54

Byrt CS, Grof CPL, Furbank RT (2011) C4 plants as biofuel feedstocks: optimising biomass production and feedstock quality from a lignocellulosic perspective. J Integr Plant Biol 53:120–135

Casler MD (1997) Breeding for improved forage quality: potentials and problems. In: Proceedings of the 18th international grassland congress. Winnipeg, Manitoba and Saskatoon, Saskatchewan (Canada), pp 323–330

Casler MD (2005) Ecotypic variation among switchgrass populations from the northern USA. Crop Sci 398:388–398

Casler MD (2010) Changes in mean and genetic variance during two cycles of within-family selection in switchgrass. Bioenergy Res 3:47–54

Casler MD (2012) Switchgrass breeding, genetics, and genomics. In: Monti A (ed) Switchgrass: a valuable biomass crop for energy. Springer, London, pp 29–54

Casler MD, Buxton DR, Vogel KP (2002) Genetic modification of lignin concentration affects fitness of perennial herbaceous plants. Theor Appl Gen 104:127–131

Casler MD, Vogel K, Taliaferro C (2004) Latitudinal adaptation of switchgrass populations. Crop Sci 44:293–303

Chamberlain SK, Paine LK, Harrison JL, Jackson RD (2012) Tradeoffs in performance of native warm-season grass cultivars and locally harvested seed managed for wildlife habitat or livestock production. Agron J 104:1383

Clark FE (1977) Internal cycling of nitrogen in shortgrass prairie. Ecology 58:1322–1333

Collins Johnson N, Wilson GWT, Bowker MA et al (2010) Resource limitation is a driver of local adaptation in mycorrhizal symbioses. Proc Natl Acad Sci USA 107:2093–2098

Costich DE, Friebe B, Sheehan MJ et al (2010) Genome-size variation in switchgrass (*Panicum virgatum*): flow cytometry and cytology reveal rampant aneuploidy. Plant Genome 3:130–141

Das MK, Fuentes RG, Taliaferro CM (2004) Genetic variability and trait relationships in switchgrass. Crop Sci 44:443–448

Eberhart SA, Russell WA (1966) Stability parameters for comparing varieties. Crop Sci 6:36–40

Emal J, Conard E (1973) Seed dormancy and germination in Indiangrass as affected by light, chilling, and certain chemical treatments. Agron J 65:383–385

Epstein H, Lauenroth W, Burke I, Coffin D (1998) Regional productivities of plant species in the Great Plains of the United States. Plant Ecol 134:173–195

Fu C, Mielenz JR, Xiao X et al (2011) Genetic manipulation of lignin reduces recalcitrance and improves ethanol production from switchgrass. Proc Natl Acad Sci USA 108:3803–3808

Gasparatos A, Stromberg P, Takeuchi K (2011) Biofuels, ecosystem services and human wellbeing: Putting biofuels in the ecosystem services narrative. Agric Ecosys Environ 142:111–128

Ge Y, Fu C, Bhandari H et al (2011) Pollen viability and longevity of switchgrass. Crop Sci 51:2698–2705

Gould FW (1975) The grasses of texas. Texas A&M University Press, College Station, TX

Gressel J (2008) Transgenics are imperative for biofuel crops. Plant Sci 174:246–263

Gustafson DJ, Gibson DJ, Nickrent DL (2004) Conservation genetics of two co-dominant grass species in an endangered grassland ecosystem. J Appl Ecol 41:389–397

Hall K, George J, Riedl R (1982) Herbage dry matter yields of switchgrass, big bluestem, and indiangrass with N fertilization. Agron J 74:47–51

Heaton EA, Flavell RB, Mascia PN et al (2008) Herbaceous energy crop development: recent progress and future prospects. Curr Opin Biotechnol 19:202–209
Hopkins AA, Vogel KP, Moore KJ (1993) Predicted and realized gains from selection for in vitro dry matter digestibility and forage yield in switchgrass. Crop Sci 33:253–258
Hull S, Arntzen J, Bleser C et al (2011) Wisconsin sustainable planting and harvest guidelines for nonforest biomass. Madison, WI
Jakubowski AR, Casler MD, Jackson RD (2010) The benefits of harvesting wetland invaders for cellulosic biofuel: an ecosystem services perspective. Restor Ecol 18:789–795
Jakubowski AR, Casler MD, Jackson RD (2011) Has selection for improved agronomic traits made reed canarygrass invasive? PLoS ONE 6:e25757
Jung C, Müller AE (2009) Flowering time control and applications in plant breeding. Trends in Plant Science 14:563–573
Jung H, Vogel K (1992) Lignification of switchgrass (*Panicum virgatum*) and big bluestem (*Andropogon gerardii*) plant parts during maturation and its effect on fibre degradability. J Sci Food Agric 59:169–176
Kausch A, Hague J, Oliver M, Li Y (2010) Transgenic perennial biofuel feedstocks and strategies for bioconfinement. Biofuels 1:163–176
Keeler KH (1992) Local polyploid variation in the native prairie grass *Andropogon gerardii*. Amer J Bot 79:1229–1232
Klironomos J (2003) Variation in plant response to native and exotic arbuscular mycorrhizal fungi. Ecology 84:2292–2301
Kwit C, Stewart C (2012) Gene flow matters in switchgrass (*Panicum virgatum* L.), a potential widespread biofuel feedstock. Ecol Appl 22:3–7
Lesica P, Allendorf FW (1999) Ecological genetics and the restoration of plant communities: mix or match? Restor Ecol 7:42–50
Lonsdale W, FitzGibbon F (2011) The known unknowns—managing the invasion risk from biofuels. Curr Opin Environ Sustain 3:31–35
Martinez-Reyna JM, Vogel KP (2008) Heterosis in switchgrass: spaced plants. Crop Sci 48:1312–1320
McKay JK, Christian CE, Harrison S, Rice KJ (2005) How local is local? A review of practical and conceptual issues in the genetics of restoration. Restor Ecol 13:432–440
Mclaughlin S, Bouton J, Bransby D, et al. (1999) Developing switchgrass as a new crop. In: Janick J (ed) Proceedings fourth national new crops symposium. ASHS Press, Alexandria, VA, pp 282–299
McMillan C (1964) Ecotypid differentiation within four North American prairie grasses. I. Morphological variation within transplanted community fractions. Amer J Bot 51:1119–1128
McMillan C (1965) Ecotypic differentiation within four North American prairie grasses. II. Behavioral variation within transplanted community fractions. Amer J Bot 52:55–65
Merigliano MF, Lesica P (1998) The native status of reed canarygrass (*Phalaris arundinacea* L.) in the Inland Northwest. USA. Nat Areas J 18:223–230
Mitchell RB, Vogel KP, Klopfenstein TJ et al (2005) Grazing evaluation of big bluestems bred for improved forage yield and digestibility. Crop Sci 45:2288–2292
Norrmann G, Quarin C, Keeler K (1997) Evolutionary implications of meiotic chromosome behavior, reproductive biology, and hybridization in 6x and 9x cytotypes of Andropogon gerardii (Poaceae). Amer J Bot 84:201–207
Parrish DJ, Fike JH (2005) The biology and agronomy of switchgrass for biofuels. Crit Rev Plant Sci 24:423–459
Parrish DJ, Casler MD, Monti A (2012) The evolution of switchgrass as an energy crop. In: Monti A (ed) Switchgrass. Springer, London, pp 1–28
Perrin R, Vogel K, Schmer M, Mitchell R (2008) Farm-scale production cost of switchgrass for biomass. Bioenergy Res 1:91–97
Perry L, Baltensperger D (1979) Leaf and stem yields and forage quality of three N-fertilized warm-season grasses. Agron J 71:355–358

Price DL, Salon PR, Casler MD (2012) Big bluestem gene pools in the Central and Northeastern United States. Crop Sci 52:189–200
Raghu S, Anderson RC, Daehler CC et al (2006) Adding biofuels to the invasive species fire. Science 313:1742
Riley R, Vogel KKP (1982) Chromosome numbers of released cultivars of switchgrass, indiangrass, big bluestem, and sand bluestem. Crop Sci 22:1082–1083
Rose LW, Das MK, Taliaferro CM (2008) Estimation of genetic variability and heritability for biofuel feedstock yield in several populations of switchgrass. Ann Appl Biol 152:11–17
Sakai AK, Allendorf FW, Holt JS et al (2001) The population biology of invasive species. Annu Rev Ecol System 32:305–332
Salehi H, Ransom CB, Oraby HF, et al. (2005) Delay in flowering and increase in biomass of transgenic tobacco expressing the Arabidopsis floral repressor gene FLOWERING LOCUS C. J Plant Physiol 162:711–717
Samson F, Knopf F (1994) Prairie conservation in North America. Bioscience 44:418–421
Sanderson MA, Schnabel RR, Curran WS et al (2004) Switchgrass and big bluestem hay, biomass, and seed yield response to fire and glyphosate treatment. Agron J 96:1688–1692
Selbo SM, Snow AA (2005) Flowering phenology and genetic similarity among local and recently introduced populations of Andropogon gerardii in Ohio. Restor Ecol 13:441–447
Simberloff D, Gotelli N (1984) Effects of insularisation on plant species richness in the prairie-forest ecotone. Biol Conserv 29:27–46
Simmonds NW (1991) Selection for local adaptation in a plant-breeding program. Theor Appl Genet 82:363–367
Smith DS, Schweitzer JA, Turk P et al (2012) Soil-mediated local adaptation alters seedling survival and performance. Plant Soil 352:243–251
Song G, Walworth A, Hancock JF (2012) Factors influencing Agrobacterium-mediated transformation of switchgrass cultivars. Plant Cell, Tiss Org Cult 108:445–453
Stewart CN, Halfhill MD, Warwick SI (2003) Transgene introgression from genetically modified crops to their wild relatives. Nat Rev Genet 4:806–817
Stubbendieck J, Hatch S, Butterfield C (1991) North American range plants. University of Nebraska Press, Lincoln, NE
Theoharides KA, Dukes JS (2007) Plant invasion across space and time: Factors affecting nonindigenous species success during four stages of invasion. New Phytol 176:256–273
Van Kleunen M, Weber E, Fischer M (2010) A meta-analysis of trait differences between invasive and non-invasive plant species. Ecol Lett 13:235–245
Ventura S, Hull S, Jackson R et al (2012) Guidelines for sustainable planting and harvest of non-forest biomass in Wisconsin. J Soil Water Conserv 67:17A–20A
Vogel KP, Mitchell RB (2008) Heterosis in switchgrass: biomass yield in swards. Crop Sci 48:2159–2164
Vogel K, Gorz H, Haskins F (1981) Heritability estimates for forage yield, in vitro dry matter digestibility, crude protein, and heading date in indiangrass. Crop Sci 21:35–38
Vogel KP, Schmer MR, Mitchell RB (2005) Plant adaptation regions: ecological and climatic classification of plant materials. Range Ecol Man 58:315–319
Vogel KP, Mitchell RB, Klopfenstein TJ, Anderson BE (2006) Registration of "Goldmine" big bluestem. Crop Sci 46:2314–2315
Vogel KP, Mitchell RB, Gorz HJ et al (2010) Registration of "Warrior", "Scout", and "Chief" Indiangrass. J Plant Registr 4:115
Warwick SI, Beckie HJ, Hall LM (2009) Gene flow, invasiveness, and ecological impact of genetically modified crops. Ann New York Acad Sci 1168:72–99
Watrud LS, Lee EH, Fairbrother A et al (2004) Evidence for landscape-level, pollen-mediated gene flow from genetically modified creeping bentgrass with CP4 EPSPS as a marker. Proc Natl Acad Sci USA 101:14533–14538
Weimer PJ, Springer TL (2007) Fermentability of eastern gamagrass, big bluestem and sand bluestem grown across a wide variety of environments. Biores Technol 98:1615–1621

Whitham TG, Gehring CA, Lamit LJ et al (2012) Community specificity: life and afterlife effects of genes. Trends Plant Sci 17:271–281

Wilsey BJ (2010) Productivity and subordinate species response to dominant grass species and seed source during restoration. Restor Ecol 18:628–637

Wu Y, Taliaferro C (2009) Switchgrass cultivar, cv "Cimarron." US Patent Number: US 2009/0300977 A1

Wullschleger SD, Davis EB, Borsuk ME et al (2010) Biomass production in switchgrass across the United States: database description and determinants of yield. Agron J 102:1158–1168

Yamada T, Forster JW, Humphreys MW, Takamizo T (2005) Genetics and molecular breeding in Lolium/Festuca grass species complex. Grassl Sci 51:89–106

Zalapa JE, Price DL, Kaeppler SM et al (2011) Hierarchical classification of switchgrass genotypes using SSR and chloroplast sequences: ecotypes, ploidies, gene pools, and cultivars. Theor Appl Genet 122:805–817

Zapiola M, Mallory-Smith C, Thompson J, et al. (2007) Gene escape from glyphosate-resistant creeping bentgrass fields: past, present, and future. In: Proceedings of the western society of weed science, p 18

Zapiola ML, Campbell CK, Butler MD, Mallory-Smith CA (2008) Escape and establishment of transgenic glyphosate-resistant creeping bentgrass *Agrostis stolonifera* in Oregon, USA: a 4-year study. J Appl Ecol 45:486–494

Zhang Y, Zalapa JE, Jakubowski AR et al (2011a) Natural hybrids and gene flow between upland and lowland switchgrass. Crop Sci 51:2626–2641

Zhang Y, Zalapa JE, Jakubowski AR et al (2011b) Post-glacial evolution of *Panicum virgatum*: centers of diversity and gene pools revealed by SSR markers and cpDNA sequences. Genetica 139:933–948

Zhang K, Johnson L, Nelson R et al (2012) Chemical and elemental composition of big bluestem as affected by ecotype and planting location along the precipitation gradient of the Great Plains. Industr Crops Prod 40:210–218

Chapter 6
Mining Genetic Diversity of Sorghum as a Bioenergy Feedstock

Cynthia M. B. Damasceno, Robert E. Schaffert and Ismail Dweikat

Abstract Sorghum is a drought-tolerant rainfed crop that requires about 30 % less nitrogen fertilizer than corn to produce equal amount of ethanol per acre under non-irrigated conditions. Excellent genetic and genomic resources exist for improvement of sorghum as a bioenergy source. We expect a huge impact on biomass yield, quality, and conversion efficiency with appropriate plant breeding and biotechnology tools in order to develop energy sorghum germplasm that allows highly efficient production of biofuel. The outlined improvement should produce benefits that include: (1) genetic improvement of a biomass crop with significantly reduced overall cost of biomass-to-ethanol conversion; (2) selection of a reliable bioenergy feedstock that is drought tolerant, inexpensive to grow, environmentally friendly and cultivated in nearly all temperate and tropical climate regions; (3) expansion of the production area for bioenergy crops by developing cold tolerance germplasm and hybrids and by offering both annual and perennial sweet sorghum types; and (4) reduction in cell wall lignin for improved efficiency in production of biofuels.

C. M. B. Damasceno · R. E. Schaffert
Embrapa Maize and Sorghum, Caixa Postal 151 Rodovia MG-424, Km 45,
34701-970 Sete Lagoas, MG, Brazil
e-mail: cynthia.damasceno@embrapa.br

R. E. Schaffert
e-mail: robert.schaffert@embrapa.br

I. Dweikat (✉)
Department of Agronomy and Horticulture, University of Nebraska, 202 Keim Hall,
Lincoln, NE 68583-0915, USA
e-mail: idweikat@unlnotes.unl.edu

M. C. McCann et al. (eds.), *Plants and BioEnergy*, Advances in Plant Biology 4,
DOI: 10.1007/978-1-4614-9329-7_6, © Springer Science+Business Media New York 2014

6.1 Introduction

Sorghum, [*Sorghum bicolor* (L.) Moench], is an excellent choice for a feedstock for bioenergy and bioproduct production as it is a very photosynthesis efficient species comparable to or more efficient than other C4 species such as sugarcane and elephant grass. Sweet sorghum can be used for first generation technology (G1) which involves the direct fermentation of the sugar produced in the stem and extracted in the juice, similar to sugarcane. Highly productive energy sorghum or biomass sorghum can be used as an efficient feedstock for second generation technology (G2) involving the hydrolysis of the cellulose and hemicellulose to simple sugars and fermentation to produce bioproducts and/or biofuels. Highly productive energy sorghum or biomass sorghum can also be used as a feedstock for third, fourth generation and beyond technologies for generating bioenergy and bioproducts. Also, a strong demand for energy sorghum is evolving in Brazil to burn the biomass for co-generation of electrical energy or the generation of steam in certain industrial processes.

The top ethanol producers in the world are the United States and Brazil. A research and development project to develop feedstock for G1 bioenergy was initiated by the Brazilian Enterprise for Agricultural Research (Embrapa) in Brazil in 1975 as a response to the demand of the "Pró-álcool" Program of Brazil, following the 1973 OPEC oil embargo. In response to the new energy demands Brazilian growing economy will require, the Brazilian government launched in 2006 the National Agroenergy Plan in order to stimulate research and production in the bioenergy sector. In the United States (US), the Energy Independence and Security Act of 2007 was established amending the renewable fuels standard (RFS)-Energy Policy Act of 2005, in order to reduce its dependence on foreign fossil energy supplies, reduce greenhouse gas emissions, and provide meaningful economic opportunity. By year 2050, the energy demands in the US alone are expected to be about 27 Terawatts per year (Lewis and Nocera 2006). Biomass feedstocks are expected to contribute to as much as 40 % of total renewable energy with expected biofuel production of 136 billion liters in the US by 2022 (EIA 2011). In response to these demands, R&D projects have been launched in recent years to develop adequate technology for the production of biofuels.

The advantage of plant-based biomass material lies in its photosynthetic ability. Nature has designed a sophisticated solar conversion system that self-assembles from water, nutrients in the soil and carbon dioxide (CO_2) in the air with energy input from the sun. Enhancing the production and conversion of plant biomass to utilizable sources of energy is critical to decrease dependency on fossil fuels, increase energy security, ameliorate negative environmental impacts associated with petroleum combustion (e.g., reducing greenhouse gas emissions and toxic pollutants) and will add significant value to domestic agricultural products. Converting plant biomass to ethanol represents a renewable form of energy that is immediately utilizable in existing combustion engines and transportation systems (Sommervile 2007; Vermerris et al. 2007).

Sorghum by origin is a short day species in its native areas of evolution in Tropical Africa or commonly called photosensitive (PS). In this case, flower induction is initiated when day length is less than 12 h and 15–20 min (Rooney 2007).

In the adaptation of sorghum to temperate environments, a mutation of the dominant Ma_1 maturity gene to ma_1 was selected to develop photo insensitive sorghum that will normally flower 60–75 days after germination. The sweet sorghum cultivars currently being developed in Brazil and the US are photo insensitive (PIS) and normally flower 60–80 days after germination and reach peak sugar concentration around physiological maturity of the grain at 120–130 days after germination or 40–70 days after flowering. The months of late September to late March in Brazil and the summer months in the US have days longer than 12 h and 20 min providing a long growing season, up to 8–9 months for energy sorghum. These months coincide with the rainy period in the regions of the South East and Central West of Brazil where the large distilleries are located and where the demand for sweet sorghum and energy sorghum is evolving. The utilization of both PS and PIS sorghum cultivars for bioenergy or bioproducts offers an array of opportunities for providing feedstock for a range of industrial processes and activities.

Critical factors in utilizing biomass as an alternative energy source will be the ability of the plant to achieve high biomass yields, grow in diverse climates under various environmental conditions, and be economically converted into a bio-based product. Synergistic improvements of biomass feedstocks, transportation and handling logistics, and the efficiency of industrial conversion into bioenergy are essential for sustainable production of bioenergy. The major requirements for sustainable biomass-based bioenergy production are high biomass yield, high energy content, lower agronomic input requirements and high suitability for the end-user defined processes, such as bio- or thermo-chemical conversion of biomass (Henry 2010).

Plant breeding is a cost-effective way to achieve an increased and stable yield. Plant breeding allows for continuous increase and release of ever more productive cultivars. In industrial terms, this increase will translate to, for example, higher sugar content and juiciness that will lower the cost of making ethanol. The development of multipurpose varieties will allow farmers to have additional markets for their product (not just ethanol, it can be a food or feed and fuel crop at the same time). The 'green revolution' for major cereals would not have been made possible without the release of outstanding varieties. A new green revolution will require also new outstanding energy crop cultivars. Important features of potential bioenergy feedstocks are; (1) the importance of deploying a non-food crop with (2) capability for high growth potential on both fertile and marginal lands with efficient water and nutrient utilization, (3) growth in a range of temperate to tropical climates, (4) amenability to production practices already in place for most growers, (5) male sterility for production of hybrid cultivars or non-flowering growth habit to maximize vegetative growth and minimize transgenic pollen escape, (6) genetic tractability and availability of genomic resources for detailed study, and (7) ability to supply carbon in a form that can be utilized within existing bioenergy production schemes.

Sweet sorghum fits in very well as an additional feedstock to sugarcane in the large distilleries in Brazil increasing the industrial processing period up to 60–90 days. Industrial processing of sugarcane begins in the month of April or May until mid-December. The limiting factors for sugarcane are low sugar content in April and high rainfall in November and December through January and February complicating harvest and transport of the crop. Sugar content also declines during the rainy season. Sweet sorghum can be planted at the beginning or the rainy season in November and December for harvest and processing in late February, March and April before the beginning of the sugarcane harvest in April and May in areas of Sugarcane renovation. Sugarcane renovation is recommended every five years in Brazil, thus generating a potential area of 20 % of the sugarcane acreage or 1.9 million hectares for sweet sorghum production between the months of October–May. Sustainable levels of ethanol production of 2,500–3,000 l ha^{-1} in these conditions have been attained in four months in pilot trials in Central Brazil (Embrapa Documentos 138). There are also other scenarios or niches for sweet sorghum, such as in the implementation of new distilleries, areas where sugarcane is restricted for environmental concerns, and in areas where rainfall is not sufficient for sugarcane production.

6.2 Sorghum: Excellent Genetic and Genomic Resources for Systems-Based Crop Improvement

Sorghum [*Sorghum bicolor* (L.) Moench] is the fifth most important cereal crop world-wide (http://apps.fao.org/default.jsp) as well as an important source of food, feed, fiber, and biofuel (Doggett 1988). Sorghum, like maize and sugarcane, carries out C4 photosynthesis, a specialization that makes these grasses well adapted to environments subject to high temperature and water limitation (Edwards et al. 2004). Different varieties of sorghum have been bred for different commodities. Sweet sorghum has traditionally been bred for high sugar content in the stem for producing molasses. These lines typically have high biomass and low seed yield. On the other hand, grain sorghum has been bred for high seed yield, has reduced plant biomass and low accumulation of sucrose in stem tissue. It is possible to cross these varieties to generate populations that segregate for the various traits. The gene for male sterility, Ms_3/ms_3 can be used for developing random mating sweet sorghum populations for use in selection and improvement. The cross of contrasting sweet sorghum lines to generate recombinant inbred lines (RILs) by single seed descent is currently being used to map traits related to sucrose accumulation and plant biomass at Embrapa.

Sorghum has cultivated varieties spread among 5 races and over 25 species of wild relatives, providing tremendous genetic diversity for crop improvement (Bhattacharya et al. 2011). Since its introduction to the Americas about 250 years ago, sorghum has been established as the third major cereal crop of both Brazil and the U.S. During the last century, conversion of sorghum genotypes to adapt to long day conditions has increased genetic diversity and greatly contributed to improved grain crop quality and productivity (Marguerat and Bahler 2010). However only about 1,000 of over 44,000 accessions of sorghum found at the USDA National Plant Germplasm System (USDA-NPGS) flower under long day conditions (photoperiod insensitive) suggesting that a vast untapped genetic diversity is available for crop improvement for bioenergy traits.

Sorghum is also an important target of genome analysis among the C4 grasses because the sorghum genome is relatively small (730 Mbp) (Paterson et al. 2009), the cultivated species is diploid (2n = 20) and the sorghum germplasm is diverse (Dje et al. 2000; Menz et al. 2004; Casa et al. 2005). As a consequence, numerous sorghum genetic, physical, and comparative maps have been constructed (Tao et al. 1998; Boivin et al. 1999; Peng et al. 1999; Klein et al. 2000; Haussmann et al. 2002; Menz et al. 2002; Bowers et al. 2003), a sorghum EST project (Pratt et al. 2005) and associated microarray analyses of sorghum gene expression have been carried out (Buchanan et al. 2005; Salzman et al. 2005). Also, a comprehensive analysis of sorghum chromosome architecture has been done (Kim et al. 2005), and an 8x draft sequence of the sorghum genome (about twice the size of rice) has been completed by the DOE Joint Genome Sequencing program (Paterson et al. 2009; http://www.phytozome.net/sorghum). In addition, genetic maps have been assembled at Texas A&M and University of Georgia (Menz et al. 2002; Bowers et al. 2003). Several projects using the method genotyping-by-sequencing (GBS) are underway at the Institute of Genetic Diversity at Cornell University, allowing large scale SNP discovery for thousands of sorghum materials (lines and populations). Embrapa has genotyped using GBS 275 F_7 RILs developed from the cross of two sweet sorghum varieties, Brandes and Wray and is in the process of phenotyping this population for both productivity and quality. In addition to this, Embrapa has phenotyped a diverse panel for biomass production and G2 quality which will be genotyped using GBS as well, allowing for rapid and high resolution QTL mapping of bioenergy traits. The whole genome sequence of sorghum BTx623 (Paterson et al. 2009) is a valuable reference for genome based discoveries in sorghum and existing and developing genomics efforts complement these genetic resources and allow genomics-based crop improvement for bioenergy traits.

6.3 Sorghum Has Been Identified as a Preferred Biomass Crop for Biofuels

Biomass can be used to generate electricity or to produce liquid transportation fuels. Among the various types of renewable fuels (such as wind, solar, and geothermal), biomass is unique because it is the only current renewable resource of liquid transportation fuel. Sweet sorghum is a type of sorghum that has a high concentration of soluble sugars in the juice, similar do sugarcane. Characteristics of high fermentable sugars, efficient nutrient use, high water use efficiency (1/3 of sugarcane and 1/2 of corn), short growing period with PIS and long growing season with PS sorghum cultivars, and the ability to adapt well to diverse climate and soil conditions make sweet sorghum a potential feedstock for ethanol production (Wu et al. 2008). While single-cut yields may be lower, an increased growing season increases cumulative yields in PS cultivars and the ratoon potential of PIS cultivars (Rooney et al. 2007).

Sorghum is among the most widely adaptable cereal grasses potentially useful for biomass and fuel production (Hons et al. 1986). The adaptation of sorghum to sub-humid and semiarid climates has extended the geographic scope of sorghum production far beyond that of other warm-cereal grains. Sorghum is a versatile crop, one grown in warm and cool climates. This annual C4, self-pollinating, highly water efficient plant of African (tropical) origin is also well adapted to sub-tropical and temperate regions. Sweet sorghum is a warm-season crop that matures earlier under high temperatures and short days. It is not only known as a "high energy crop" for its high photosynthetic rate, but it is also the "camel among crops" for its drought and aluminum toxicity tolerance. Sweet sorghum is adapted to widely differing climatic and soil conditions. As a consequence, sorghum plays a vital role in the global food economy and is the fifth most important cereal crop—following wheat, rice, corn, and barley—with more than 43 million hectares planted to sorghum worldwide each year (FAO 2011). Sweet sorghum varieties can grow up to 5 m tall and produce 50–120 tons of biomass (fresh weight) per hectare. Biomass, sugar extraction, and ethanol production from traditional and new varieties of sweet sorghum ranged from 45 to 52 Mg ha^{-1}, 72 to 109 kg Mg^{-1} and 1,800 to 2,600 L ha^{-1}, respectively in Brazil (Schaffert et al. 1986). Sweet sorghum requires less than 50 % total nitrogen to produce similar ethanol yields as corn (Anderson et al. 1995) and is capable of removing 62 % of total nitrogen with no difference in DM yield (Bean et al. 2008).

Reports have shown that sweet sorghum yielding 11–16 Mg ha^{-1} will remove nitrogen, phosphorus and potassium at the rate of 112, 45, and 202 kg ha^{-1}, respectively (Undersander et al. 1990). Under favorable conditions, sweet sorghum is capable of producing up to 13.2 metric tons per hectare of total sugars, which is equivalent to 7,682 l of ethanol per hectare (Murray et al. 2009). Sorghum stalks can reach a height of 5 m with diameters ranging from 1 to 5 cm. Sorghum's small leaf surface and very developed root structure (twice that of corn) are likely responsible for the plant's exceptional drought tolerance. Sorghum has also been reported to tolerate fungal disease, viruses, herbicides, heat, insects, weeds, and poor quality (alkali or acid) and water logged soils. Most of the sugars (sucrose, fructose, and glucose) are uniformly distributed in the stalk, while about 2 % are in the leaves and inflorescences (Vietor and Miller 1990). Even in dry climates, sorghum can yield high levels of fermentable sugar, grain, and lignocellulose (Gnansounou et al. 2005).

6.4 Sweet Sorghum is Ideal for Both Sugar- and Cellulosic-Based Ethanol

Sorghums can be classified into four main groups depending on their production characteristics: grain sorghum, forage sorghum, high-tonnage sorghum or energy sorghum, and sweet sorghum. Although sweet sorghum, as a crop, meets the needs of US Midwest growers, grower acceptability of dedicated energy crops will be greatest if the crops can be planted and harvested with the machinery used for current crops, if they can be easily eradicated should landowners want to change land use, and if they can provide harvestable material in a short period of time. Sweet sorghum fits all of these requirements. Sorghum's high water and N use efficiency will further enhance farmer acceptability.

Sweet sorghum is characterized by high fermentable sugar content. At the soft dough stage, the sugar is composed mainly of sucrose (60–74 %) (Table 6.1), fructose and glucose that can be easily fermented to produce ethanol. Sweet sorghum total fresh weight yields vary considerably from 12 to 60 Mt ha^{-1} with arrange from depending on the cultivars/hybrids used, the location, inputs, and production practices (Dweikat et al. 2012; Jackson et al. 1980; Reddy et al. 2007; Propheter et al. 2010; Zhao et al. 2009). Sweet sorghum sugar yields range between 1.6 and 13.2 Mg ha^{-1}, with significant variation observed across different years and regions (Jackson et al. 1980; Reddy et al. 2007; Propheter et al. 2010; Zhao et al. 2009).

Cellulosic materials are generally thought to be the most likely feedstock for large-scale ethanol production from biomass in the long-term, due to their potentially larger supply and lower price compared to other carbohydrate sources (Perlack et al. 2005). However, low cost, plentiful supply, and ease of conversion have made readily fermentable carbohydrates (FC) the preferred feedstocks for bioethanol production. Starch-rich materials, such as grains, have the advantage of established feedstock and processing infrastructure in the US, and a more homogenous and reactive form of carbohydrate than that found in cellulosic materials. An advantage of both starch and sugar-rich materials over cellulosic materials is that they can be processed to sugar streams of sufficient purity to accommodate production of high-value products such as food, pharmaceuticals, and fiber-grade polymers. Plant materials high in soluble sugars yield the most readily converted form of carbohydrate, requiring lower inputs of chemicals and energy for processing, and the technology for the extraction of sugars is fully mature and highly efficient, reducing processing costs. Sugar is the preferred carbohydrate feedstock for many high-value products and is also used to produce around half of the world's ethanol, the largest bio-commodity (Murray 2005).

Sweet sorghum is of particular interest because of the large volume of readily fermentable juice that can be extracted. Hunter and Anderson (1997) indicated

Table 6.1 Properties of sweet sorghum at the soft dough stage, including stalk, leaves, panicle and grain, for two varieties. Except for moisture, % is in terms of dry mass. Sampling was done field planted at 70,000 plants/acre

Character	Variety	Variety
Cultivar	Simon	M 81E
Moisture content	65–70	65–70
Sugar	35–42	35–40
Fiber	13–15	13–15
Cellulose	33–37	32–37
Hemicellulose	20–24	20–24
Protein	4–7	5–8
Aconitic acid	2–4	2–4
Starch (Juice)	0.5–4	0.5–4
Oil and wax	3–4	3–4
Ash	2–4	2–4
Total	100 % of dry matter	100 % of dry matter

that the sugar produced in sweet sorghum has a potential ethanol yield up to 8,000 L ha^{-1}, or about twice the ethanol yield potential of maize grain, In addition to producing large amounts of sugar-rich biomass, hybrids can be developed from crosses between grain-type seed parents and sweet-type pollen parents (Hunter and Anderson 1997). The product of these crosses typically increase biomass yields and sugar content when compared to the original grain-type seed parents, but are inferior in sugar quality compared to the pollinator parent. Such hybrids can co-produce grain at levels approaching the yields of the grain-type seed parent (Miller and McBee 1993). Sweet sorghum has been found to be competitive with corn for theoretical ethanol yield with less energy invested (Smith et al. 1987; Smith and Buxton 1993; Hunter and Anderson 1997). Ethanol production from sugar does not require energy to depolymerize carbohydrates such as is required for grain starch or cellulosic ethanol. Smith et al. (1987) reported total sugar yield ranging from 4 to 10.7 Mg ha^{-1} for the continental USA and up to 12 Mg ha^{-1} for Hawaii while Smith and Buxton (1993) reported sugar yields at 6 Mg ha^{-1} in Iowa and Colorado. Vermerris et al. (2007) reported total sugar concentrations of the juice ranging from 9 to 15 %. Ricaud et al. (1979) found sugar concentration in juice at the soft dough stage to range from 12.8 to 16.6 %. Theoretical ethanol yield estimates for SS have ranged from 3,850 to 4,410 L ha^{-1} (Lueschen et al. 1991; Hunter 1994) although the crop has been estimated to have the potential of 8,000 L ha^{-1} of ethanol (Hunter and Anderson 1997), equivalent to ethanol produced from approximately 20 Mg ha^{-1} of corn.

The conversion efficiency of lignocellulosic biomass to liquid fuels like ethanol is strongly influenced by cell wall composition. Lignin has been shown to hamper saccharification by physically shielding the cellulose from degradation, making it difficult to convert into ethanol and increasing the energy requirement for processing. Sorghum stover, containing lignin, hemi-cellulose and cellulose (Table 6.1), can serve as an excellent feedstock for ethanol production. A set of mutation stocks, developed by the USDA Plant Stress and Germplasm Development Unit in Lubbock, Texas, USA (Xin et al. 2008), is sufficiently extensive to allow identification of mutations in virtually every sorghum gene. The best known such mutations are the brown midrib (BMR) mutants, which were first discovered in maize in 1926. Early studies revealed the trait resulted in lower fiber and lignin within the plant and could increase the conversion efficiency of sorghum biomass for lignocellulosic bioenergy. In sorghum, more than 19 *bmr* mutants were discovered by Porter et al. (1978). The *bmr* mutants are characterized by the reddish-brown coloration of the vascular tissue of the leaf blade, leaf sheath and stem, which is associated with alteration of secondary cell wall composition, especially lignin.

The *bmr* mutant sorghum, pearl millet (*Pennisetum glaucum*) and maize lines have significantly lower levels of lignin content (51 % less in their stems and 25 % less in their leaves). Purdue University research showed 50 % higher yield of fermentable sugar from the stover of certain sorghum *bmr lines* after enzymatic hydrolysis (Vermerris 2011). Therefore, the use of *bmr* cultivars would reduce the cost of biomass-based ethanol production. In addition, The *bmr* crop residues have higher rumen digestibility and palatability, making them good for fodder as well.

6.5 Sorghum Production

Sorghum is relatively inexpensive to grow to high yields, and can be used to produce high value-added products like ethanol and distillers' dried grains (Chiaramonti et al. 2004). Due to its high productivity and rapid growth cycle (120–150 days), sweet sorghum has an impressive capacity to absorb a large amount of CO_2 from the atmosphere during its growing cycle. When compared to the input requirements of other crops, sorghum requires half of those needed by sugar beets, and one-third of the requirements of sugar cane or corn (Soltani and Almodares 1994; Renewable Energy World 2000). The cost of production per unit area e.g. one hectare is about $250 less than corn due to seed cost and lower requirement of fertilizers. Bennett and Anex (2009) compared the production, milling of sweet sorghum to corn and concluded that when combustion credits are $6–8 GJ^{-1}, sweet sorghum costs are in the range of $91–149 Mg^{-1} compared to $171–258 Mg^{-1} for corn. Sweet sorghum juice is ideally suited for ethanol production given its higher content of total reducing sugars compared to the content of other sources, including sugarcane juice. Further, following the extraction of juice, sorghum bagasse can be burned to generate steam for ethanol distillation and co-generation of electricity. Remaining bagasse can be used as fodder for animals or for additional ethanol production through lignocellulose conversion.

Also important is the amount of energy used to produce ethanol. Historically, for each unit of energy it took to plant and harvest a crop and process it into ethanol, the fuel returned 0.92 units of energy. Ethanol had a negative "energy balance" of one unit in for 0.92 out (1:0.92). However, steady improvements have been made in corn yield and harvesting and in ethanol processing efficiency. The latest studies show corn ethanol boasts a positive energy balance of 1:1.25; a 25 % net increase in net energy (Farrell et al. 2006). Today, corn ethanol is made by converting the starch in corn to sugars and then into alcohol by a fermentation process. Sugar beets (*Beta vulgaris* L.) are a better ethanol source, producing nearly two units of energy for every unit used in production. Sugarcane, though, is by far the most efficient of the current feedstocks, yielding more than three units as much energy as is needed to produce the ethanol derived from it (Hopkinson and Day 1980). Sweet sorghum's positive energy balance, with a ratio of 1:8, is comparable to that of sugarcane (Worley et al. 1992). Given their positive energy balances and higher yields, it makes more sense to produce ethanol from sugar crops than from starchy grains.

6.6 Sweet Sorghum Does Not Compromise Food Security

Corn is currently the feedstock of choice for U.S. ethanol producers. Increasing ethanol production led to higher domestic corn utilization, as it is also widely used in the food and livestock sectors. This, coupled with other factors such as the value of the dollar and investment markets, has contributed to corn prices rising to some of the highest levels in U.S. history. Farmers responded to high corn prices by shifting planted acres to corn, which has caused ripple effects across other crops, contributing to higher price levels of competing crops. As a result, public and political interest has escalated for the production of ethanol from sources other than corn. Economic research has explored various alternative ethanol production technologies. Progress has been made with respect to biochemical and thermo-chemical technologies for cellulosic ethanol, yet the ability to reach commercial viability continues to elude the industry. Herbst (2003), Shapouri et al. (2006), Ribera et al. (2007), Salassi (2007), and Outlaw et al. (2007) have examined the economic feasibility of ethanol production from grain sorghum and corn, sugarcane juice and molasses, respectively. Studies by Epplin (1996), Graham et al. (2000), and Mapemba et al. (2007) have explored transportation, harvest, and delivered feedstock cost components of biomass used for cellulosic ethanol. Outlaw et al. (2007) conclude ethanol production from sugarcane juice, a predominant production method in Brazil, would be economically feasible in certain regions of the United States. However, sugar policy has left little opportunity for this method to gain traction in the United States.

6.7 Sweet Sorghum is Ideal for Double Cropping

Because sorghum grown for biomass can be harvested before it is fully mature, it is possible to grow it in a double crop sequence with a winter annual. Winter annuals are planted in the fall, grow rapidly in the spring, and reach harvest anytime during late spring to early summer. Sorghum, which is well adapted to germination under limited moisture, can be no-till planted into the stubble of a winter annual crop. The primary advantages of a double crop sequence are to maximize use of solar radiation, provide winter cover against wind and water erosion, and increase biomass/ha. Because sorghum and winter annuals have differing cardinal temperatures for growth, the double crop sequence can take advantage of a longer growing season than either crop alone.

Sweet sorghum yield has not been found to be sensitive to plant density (Lueschen et al. 1991; Ferraris and Charles-Edwards 1986), but sucrose content, sugar yield, and juice content have been decreased with high plant density (Broadhead and Freeman 1980; Martin and Kelleher 1984; Kuepper 1992). Efficient nitrogen (N) use is important for net energy yield (energy yield relative to energy invested), and associated life cycle greenhouse gas (GHG) emissions (Liska and Cassman 2008). Response to applied nutrients has varied with location. Nitrogen application did not affect fermentable sugar yield (Smith and Buxton 1993), total and stalk dry matter yield at harvest (Barbanti et al. 2006), or fermentable carbohydrate and ethanol yield (Lueschen et al. 1991). Biomass yield has been shown to increase in Louisiana by 140 % by applying 100 kg ha^{-1} N, but no further increase was observed with an additional 100 kg ha^{-1} N (Ricaud and Arenneaux 1990). Total sugar yield was also increased by 150 % in the same study by applying 100 kg ha^{-1} N, with a 4 % increase from an additional

100 kg ha^{-1} N. Sweet sorghum has been found to require roughly 36 % of the fertilizer N required for similar yield levels in corn (Geng et al. 1989). Total dissolved solid concentration in stalk juice was shown to decrease with increased N rate (Wiendenfeld 1984). Consequently, some priority areas of research for sweet sorghum as an ideal bioenergy crop are the annual nature of the crop in the temperate regions, sensitivity to cool temperatures at the early growing stages, weed controls for industrial scale plantation, sustainable production systems.

6.8 Development of Perennial Sorghum

Perennial plants are highly efficient and responsive micro-managers of soil, nutrients, and water. In contrast, annual crops require seedbed preparation, precisely timed inputs and management, and good weather during narrow time windows. With shorter growing seasons and less extensive root systems, annual crops provide less protection against soil erosion, manage water and nutrients less effectively, store less carbon below ground, and are less resilient to pests and abiotic stresses than are perennial plant communities (Glover 2005). Perennials generally yield more above ground biomass than do annuals, and some of the carbon that goes into the biomass might be reallocated to above ground section of the plant production through breeding. Although those species currently being domesticated as perennial grain crops have low seed yields, their total aboveground productivity is often higher than that of annual crops with long breeding histories (DeHaan et al. 2005). Piper and Kulakow (1994), for example, reported a mean aboveground biomass for self-pollinated progeny of annual X perennial sorghum hybrids that was 62 % higher than that of their annual parent.

The perennial species *Sorghum halepense* is a tetraploid with 20 pairs of chromosomes, 10 of them similar to those of *S. bicolor* and 10 similar to those of the diploid perennial *S. propinquum* (Paterson et al. 1995). *Sorgum halepense* as migrated throughout much of the United States and into Canada as a highly successful, rhizomatous weed known as johnsongrass. Because the two species differ in chromosome number, those early hybrids were produced using artificially chromosome-doubled (tetraploid) *S. bicolor* plants as female parents. The tetraploid F_1 hybrids were fertile, and F_2 plants derived by self-pollination of the hybrids varied widely in rhizome production. A majority of those plants were perennial; that is after harvest, their rhizomes survived through the winter to produce new shoots the following spring. Advance sorghum lines with rhizomes have been generated as a result of crossing grain sorghum nuclear–male sterile lines to Johnsongrass materials (Dweikat 2005, Fig. 6.1). The sorghum lines were backcrossed five generations, and grain types with rhizomes habit were selected. Our goal is to use these sorghum lines for crosses with elite sweet sorghum cultivars. The F1 plants were backcrossed to sweet sorghum to generate sweet sorghum advanced lines with rhizomes. The advanced materials have been planted in the field and have been allowed to over-winter to select for perennial biotypes.

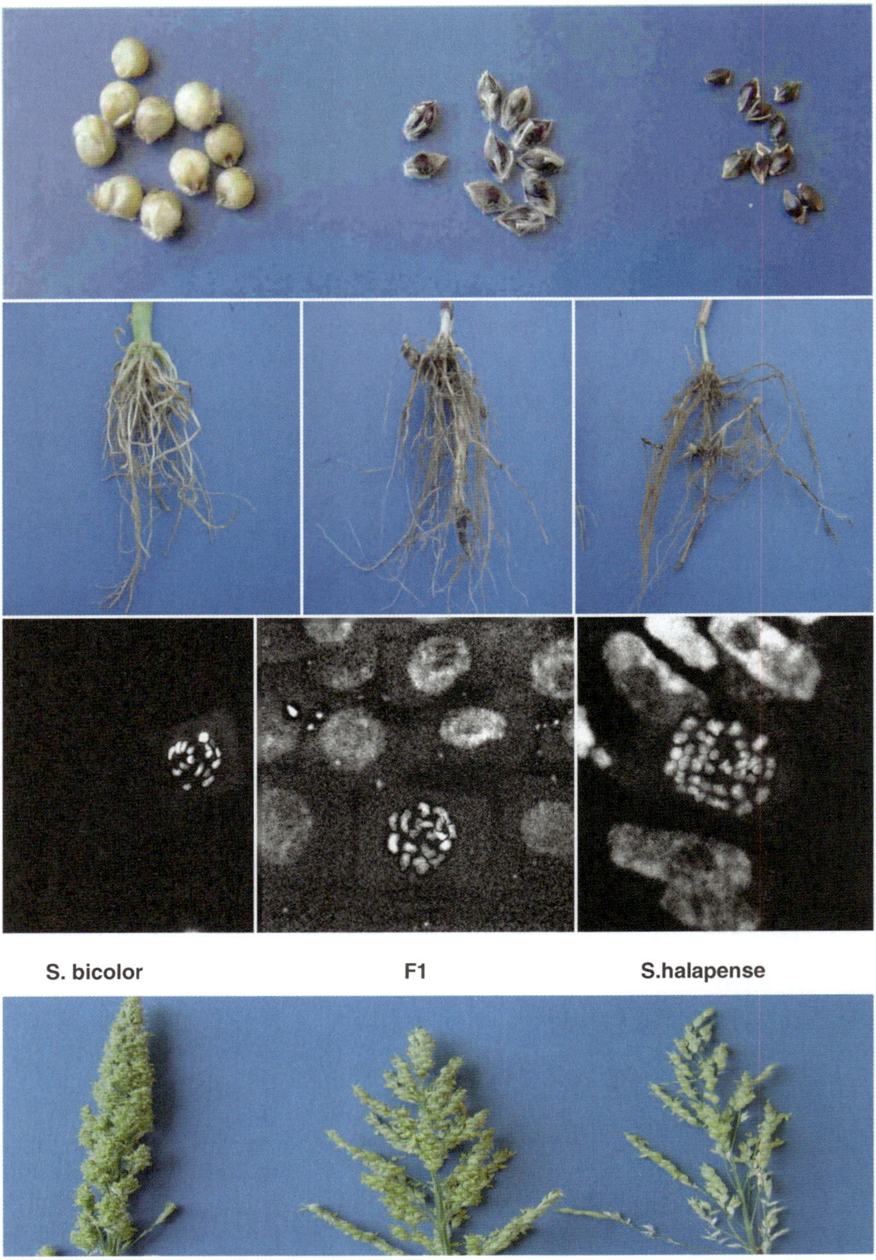

Fig. 6.1 **a** Chromosome analysis of root-tip cells collected from Johnsongrass (*right*), cultivated sorghum (*left*) and the putative F1 plant (*middle*) confirmed that Johnsongrass contained 40 chromosomes while both the cultivated sorghum and the F1 contained 20 chromosomes. **b** Phenotypic characterization of the seed (*top panel*), panicles (*middle panels*) and roots (*bottom*) of sorghum (*left*), Johnsongrass (*right*), and the F1 hybrid (*middle*)

6.9 Breeding for Cold Tolerance as a Mean to Extend Sorghum Production Regions

Sorghum originated from tropical regions and is a cold and frost-sensitive plant. Low-temperature stresses, including chilling and frost, greatly affect the germination and growth of the plant, and limit the geographical distribution of the crop. This poses a major problem in temperate environments when sorghum is planted early. Cool air and soil temperatures result in poor seedling establishment of sorghum because of slow emergence rate, reduced emergence percentage, and reduced growth rate after emergence (Pinthus and Rosenblum 1961; Singh 1985). Early planting of sorghum hybrids in temperate climates confers numerous advantages, but requires genetic improvement, both in lines and in parental effects for germination and seedling tolerance to low soil temperatures. Analysis of the phenotypic data showed a high degree of heritability for all traits measured (Gunaratna 2002), suggesting that gains from selection for seedling cold tolerance should be significant. The development of sorghum hybrids with increased cold tolerance would help to expand and stabilize sorghum production under these conditions. Sorghum hybrids with early-season cold tolerance have several advantages. First, early canopy development reduces weed competition and soil moisture evaporation. Second, sorghum primarily is grown in hot and dry environments, and early planting allows the crop to better utilize spring rains and reach flowering prior to drought conditions that are more prevalent during the mid-summer period (Pendleton et al. 1965). Finally, improved cold tolerance would be important in no-tillage production systems. No-tillage systems are becoming increasingly popular due to their numerous advantages, but they also reduce soil temperature by 1–5 °C at typical planting dates (Carter and Barnett 1987; Graven and Carter 1991).

6.10 Development of Herbicides Tolerance

Sorghum has the ability to tolerate short-term drought and a late summer sorghum crop may follow an early-season corn crop. Achieving good control of grassy weeds has been identified by producers as a significant management challenge that must be addressed in order for the crop to be economically sustainable. New herbicides are not developed specifically for grain sorghum. If a herbicide developed for corn or wheat does not cause severe phytotoxicity in grain sorghum, then grain sorghum may be added to the label. In fact, because of the high cost of developing, testing, and registering herbicides, there are very few new herbicides being developed now even for the major crops. Weed control in sorghum is essential if high yields and efficient harvest are to be achieved; however, good weed control in sorghum is often difficult to achieve. Sorghum is a small seeded grass and is relatively slow growing in the first few weeks after emergence. The slow seedling growth combined with the limited number of herbicides and the low rates which must be used creates a problem in sorghum weed control.

The most troublesome weeds for grain sorghum include morning glory, pigweed, broadleaf signal grass, barnyard grass, prickly sida, crabgrass and sicklepod (references—this is not uniform in all sorghum growing areas). There are fewer control options for weed control in grain sorghum than in corn, cotton and soybeans. Preemergence and POST ALS-inhibiting herbicides are used effectively to control weeds in corn (*Zea mays* L.) and other crops. Unfortunately, sorghum is susceptible to grass control ALS-inhibiting herbicides such as nicosulfuron and rimsulfuron, which makes it impossible to use these herbicides in sorghum. Recently, researchers at both Kansas State University and the University of Nebraska have developed different types of sorghum that is resistant to several ALS-inhibiting herbicides by transferring resistance genes from a wild sorghum relative (Tuinstra and Al-Khatib 2007; Tuinstra et al. 2009; Hennigh et al. 2010; Gelli et al. "unpublished").

Shattercane (*Sorghum bicolor*) is a monocot weed in the *Poaceae* family. It occurs as a weed in wherever cultivated grain sorghums and their wild relatives grow in the same region. All races of the subspecies *bicolor* and all wild kinds of *S. bicolor* can hybridize into the weedy shattercanes. Four different shattercanes resistant to ALS-inhibiting herbicides were collected between 1992 and 1996 from plants in 16 fields located in southeastern and south central Nebraska. The plants were randomly selected from those fields which had been treated for three consecutive years with different classes of ALS inhibiting herbicides. Greenhouse experiments will be conducted to evaluate the response of the four different putative resistant shattercane biotypes to ALS-inhibiting herbicides. The resistant and the susceptible biotypes will be tested against the four classes of ALS-inhibiting herbicides SU, PTB, TP, and SCT (Gelli et al. "unpublished").

6.11 Breeding Strategies for Sweet Sorghum

Historically, sweet sorghum cultivars have been photo insensitive varieties. Sweet sorghum varieties were initially introduced into the USA in the 1850s and developed in the United States in the 1880s and 1890s to develop sweet syrup, especially when crystal sugar was unavailable. Research investment continued with the development of improved varieties through the 1970s and 1980s. The sweet sorghum varieties developed for high quality syrup production were selected for high Brix and high total sugars in the juice, but reduced levels of sucrose to avoid crystallization of the syrup. Sweet sorghum syrup that crystallized was considered to have reduced quality. High sucrose sweet sorghum varieties such as Rio, Roma, Ramada, Keller and Wray were developed and released in the 1970s and 1980s with the objective of producing crystal sugar in the sugar mills of Southern Texas and Mexico. The high level of starch (up to 5,000 ppm) in the juice in high sucrose varieties interferes in the sucrose crystallization process.

Table 6.2 Sweet sorghum minimum yield and quality goals

Trait	Target 1975	Target 2013
Minimum biomass yield (Genetic potential × production system)	40 t ha^{-1} (10 t ha^{-1} month)	50–60 t ha^{-1} (12–15 t ha^{-1} month)
Minimum brix[a]	15–17°	16–19°
Peak brix[a]	21°	23°
Minimum total sugar extracted[a] (kg t^{-1} biomass)	80	100–120
Sucrose (POL)		10–18
Purity (% Sucrose)		70–90 %
Juice extraction efficiency	60–65 %	90–95 %
Minimum total sugar (ART) content in juice	12.5 %	14 %
Minimum alcohol yield	40 l t^{-1} biomass	60–70 l t^{-1}
Minimum alcohol yield	2,500 l ha^{-1}	3,500 l ha^{-1}
Fermentation efficiency (%)	90	95
Distillation efficiency (%)	90	95
Industrial efficiency (%)	81	90
Period of industrial utilization (PIU)	30 days	30 days
Panicle size		Small
Tillering		Non-tillering
Cultivar type	Variety	Variety/hybrid

[a] Based on standard hydraulic press extraction of 500 g sample with 245 kg cm^2 for 60 s

Embrapa conducted a sweet sorghum breeding program for ethanol production between 1975 and 1987 and reinitiated this R&D activity in 2008. The original and updated breeding objectives of Embrapa for developing sweet sorghum cultivars for ethanol production in Brazil are summarized in Table 6.2. The biomass yield is a function of the genetic potential of the cultivar used and the production system and level of production technology utilized. Experimental yields of total biomass of sweet sorghum varieties developed by Embrapa have surpassed 100 Mg ha^{-1}. The Brix is a measure of soluble solids in solution including sugars and soluble salts. In sweet sorghum with high levels of Brix (16–21), the total sugar content is normally approximately 1–1.5 units less than Brix and at lower levels of Brix (8–12), the sugar content is several units less than Brix, frequently 4–6 % sugar. The sugar extracted is a function of biomass yield, fiber content (related to juice extraction) and juice quality. The presence of sucrose in the total sugar content is not essential for G1 ethanol production. However, higher levels of sucrose appear to be correlated (Non-published data Embrapa) with both higher total sugars and longer periods of industrial utilization (PIU), the minimum period of adequate quality for efficient and sustainable industrialization (PIU is discussed in more detail below). Consequently, a breeding strategy for G1 sweet sorghum should include selection for high sucrose (POL) and high purity (sucrose as a percent of total sugars). The juice extraction efficiency is based on the fiber content of the sorghum stalks and the efficiency of the equipment used in processing the feedstock and may vary depending upon the size and efficiency of the distillery. The minimum level of total sugars (12.5 %) in the juice is governed by the efficiency of the yeast and tolerance of the yeast to the level of ethanol and may be increased as progress is made in selecting more efficient yeast strains more tolerant to higher ethanol concentration. Juice with total sugar content higher than 12.5–14.0 % sugar content can be diluted with water for efficient fermentation. The alcohol yield per ton of stalks is a function of extraction efficiency, sugar content in the juice, and industrial efficiency, and alcohol yield per hectare is a function of biomass yield, extraction efficiency, sugar content in the juice and industrial efficiency.

The above discussion is based on a standardized process of determining Brix, sugar content of the juice and total sugar extracted. We have used the hydraulic press (http://www.pontalmaq.com.br/links/sub_link/prensa_hidraulica.html) developed and utilized by the sugarcane industry to characterize experimental germplasm and commercial cultivars. The Brix measurement from a field sample can vary significantly depending upon how the juice was extracted. The Brix reading of a few drops of juice (easy juice) on a digital refractometer in the field will be several units higher than the Brix reading of the juice extracted with a hydraulic press (minimum of 8–10 stalks). Consequently, one must beware and knowledgeable of how the juice was extracted for proper interpretation. Economic ethanol yield projections based on easy juice samples will be very misleading by projecting exaggerated ethanol yield productions. Also, one must keep in mind that Brix is not sugar and cannot be used directly to substitute total sugar levels in estimating ethanol production as has been reported in several publications [Degrees Brix—symbol °Bx—is the sugar content of an aqueous solution (http://www.engineeringtoolbox.com/degrees-brix-d_1828.html)].

6.12 Industrial Management

The period of industrial utilization (PIU) is terminology borrowed from the sugarcane industry to describe the time interval that a cultivar of sweet sorghum has the minimum levels of sugar extraction and minimum level of sugar in the juice for economic production of ethanol. Experience at Embrapa, based on data utilizing the hydraulic press, indicates that a minimum total sugar of 12.5 % in the juice and extracted sugar of 80 kg t^{-1} biomass over a period of 30 days are the minimum parameters for determining the PIU. Normally total sugar of 12.5 % and extracted sugar of 80 kg t^{-1} biomass occur simultaneously. The PIU of a cultivar can vary at different planting times and at different locations and should be determined for each eco-region where the cultivar will be utilized. This information is necessary for industrial planning in order to provide the necessary amount of biomass daily for the programmed harvest period. The logic of using a minimum period of 30 days is so that the industrial manager can program each cultivar and planting period for a 15 day harvest window, thus allowing a 15 day cushion for providing the necessary quantity of daily feedstock for milling. A maturity curve, the interaction of juice extraction, fiber, and sugar content of the juice, is developed by weekly sampling of a cultivar beginning a few days (15) after flowering for a period of 8–10 weeks. An example of a sweet sorghum maturity curve is presented in Fig. 6.2.

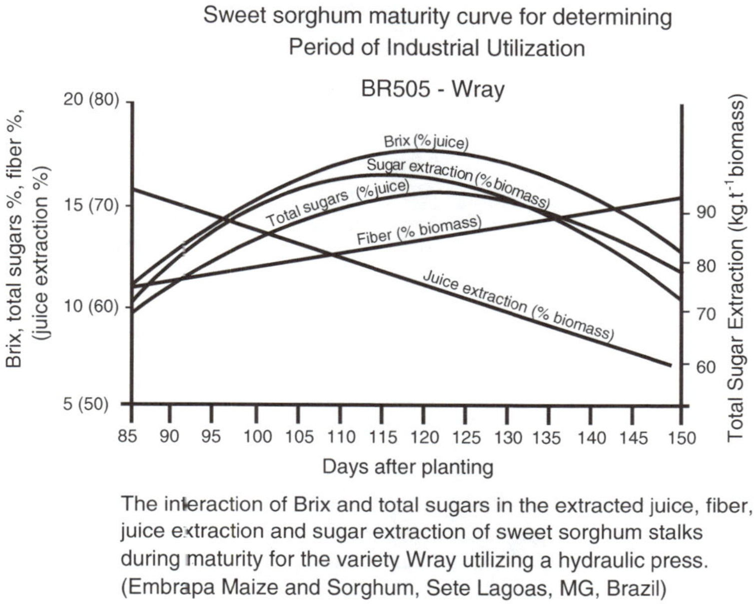

Fig. 6.2 Sweet sorghum maturity curve for determining period of industrial utilization

6.13 Sweet Sorghum Idiotype

The ideal sweet sorghum idiotype is a non-tillering cultivar with a small panicle (Fig. 6.3) with the parameters in Table 6.2. The number of stems per hectare influences biomass yield, biomass quality, and lodging. The only way to control stem population is by controlling the seeding rate of non-tillering cultivars. Small panicles are desirable as this reduces lodging, provides less competition for photosynthates and reduces harvesting and processing complications.

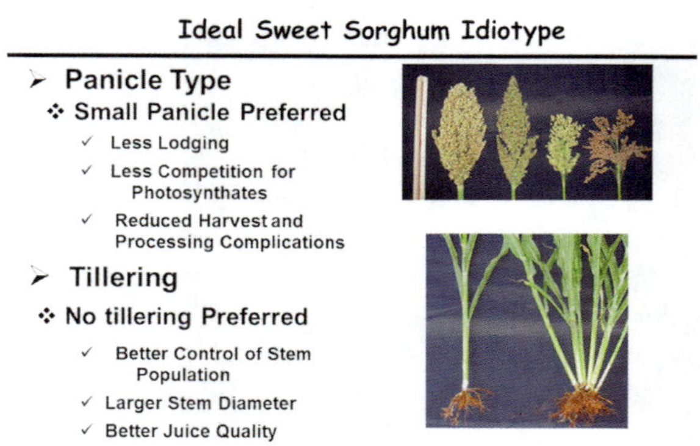

Fig. 6.3 Ideal sweet sorghum idiotype

Table 6.3 Minimum thresholds recommended for sweet sorghum in Brazil

	1975–2011	2012/2013	Ideal[a]
Biomass production (t ha^{-1})	40	50	60–70
Total sugar in juice (%)	12.5	12.5	>14.5
Ethanol production (l t^{-1})	60	60	70
Ethanol production (l ha^{-1})	2,500	3,000	4,200–4,900
Period of industrial utilization—PIU (days)	30	30	30

[a] Ideal threshold of a non-photosensitive (PIS) sweet sorghum cultivar in Brazil (Embrapa maize and sorghum)

6.14 Varieties Versus Hybrids

Both hybrids and varieties of sweet sorghum cultivars have been evaluated in experimental trials and pilot evaluations in Brazil. The varieties were developed by Embrapa and the hybrids were developed and provided by several private seed companies. For the most part, no differences were observed in biomass productivity, with both varieties and hybrids producing acceptable levels of biomass of more than 40 t ha^{-1}. However, for the most part, the sugar quality of the hybrid biomass has not reached the minimum thresholds (Table 6.3) of sugar quantity, for both sustainable processing and sustainable ethanol yields. The varieties, on the other hand, produced both adequate biomass quantity and quality when adequate production systems are used, reaching the minimum ethanol production of 2,500–3,000 l ha^{-1}. Research at Embrapa has indicated that the lower industrial quality of hybrids can be contributed to the low sugar level of the female lines. No differences have been observed in the quantity of juice extracted between varieties and hybrids, but the Brix and sugar levels in the juice of the hybrids is only slightly superior to the average or mid-parent of the male and female parents. This has also been reported by several authors (Reddy et al. 2011).

The presence and absence of juice in the stem of sorghum is simply inherited (one recessive gene) and many juicy stem female lines are used in producing both grain and forage sorghum hybrids. Sugar in the juice, on the other hand, is a more complexly inherited trait with several genes being involved. There is a very limited number of short statured three dwarf sweet sorghum female lines available to produce sweet sorghum hybrids. Another complicating factor to develop ideal sweet sorghum lines is that all female lines developed since the 1950s have been selected for high general combining capacity (GCC) for grain production, whereas in sweet sorghum the ideal type is minimum grain production. The rational for this is to reduce lodging, to simplify the mechanical harvest of sweet sorghum biomass for producing ethanol using G1 technology, and to reduce possible competition for sugars in the photosynthesis process. We have observed that in experimental hybrids, both an increased amount of stalk lodging occurs and lower levels of sugar in the juice resulting in an inadequate and shorter PIU.

The most recently released cultivars (USA), cultivars released in the 1970s and 1980s such as Topper, M81E, Wray, Theis, Rio, etc. were improved varieties.

A small emphasis was placed on developing cytoplasmic male-sterile female A and B lines for the A1 cytoplasm by some public institutions in the USA, but desirable results with hybrids have not been documented using these female lines. The experience of Embrapa with sweet sorghum hybrids in Brazil is very similar to the released and experimental hybrids of the private sector in that the sugar quality has not met the minimum standards of the distilleries. In collaboration with the distilleries, Embrapa has established the concept of minimum ethanol yield per hectare. The private sector distilleries initially established a minimum ethanol yield of 2.500 l ha^{-1} in 2011 and have suggested in 2012 and 2013 that an ideal minimum ethanol yield of 3,000 l ha^{-1} provides a more adequate profit scenario. The first variety released by Embrapa, BR506 has met these minimum requirements in pilot operations at large distilleries. The varieties released by Embrapa in 2012, BRS508, BRS509 and BRS511, also meet these minimum requirements.

Sweet sorghum varieties are the preferred cultivars; however, hybrids will be the cultivars of the future because of the possibility of mechanized seed production to produce the large quantity of seed necessary to meet the projected demand. Also, the private sector prefers hybrids to protect their intellectual property. The path to successful development of high yielding, high quality sweet sorghum hybrids is to develop short statured three or two dwarf sweet sorghum high quality male sterile lines suitable for mechanical harvesting.

6.15 Strategy for Developing Cytoplasmic Male Sterile Sweet Sorghum Lines

The principal cytoplasmic male sterility utilized in developing male sterile (A and B lines) and restorer (R) lines is the A1 cytoplasm. The fertility restorer lines (R-lines) depend upon at least one of three dominant genes, Rf_1/rf_1, Rf_2/rf_2 and Rf_5/rf_5. The recovery of B-lines from B × R crosses that give 100 % male sterility when backcrossed into the A1 sterile cytoplasm is much lower than from B × B crosses. The great majority of the sweet sorghum varieties developed over the past several decades are R-lines. Embrapa has used the elite R-line, Wray, and Wray derivatives in crosses with elite juicy stem B-lines to develop high sugar, high sucrose, non-tillering, small panicle size B-lines for use in developing ideal sweet sorghum male sterile A-lines for use in developing high quality sweet sorghum hybrids with the desirable characteristics described in Table 6.2. The first set of high sugar female lines is currently being evaluated at Embrapa for 100 % male sterility in the A lines and in hybrid combinations.

Over the past 2 years, Embrapa has identified 12 sweet sorghum non-restorer cultivars from its sweet sorghum germplasm collection. These lines have been characterized and the best lines are being utilized in B × B crosses with elite multiple stress tolerant juicy stem B-lines to develop a broad based array of male sterile lines for developing sweet sorghum hybrids.

6.16 Breeding Strategies for Energy Sorghum

The basic objective for developing energy or biomass sorghums is to maximize the productivity. This is best achieved by using photosensitive sorghum hybrids and it can be achieved in Brazil using two different models. The first and most efficient model is using photo insensitive (PIS) female lines with the recessive maturity gene ma_1ma_1 and PS restorer lines with dominant maturity gene Ma_1Ma_1 as the male parent. These two types of sorghum will have synchronized flowering when planted in March and April, facilitating commercial seed production. Total biomass production of 60 t ha^{-1} dry weight has been observed for experimental PS hybrids in experimental biomass trials. Pilot production of PS experimental hybrids has been over 50 t ha^{-1} dry matter.

6.17 Breeding Strategies for Modifying Lignin Content in Sorghum for Cellulosic Ethanol Production and Electrical Energy

Chemical composition of lignocellulosic feedstocks is a key factor affecting efficiency of cellulosic ethanol production during the biomass conversion process. Biomass chemical composition and structural characteristics can be affected by several factors including plant genetics, growth environment, developmental stage, harvesting method, storage, and others. Since these sources of variation are difficult to control; tailoring of feedstock chemical composition can be used to attain high efficiency and optimal biomass conversion.

Currently in Brazil, the abundant and low cost sugarcane bagasse is used as burning fuel in boilers to produce electricity, making sugarcane mills energy self-sufficient. However, a considerable amount of bagasse is still not utilized, resulting in a waste problem. Therefore, production of cellulosic ethanol could increase ethanol production in Brazil by up to twofold and also eliminate an environmental waste problem.

Cellulosic ethanol production is achieved by hydrolyzing cellulose and hemicellulose fractions of biomass in order to release fermentable sugars. There are several obstacles to biomass conversion that makes developing an economically feasible technology very difficult (Lynd et al. 2005; Vermerris 2011). In addition to cellulose recalcitrance, the presence of lignin, an important component of cell walls, is one of the major problems in cellulosic ethanol production. It has been shown that biomass conversion yields were highly improved by reducing lignin content in sorghum lines by carrying the *bmr* mutations (Dien et al. 2009).

However, in the case of using bagasse as feedstock for water vapor generation, varieties with higher lignin content are desirable since their biomass have a higher caloric value.

Therefore, it is important to consider the structural and biochemical characteristics that can be enhanced in energy crops, allowing achievement of set goals for liquid biofuels production. Ideal energy crops will maximize yield per hectare with minimum inputs and exhibit value-added traits that enhance their use as biofuel feedstock. In addition to enabling agronomic traits, many argue that new energy crops must focus on cell wall production since the bulk of photosynthetic free energy is found in the polymers of this complex matrix (U.S. DOE 2006).

At Embrapa, our objective in developing biomass sorghums is to tailor the biomass composition to increase efficiency of biomass conversion. Sorghum with a higher level of lignin is desirable for burning, and with a lower level of lignin is desirable for biomass hydrolysis to fermentable sugars. There are already several techniques to utilizing sweet sorghum bagasse for burning, as well as methane and ethanol conversion (Gnansounou et al. 2005; Rooney et al. 2007; Xin and Wang 2011).

Embrapa Maize and Sorghum is developing very high yielding (50–60 t ha^{-1} dry matter) high energy photosensitive hybrids as a feedstock for cellulosic ethanol. Some of these hybrids have *bmr* (*brown midrib*) mutant genes introgressed in different genetic backgrounds, which allows for selection of materials showing low lignin content without lodging problems. Several of these hybrids are already being evaluated under different field conditions. Strategies for studying biomass composition and its accumulation differences, as well as sugar content and juiciness, are under way and are great resource for high resolution mapping of desirable QTLs and/or genes. This is being carried out in a PS sweet sorghum RIL population (275 F7 individuals) using the genotyping by sequencing (GBS) method (Elshire et al. 2011). We expect this molecular approach to significantly contribute to the improvement of sorghum as an energy dedicated crop.

In a similar approach, Embrapa has also developed a diverse sorghum association panel composed of approximately 200 photosensitive and insensitive materials (high biomass). The panel has been fully characterized for biomass compositional traits, including lignin content, using Near-infrared spectroscopy technology and biochemical assays. This association panel was put together using accessions from Embrapa's germplasm collection and breeding program, as well as accessions from the CIRAD core collection. The data showed great variability of cell wall composition among the genotypes. All individuals of this panel will be sequenced by GBS, allowing us to genotype the panel with thousands of SNPs across the sorghum genome. This will give us an opportunity for genome-wide association studies on traits related to biomass quality and yield. Genotyping this panel using the GBS method will also enable genetic diversity characterization of the panel, which will be useful in the selection of materials and development of new mapping populations for biomass accumulation and composition studies in sorghum.

Together, these different approaches will contribute to the development of sorghum materials with better biomass conversion properties, adding an aggregated value for producing bagasse/biomass that produce ethanol or electricity.

References

Anderson IC, Buxton DB, Hallam A, Hunter E (1995) Biomass production and ethanol potential from sweet sorghum. Leopold Center for Sust Agric, Iowa State Univ Ames, IA 4:97–101

Barbanti L, Grandi S, Vecchi A, Venturi G (2006) Sweet and fiber sorghum (*Sorghum bicolor* (L) Moench), energy crops in the frame of environmental protection from excessive nitrogen loads. Europ J Agron 25:30–39

Bean B, Bronson KF, Schwartz R, Malapati A (2008) Nitrogen requirements of sorghums for biofuel feedstock production in the southern high plains. In: Proceedings of great plains soil fertility conference, Denver, CO. 4–5 March 2008. Intern Plant Nutrition Inst, Norcross, GA, pp 15–18

Bennett AC, Anex RP (2009) Production, transportation and milling costs of sweet sorghum as a feedstock for centralized bioethanol production in the upper Midwest. Bioresources Technol 100:1595–1607

Bhattacharya A, Rice N, Shapter FM, Norton SL, Henry RJ (2011) Sorghum. In: Kole C (ed) Wild crop relatives: genomic and breeding resources (cereals). Springer, Berlin, pp 397–406

Boivin K, Deu M, Rami JF, Trouche G, Hamon P (1999) Towards a saturated sorghum map using RFLP and AFLP markers. Theor Appl Genet 98:320–328

Bowers JE, Abbey C, Anderson S, Chang C, Draye X, Hoppe AH, Jessup R, Lemke C, Lennington J, Li Z, Lin Y-R, Liu S-C, Luo L, Marler BS, Ming R, Mitchell SE, Qiang D, Reischmann K, Schulze SR, Skinner DN, Wang Y-W, Kresovich S, Schertz KF, Paterson AH (2003) A high-density genetic recombination map of sequence-tagged sites for *Sorghum*, as a framework for comparative structural and evolutionary genomics of tropical grains and grasses. Genetics 165:367–386

Broadhead DM, Freeman KC (1980) Stalk and sugar yield of sweet sorghum as affected by spacing. Agron J 72:523–524

Buchanan C, Lim S, Salzman RA, Kagiampakis I, Klein RR, Pratt LH, Cordonnier-Pratt M-M, Klein PE, Mullet JE (2005) *Sorghum bicolor*'s transcriptome response to dehydration, high salinity and ABA. Plant Mol Biol 58:699–720

Carter PR, Barnett KH (1987) Corn-hybrid performance under conventional and no-tillage systems after thinning. Agronomy J 79:919–926

Casa AM, Mitchell SE, Hamblin MT, Sun H, Bowers JE, Paterson AH, Aquadro CF, Kresovich S (2005) Diversity and selection in sorghum: concurrent studies with simple sequence repeats (SSRs). Theor Appl Genet 111:23–30

Chiaramonti D, Grassi G, Nardi A, Grimm HP (2004) ECHI-T: large bio-ethanol project from sweet sorghum in China and Italy. Energia Trasporti Agricoltura, Florence

DeHaan LR, Van Tassel DL, Cox TS (2005) Perennial grain crops: a synthesis of ecology and plant breeding. Renewable Agric Food Syst 20:5–14

Dien BS, Sarath G, Pedersen JF, Sattler SE, Chen H, Funnell-Harris DL, Nichols NN, Cotta MA (2009) Improved sugar conversion and ethanol yield for forage sorghum (*Sorghum bicolor* L. Moench) lines with reduced lignin contents. BioEnerg Res 2:153–164

Dje Y, Heuertz M, Lefebvre C, Vekemans X (2000) Assessment of genetic diversity within and among germplasm accessions in cultivated sorghum using microsatellite markers. Theor Appl Genet 100:918–925

Doggett H (1988) Sorghum, 2nd edn. Wiley, New York

Dweikat I (2005) A diploid, interspecific, fertile hybrid from cultivated sorghum, *Sorghum bicolor*, and the common Johnsongrass weed *Sorghum halepense*. Molec Breed 16:93–101

Dweikat I, Weil CF, Moose SP, Kochian L, Mosier NS, Ileleji KE, Brown PJ, Peer WA, Murphy AS, Taheripour F, McCann MC, Carpita NC (2012) Envisioning the transition to a next-generation biofuels industry in the midwest. Biofuels Bioprod Bioref 6:376–386

Edwards GE, Franceschi VR, Voznesenskaya EV (2004) Single-cell C_4 photosynthesis versus the dual-cell (Kranz) paradigm. Annu Rev Plant Biol 55:173–196

EIA (2011) Annual energy outlook 2011 with projections to 2035. Energy Information Administration, Washington, D.C

Elshire RJ, Glaubitz JC, Sun Q, Poland JA, Kawamoto K, Buckler ES, Mitchell SE (2011) A robust, simple genotyping-by-sequencing (GBS) approach for high diversity species. PLoS ONE 6:e19379. doi:10.1371/journal.pone.0019379

Epplin FM (1996) Cost to produce and deliver switchgrass biomass to an ethanol conversion facility in the southern plains of the U.S. Biomass Bioenergy 11:459–467

FAO (2011) http://faostat.fao.org/site/339/default.aspx (last time accessed: 06/28/2013)

Ferraris R, Charles-Edwards DA (1986) A comparative analysis of the growth of sweet and forage sorghum crops II. Accumulation of soluble carbohydrates and nitrogen. Aust J Agric Res 37:513–552

Geng S, Hills FJ, Johanson SS, Sah RN (1989) Potential yields and on-farm ethanol cost of corn, sweet sorghum, fodder beet and sugarbeet. J Agron Crop Sci 162:21–29

Glover J (2005) The necessity and possibility of perennial grain crops. Renewable Agric Food Syst 20:1–4

Gnansounou E, Dauriat A, Wyman CE (2005) Refining sweet sorghum to ethanol and sugar: economic trade-offs in the context of North China. Bioresour Technol 96:985–1002

Goodstein DM, Shu S, Howson R, Neupane R, Hayes RD, Fazo J, Mitros T, Dirks W, Hellsten U, Putnam N, Rokhsar DS (2012) Phytozome: a comparative platform for green plant genomics. Nucleic Acids Res 40:D1178–D1186

Graham R, English B, Noon C (2000) GIS-based modeling system for evaluating the cost of delivered energy crop feedstock. Biomass Bioenergy 18:309–329

Graven LM, Carter PR (1991) Seed quality effect on corn performance under conventional and no-tillage systems. J Prod Agric 4:366–373

Gunaratna N (2002) Early season cold tolerance in sorghum. MS thesis, Purdue University, West Lafayette, IN

Haussmann BIG, Hess DE, Seetharama N, Welz HG, Geiger HH (2002) Construction of a combined sorghum linkage map from two recombinant inbred populations using AFLP, SSR, RFLP, and RAPD markers, and comparison with other sorghum maps. Theor Appl Genet 105:629–637

Hennigh D, Al-Khatib K, Tuinstra M (2010) Response of acetolactate synthase-resistant grain sorghum to nicosulfuron plus rimsulfuron. Weed Technol 24:411–415

Henry R (2010) Evaluation of plant biomass resources available for replacement of fossil oil. Plant Biotechnol J 8:288–293

Herbst BK (2003) The feasibility of ethanol production in Texas. MS Thesis, Texas A&M University

Hons FM, Moresco RF, Wiedenfeld RP, Cothren JT (1986) Applied nitrogen and phosphorus effects on yield and nutrient uptake by high-energy sorghum produced for grain and biomass. Agron J 76:1010–1069

Hopkinson CS Jr, Day JW Jr (1980) Net energy analysis of alcohol production from sugarcane. Science 207:302–304

Hunter EL (1994) Development, sugar yield, and ethanol potential of sweet sorghum. MS thesis, Iowa State University, Ames, Iowa

Hunter EL, Anderson IC (1997) Sweet sorghum. Hortic Rev 21:73–104

Jackson DR, Arthur MF, Davis M, Kresovich S, Lawhon WT, Lipinsky ES, Price M, Rudolph A (1980) Research report on development of sweet sorghum as an energy crop. Volume 1: Agricultural task

Kim J-S, Klein PE, Klein RR, Price HJ, Mullet JE, Stelly DM (2005) Chromosome identification and nomenclature of *Sorghum bicolor*. Genetics 169:1169–1173

Klein PE, Klein RR, Cartinhour SW, Ulanch PE, Dong J, Obert JA, Morishige DT, Schlueter SD, Childs KL, Ale M, Mullet JE (2000) A high-throughput AFLP-based method for constructing integrated genetic and physical maps: progress toward a sorghum genome map. Genome Res 10:789–807

Kuepper G (1992) Sweet sorghum: production and processing. Kerr Center for Sustainable Agriculture, Poteau

Lewis NS, Nocera DG (2006) Powering the planet: chemical challenges in solar energy utilization. Proc Natl Acad Sci USA 103:15729–15735

Liska AJ, Cassman KG (2008) Towards standardization of life-cycle metrics for biofuels: greenhouse gas emissions mitigation and net energy yield. J Bio-based Mater Bioenergy 2:187–203

Lueschen WE, Putnam BK, Kanne BK, Hoverstad TR (1991) Agronomic practices for production of ethanol from sweet sorghum. J Prod Agric 4:619–625

Lynd LR, van Zyl WH, McBride JE, Laser M (2005) Consolidated bioprocessing of cellulosic biomass: an update. Curr Opin Biotechnol 16:577–583

Mapemba LD, Epplin FM, Taliaferro CM, Huhnke RL (2007) Biorefinery feedstock production on conservation reserve program land. Rev Agric Econ 29:227–246

Marguerat S, Bahler J (2010) RNA-seq: from technology to biology. Cell Mol Life Sci 67:569–579

Martin PM, Kelleher FM (1984) Effects of row spacing and plant population on sweet sorghum yield. Aust J Exp Agric Husb 24:386–390

Menz MA, Klein RR, Mullet JE, Obert JA, Unruh NC, Klein PE (2002) A high-density genetic map of *Sorghum bicolor* (L.) Moench based on 2926 AFLP, RFLP and SSR markers. Plant Mol Biol 48:483–499

Menz MA, Klein RR, Unruh NC, Rooney WL, Klein PE, Mullet JE (2004) Genetic diversity of public inbreds of sorghum determined by mapped AFLP and SSR markers. Crop Sci 44:1236–1244

Miller FR, McBee GG (1993) Genetics and management of physiologic systems of sorghum for biomass production. Biomass Bioenergy 5:41–49

Murray D (2005) Ethanol's potential: looking beyond corn. Earth Policy Institute, Washington, D.C. http://www.xmarks.com/site/www.earth-policy.org/Updates/2005/Update49.htm. Last accessed 06/28/2013

Murray SC, Rooney WL, Hamblin MT, Mitchell SE, Kresovich S (2009) Sweet sorghum genetic diversity and association mapping for brix and height. Plant Genome 2:48–62

Outlaw JL, Ribera LA, Richardson JW, da Silva J, Bryant H, Klose SL (2007) Economics of sugar-based ethanol production and related policy issues. J Agric Appl Econ 39:357–363

Paterson AH, Lin YR, Li Z, Schertz KF, Doebley JF, Pinson SRM, Liu SC, Stansel JW, Irvine JE (1995) Convergent domestication of cereal crops by independent mutations at corresponding genetic loci. Science 269:1714–1718

Paterson AH, Bowers JE, Bruggmann R, Dubchak I, Grimwood J, Gundlach H, Haberer G, Hellsten U, Mitros T, Poliakov A, Schmutz J, Spannagl M, Tang H, Wang X, Wicker T, Bharti AK, Chapman J, Feltus FA, Gowik U, Grigoriev IV, Lyons E, Maher CA, Martis M, Narechania A, Otillar RP, Penning BW, Salamov AA, Wang Y, Zhang L, Carpita NC, Freeling M, Gingle AR, Hash CT, Keller B, Klein P, Kresovich S, McCann MC, Ming R, Peterson DG, Mehboob-ur R, Ware D, Westhoff P, Mayer KF, Messing J, Rokhsar DS (2009) The Sorghum bicolor genome and the diversification of grasses. Nature 457:551–556

Pendleton J, Brown CM, Weibel RO (1965) Effect of reflected light on small grain yields. Crop Sci 5:373

Peng Y, Schertz KF, Cartinhour S, Hart GE (1999) Comparative genome mapping of *Sorghum bicolor* (L.) Moench using an RFLP map constructed in a population of recombinant inbred lines. Plant Breed 118:225–235

Perlack RD, Wright LL, Turhollow AF, Graham RL, Stokes BJ, Erbach DC (2005) Biomass as feedstock for bioenergy and bioproducts industry: the technical feasibility of a billion-ton annual supply. DOE/GO-102995-2135, ORNL/TM-2005/66. Oak Ridge National Laboratory, Oak Ridge, Tenn. http://feedstockreview.ornl.gov/pdf/billion_ton_vision.pdf. Last accessed: 06/28/2013

Pinthus MJ, Rosenblum J (1961) Germination and seedling emergence of sorghum at low temperatures. Crop Sci 1:293–296

Piper JK, Kulakow PA (1994) Seed yield and biomass allocation in *Sorghum bicolor* and F1 and backcross generations of *S. bicolor* x *S. halepense* hybrids. Can J Bot 72:468–474

Porter KS, Axtell JD, Lechtenberg VL, Colenbrandu VF (1978) Phenotype fiber composition and in vitro dry matter disappearance of chemically induced brown-midrib (*bmr*) mutants of sorghum. Crop Sci 18:205–208

Pratt LH, Liang C, Shah M, Sun F, Wang HM, Reid SP, Gingle AR, Paterson AH, Wing R, Dean R, Klein R, Nguyen HT, Ma HM, Zhao X, Morishige DT, Mullet JE, Cordonnier-Pratt MM (2005) Sorghum expressed sequence tags identify signature genes for drought, pathogenesis, and skotomorphogenesis from a milestone set of 16,801 unique transcripts. Plant Physiol 139:869–884

Propheter JL, Wang D, Wu X, Staggenborg SA (2010) Performance of annual and perennial biofuel crops: yield during the first two years. Agron J 102:806–814

Reddy BVS, Kumar AA, Ramesh S (2007) Sweet sorghum: a water saving bio-energy crop. In: International conference on linkages between energy and water management for agriculture in developing countries, IWMI, ICRISAT, Hyderabad, India

Reddy BVS, Layaoen H, Dar WD, Rao PS, Eusebio JE (2011) Sweet sorghum in the Philippines: status and future. ICRISAT, Patancheru 502 324, Andhra Pradesh, India

Renewable Energy World (2000) Bioethanol-industrial world perspective. Website www.jxj.com/magsandj/rew/200_03/bioethanol.html

Ribera LA, Outlaw JL, Richardson JW, Silva JD, Bryant H (2007) Mitigating the fuel and feed effects of increased ethanol production utilizing sugarcane. St. Louis, MO, Farm Foundation Conference

Ricaud R, Arceneaux A (1990) Sweet sorghum research on biomass and sugar production in 1990. Manuscript report from the St. Gabriel Experiment Station

Ricaud RB, Cochran B, Arceneaux A, Newton G (1979) Sweet sorghum for sugar and biomass production in Louisiana. Manuscript report from the St. Gabriel Experiment Station

Rooney WL, Blumenthal J, Bean B, Mullet JE (2007) Designing sorghum as a dedicated bioenergy feedstock. Biofuels, Bioprod Biorefin 1:147–157

Salassi ME (2007) The economic feasibility of ethanol production from sugar crops. Louisiana Agric 50:1

Salzman RA, Brady JA, Finlayson SA, Buchanan CD, Sun F, Klein PE, Klein RR, Pratt LH, Cordonnier-Pratt M-M, Mullet JE (2005) Transcriptional profiling of sorghum induced by methyl jasmonate, salicylic acid, and aminocyclopropane carboxylic acid reveals cooperative regulation and novel gene responses. Plant Physiol 138:352–368

Schaffert RE, Borgonovi RA, Casela CR (1986) Metodologia usada para desenvolver novas cultivares de sorgo sacarino. Relatorio Tecnico Anual do Centro Nacional de Pesquisa de Milho e Sorgo 1980–1984. Sete Lagoas, Minas Gerais, EMBRAPA/CNPMS, Brazil

Shapouri H, Salassi M, Fairbanks JN (2006) The economic feasibility of ethanol production from sugar in the United States. USDA-OEPNA and Louisiana State University

Singh SP (1985) Sources of cold tolerance in grain sorghum. Can J Plant Sci 65:251–257

Smith GA, Buxton DR (1993) Temperate zone sweet sorghum ethanol production potential. Bioresour Technol 43:71–75

Smith MS, Frye WW, Varco JJ (1987) Legume winter cover crops. Adv Soil Sci 7:95–139

Soltani A, Almodares A (1994) Evaluation of the investments in sugar beet and sweet sorghum production. In: National convention of sugar production from agriculture products, 13–16 Mar 1994, Shahid Chamran University, Ahwaz, Iran

Somerville C (2007) Biofuels. Curr Biol 17:R115–R119

Tao YZ, Jordan DR, Henzell RG, Mcintyre CL (1998) Construction of a genetic map in a sorghum RIL population using probes from different sources and its comparison with other sorghum maps. Aust J Agric Res 49:729–736

Tuinstra MR, Al-Khatib K (2007) New herbicide tolerance traits in sorghum. In: Proceedings of the 2007 corn, sorghum, and soybean seed research conference and seed expo, Chicago, IL. 5–7 Dec 2007. Am Seed Trade Assoc, Alexandria, VA

Tuinstra MR, Soumana S, Al-Khatib K, Kapran I, Toure A, van Ast A, Bastiaans L, Ochanda N, Salami I, Kayentao M, Dembele S (2009) Efficacy of herbicide seed treatments for controlling striga infestation of sorghum. Crop Sci 49:923–929

Undersander DJ, Lueschen WE, Smith LH, Kaminski AR, Doll JD, Kelling KA, Oplinger ES (1990) Alternative field crops manual: sorghum syrup. http://www.hort.purdue.edu/newcrop/afcm/syrup.html. Last accessed: 06/28/2013

US DOE (2006) Breaking the biological barriers to cellulosic ethanol: a joint research agenda. Report from the December 2005 workshop, DOE/SC-0095. U.S. Department of Energy Office of Science. http://www.genomicscience.energy.gov/biofuels/. Last accessed: 06/28/2013

Vermerris W (2011) Survey of genomics approaches to improve bioenergy traits in maize, sorghum and sugarcane. J Integr Plant Biol 53:105–119

Vermerris W, Saballos A, Ejeta G, Mosier NS, Ladisch MR, Carpita NC (2007) Molecular breeding to enhance ethanol production from corn and sorghum stover. Crop Sci 47:S142–S153

Vietor DM, Miller FR (1990) Assimilation, partitioning and nonstructural carbohydrates in sweet compared with grain sorghum. Crop Sci 30:1109–1115

Wiendenfeld R (1984) Nutrient requirement and use efficiency by sweet sorghum. Energy Agr 3:49–59

Worley JW, Cundiff JS (1991) System analysis of sweet sorghum harvest for ethanol production in the Piedmont. Trans Amer Soc Agr Eng 34:539–547

Wu X, Staggenborg S, Propheter JL, Rooney WL, Yu J, Wang D (2008) Features and fermentation performance of sweet sorghum juice after harvest. ASABE Paper No. 080037. ASABE, St Joseph, MI

Xin Z, Wang ML (2011) Sorghum as a versatile feedstock for bioenergy production. Biofuels 2:577–588

Xin Z, Wang ML, Barkley NA, Burow G, Franks C, Pederson G, Burke J (2008) Applying genotyping (TILLING) and phenotyping analyses to elucidate gene function in a chemically induced sorghum mutant population. BMC Plant Biol 8:103

Zhao YL, Dolat A, Steinberger Y, Wang X, Osman A, Xie GH (2009) Biomass yield and changes in chemical composition of sweet sorghum cultivars grown for biofuel. Field Crops Res 111:55–64

Chapter 7
Genetics, Genomics and Crop Modelling: Integrative Approaches to the Improvement of Biomass Willows

Angela Karp, Goetz M. Richter, Ian F. Shield and Steven J. Hanley

Abstract Willows (*Salix spp.*) grown as short rotation coppice (SRC) are among the leading commercially grown biomass crops in temperate regions, however, compared with major arable crops they are relatively undomesticated. Initial advances in improving the crop were made by selecting stem characteristics (height, diameter, straightness) and coppicing response (shoot number, shoot vigour), as well as resistance to pests, diseases and environmental stress. Selections were achieved purely on the basis of phenotype, with little understanding of the genetics for many of these important traits, or how they interact with each other and the environment. To enhance yields further, and to adapt the crop to future climates and more marginal environments where biomass crops will be encouraged, a more holistic understanding is needed of the key traits to target and expected gene-environment interactions. In this chapter we begin by reviewing what is known about growth in willow in relation to the parameterisation of process-based models and the advances made in willow genetics and genomics. We finish by considering an integrative approach which feeds genotypic information into phenotypic models of source-sink interaction to identify target traits for crop improvement.

Keywords Willow • Salix • Biomass • Growth • Genetics • Process models • Breeding

A. Karp (✉) · I. F. Shield · S. J. Hanley
AgroEcology Department, Rothamsted Research,
Harpenden AL5 2JQ, Herts, UK
e-mail: angela.karp@rothamsted.ac.uk

G. M. Richter
Dept. of Sustainable Soils and Grassland Systems, Rothamsted Research,
Harpenden AL5 2JQ, Herts, UK

M. C. McCann et al. (eds.), *Plants and BioEnergy*, Advances in Plant Biology 4,
DOI: 10.1007/978-1-4614-9329-7_7,

7.1 Introduction

Willows are catkin-bearing trees of the genus *Salix*. They occupy a wide variety of ecological niches in temperate zones and are extremely variable in growth characteristics. An exception is the enclosing of buds within a single scale; one of the characteristics that distinguish *Salix* from the related genus *Populus* (poplars). The *circa* 400 species are broadly grouped into the tree willows (sub-genus *Salix*), the dwarf and alpine willows (sub genus *Chamaetia*) and the shrubby willows (sub-genus *Vetrix*). The latter are particularly suited as biomass crops due to their propensity for fast, vigorous growth in coppicing cycles, ease of vegetative propagation and low fertilisation requirements (an average of 20–30 kg N ha^{-1} $year^{-1}$ depending upon site) (Karp et al. 2010a).

Systematic breeding of willow for biomass has only been pursued since the late 1980s in response to a rise in the oil prices, and then in just a few places worldwide. The potential for further crop improvement is therefore huge. Central to realising this potential are the answers to two inter-related questions: what is growth in willow and what determines willow growth? These questions are being tackled indirectly and directly by many groups, including the BSBEC-BioMASS consortium, which is part of the UK BBSRC Sustainable Bioenergy Centre (BSBEC). BSBEC-BioMASS planted a trial of four different genotypes (Tora, Endurance, Resolution and Terra Nova) at two sites in the UK (Cerasuolo et al. 2013). A large number of non-destructive and destructive measurements were made on this trial, which, together with studies of willow mapping populations, have led to an improved understanding of many aspects of willow growth. In this review we draw on these results, and other studies, and show that the answers to the questions posed above lie in understanding how the plant integrates across many processes.

7.2 Growth in Willow

Growth is as an increase in size due to cell division and subsequent cell expansion. Growth processes have been most thoroughly studied in the model plant *Arabidopsis thaliana* (Thale Cress) and knowledge is accumulating in poplar as a model tree. We start by considering willow growth in comparison with Arabidopsis and poplar, as differences become relevant later when processes and candidate genes derived from these model plants are considered.

7.2.1 Perenniality and Seasonal Growth

Arabidopsis thaliana is a self-fertilising hermaphrodite annual. The primary shoot apical meristem grows monopodially, remaining active throughout the life span of the plant and continuously producing new lateral organs. Growth is primary and vegetative until environmental cues (Sect. 7.3.1) result in a developmental switch to flowering. An inflorescence is produced, seed is set and the plant dies. New plants develop from the seed.

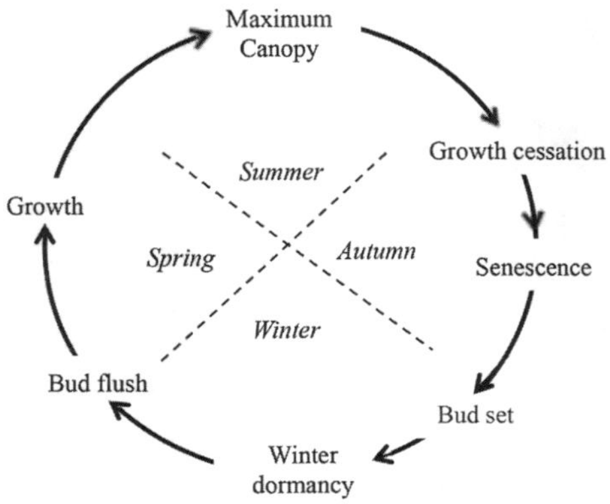

Fig. 7.1 The perennial cycle of willow and poplar depicting the main growth transitions. Although some stages overlap, or occur concurrently, they involve separate developmental processes and have been shown distinctly for this reason

Willows and poplars are obligate (diocecious) outcrossing perennials. They show sympodial primary and secondary growth. There is a distinct period in the year, from early spring to late summer when they are actively growing and the remainder when growth has ceased (Fig. 7.1). Growth cessation marks the beginning of several developmental phases which include senescence, leaf fall, the laying down of reserves, induction of cold hardiness and, like Arabidopsis, a developmental shift from leaf to bud formation. Unlike *A. thaliana*, however, willows and poplars form floral and vegetative buds in different positions on the stem and the end of the season is not demarked by flowering and seed production but by bud set and senescence. In poplar a terminal bud is formed at the shoot apex but in willow the shoot tip degenerates, or even abscises. The plants over-winter in a state of dormancy. New spring growth begins with the activation of vegetative buds (bud break or bud flush) to form leaves and floral buds (flowering) to form catkins. In willow there is considerable variation in catkin morphology and whether they appear before (precocious—e.g. *S. daphnoides* and *S. viminalis*), at the same time (coetaneous, e.g. *S. alba* and S. *babylonica*), or after leaves form (serotinous, e.g. *S. pentandra* and *S. triandra*). Bud break is closely followed by growth from the new apical meristem in the stem tip and the cambium (Sect. 7.2.3). In addition, adventitious buds may arise during the growing season to form sylleptic branches. In poplars these contribute to biomass yield, but not in willow (Sennerby-Forsse 1995).

Flowering is delayed in poplar (for around seven years) but most willows flower in the first or second year of growth from a seed or vegetative cutting, an important advantage for genetics and breeding. The tiny seed has little endosperm and needs to germinate immediately. Fast seedling growth rates (1.0–1.4 g per week), which exceed standard woody plants, have been reported (Grime et al. 1988). The seed comprises genetically distinct individuals and is difficult to handle. Consequently, commercial biomass willows are planted as stem cuttings of ~20 cm length. Growth rates from cuttings are also high, for example 1.20 g per week for *S. viminalis* under continuous light (MacDonald 1989).

7.2.2 Coppicing: The Stem and the Stool

In short rotation coppice (SRC), stem cuttings are planted in spring and grown for one year, when they can reach 2–3 m in height. After leaf fall the stems are cut back to induce a coppicing response, in which multiple shoots re-sprout from the cut base (stool) in the following spring. Further growth is allowed, typically for three more years, by which time the stems are ~7 m tall. Specialist machinery is used to harvest the stems to produce chips or billets for storage and use. Resprouting follows in spring and the cycle is continued for *circa* 20 years.

The SRC cycle has interesting features in terms of growth (Fig. 7.2). The stem cutting has everything required to re-grow a tree: vegetative buds to form leaves, adventitious nodes/lenticels to develop roots and reserves for initial growth (Sect. 7.3.3). Growth is similar to that from the intact stem in inter-harvest years. New growth from cut stools, however, relies on stool and root reserves. It originates from axillary buds in the stool that may have been dormant for successive years until stem removal at harvest releases them from apical dominance. The stool starts as a small swelling and increases in size with each successive harvest. The number of re-sprouting shoots differs among species (Sennerby-Forsse and Zsuffa 1995). Axillary buds contain three primordia, each of which has a fixed number of initiation nodes from which leaves develop to give a flush of "fixed growth". Additional "free growth" arises from new nodes initiated during the year (Sennerby-Forsse et al. 1984). Many buds sprout simultaneously but the resultant stems are progressively thinned, a process involving differential growth rates, dominance and

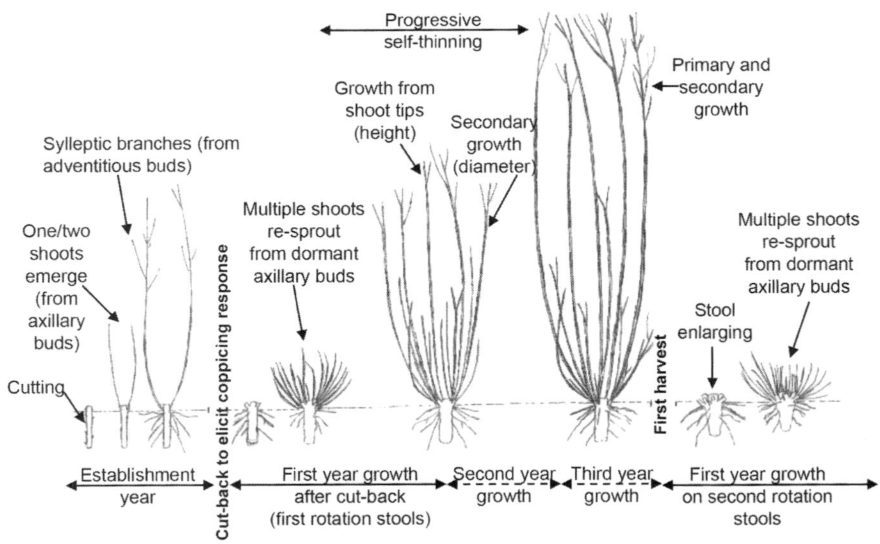

Fig. 7.2 The main growth characteristics of the short rotation coppice cycle in willow. Events are shown for the establishment year through the first rotation, including initial re-growth after the first harvest

suppression of stems and re-allocation of resources (Sennerby-Forsse et al. 1984). Coppicing re-invigorates growth and its' timing is a compromise between the diminishing yield gained per successive year, harvesting costs, and the need to replenish sufficient reserves for vigorous re-sprouting (Verwijst 1996a, b).

Root growth in *S. viminalis* is initiated at least one week before bud break. Several pulses of increased growth, followed by increased decay occur throughout the season, suggesting that the roots continually turn-over during shoot growth (Rytter and Hansson 1996; Rytter 1999, 2001). Root turn-over rate in planted willow stands was 4.9–5.8 times year^{-1}, and similar to that for willows grown in lysimeters (4.8–8.1 year^{-1}; Rytter and Rytter 1998). In *S. triandra* (Stott 1961) and *S. dasyclados* (Ericsson 1984) 90 % of roots occurred within the upper 12–20 cm. A mean rooting depth of 25–30 cm was later reported in *S. viminalis* (Rytter and Hansson 1996) but roots have been detected at 1.3 m. Estimates of the standing fine root ($\leq$2 mm) biomass, to a depth of 50 cm, average 1,320 kg ha^{-1} (Rytter 1999).

A relationship between maximum root and stem diameters in poplar and willow accounted for 7.8 and 10.9 % of the variation, respectively (Crow and Houston 2004). Older stools showed no evidence of larger roots, suggesting that regular coppicing may slow down root development or remove the need for larger roots (Crow and Houston 2004). Rytter found below-ground allocation was highest in the first year (49 and 58 % of total biomass for plants grown in lysimeters on clay and sand, respectively), and dropped to 36–38 % (clay) and 33–40 % (sand) in the second and third years (Rytter 2001). There is also a general relationship between stool size and above-ground biomass and both are affected by inter-plant competition. If planting densities are too high, excessive self-thinning occurs and stool mortality rates increase as smaller stools are out-competed (Verwijst 1996a, b). BSBEC-BioMASS has shown that allocation is genotype-dependent. All four genotypes studied showed high biomass yields but Resolution allocated proportionally more above ground (J. Cunniff, Rothamsted Research UK (unpublished results).

7.2.3 Primary and Secondary Growth

The willow shoot tip contains the shoot apical meristem (SAM) and comprises a stem portion with overlapping upturned leaves that extend above the tip. Shoot tips differ in length (e.g. 55, 15–30 and 25–40 mm, for *S. viminalis, S. caprea,* and their hybrid, respectively) but are generally very small (Sennerby-Forsse et al. 1984). Leaf initiation originates in the SAM and is alternate along the stem. The SAM typically contains an outermost layer, which produces the epidermis, and an innermost layer which gives rise to the rest of the shoot. Below this is the zone of central mother cells, beneath which cell divisions produce vertical cell files that become the pith and the pro-cambium. A few cm lower the vascular cambium is already visible as a ring. Primary growth is continuous throughout the season, although growth rate is affected by temperature and stress (Sennerby-Forsse et al. 1984).

Secondary growth, leading to an increase in stem thickness, results from mitotic activity in the cambium which produce sapwood (or secondary xylem) to the inside and bark (or secondary phloem) to the outside. In most trees large cohorts of cells within the cambium undergo divisions over discrete time intervals in the growing season, resulting in growth rings. Willows are diffuse or semi-ring porous, and growth rings are not always distinct although, the larger thinner walled vessels formed first (earlywood) can normally be distinguished from the thicker, smaller latewood produced later (Arihan and Guvenc 2011). The four genotypes of the BSBEC trial showed similar stem anatomy, although differences were present in the size of the xylem vessels and in the timing of cambial activity.

Wood formation is a dynamic process and many trees make Reaction Wood (RW) to structurally reinforce and redirect growth towards the vertical in response to mechanical or gravitational stress. In willow and poplar RW comprises Tension Wood (TW) on the 'upper' (tension) stem and Opposite Wood (OW) on the 'lower' stem. TW is often characterised by the formation of a gelatinous layer, composed mainly of cellulose, within the fibre cells of the secondary xylem. TW is problematic for the pulp and paper industry as it reduces the quality of the wood but the high cellulose content means that sugars can be released more easily in biomass processing, making it attractive for biofuel production (Liu and Liu 2010). BSBEC-BioMASS has shown that final glucose yields can be increased by TW induction without a detrimental impact on biomass yield *per se* (Brereton et al. 2011).

7.3 Determinants of Willow Growth

7.3.1 Environmental Cues: Light and Temperature

In poplar, growth cessation occurs principally in response to a change in day length from long days (LD) to short days (SD) although light quality and temperature are also important, depending on the latitudinal origin of species (Rohde et al. 2002; Druart et al. 2007; Olsen 2010). Studies of ecotypes of *S. pentandra* L indicate that similar environmental cues operate in willow (Junttila and Kaurin 1990).

Winter dormancy can be considered as an endogenous repression that is maintained, even if conditions become permissive to growth, until a specific cue triggers release (Jimenez et al. 2010). The cue used by many perennial species is quantitative accumulation of cold exposure, or chilling requirement. Evidence suggests this is also true for willow (Cannell et al. 1987). Moreover, willow shows genetic variation for this trait (Lennartsson and Ogren 2004; Weih 2009). Flushing early has been shown to result in increased yields but is not without risk, as once the buds burst they became frost sensitive (Lennartsson and Ogren 2004).

7.3.2 Endogenous Cues: Phytohormones

Cytokinin activity increases abruptly coinciding with floral and leaf bud burst in willow after which levels fall and remain low throughout the summer (Bowen and Hoad 1968; Alvim et al. 1976). In contrast, ABA levels increase over the winter following leaf abscission, then decrease before growth initiation in the spring although a further increase in ABA occurs in July just before growth cessation (Alvim et al. 1976). No correlations between ABA levels and photoperiodicity have been detected (Alvim et al. 1979).

Regulation of axillary lateral SAMs is a key way in which plant architecture is determined. The polar auxin transport stream acts as an inhibitor of axillary bud outgrowth, an effect opposed by cytokinins and strigolactones (Domagalska and Leyser 2011; Mueller and Leyser 2011). Root apical meristems are controlled negatively by cytokinins and positively by auxin. PIN proteins regulate auxin flow by localizing it asymmetrically in the vessels. We have demonstrated that hormone physiology assays developed in Arabidopsis can be directly transferred to study control of bud activation in biomass willow and that bud hormone response was qualitatively remarkably similar. Arabidopsis hormone mutants were used to assess allelic variation in the cognate willow hormone genes and allelic differences in willow strigolactone genes were observed using this approach (Ward et al. 2013).

Gibberellins (GAs) are regulators of stem or leaf elongation, and together with brassinosteroids, are considered to be the main factors influencing plant height. In poplar cessation of cell division and cell elongation in response to SD is associated with down regulation of GAs and can be reinitiated by GA application (Olsen 2010). Similarly, in *S. pentandra* GAs influence shoot elongation and cessation of apical growth can be prevented by exogenous GAs (Davies et al. 1985). In elongating shoots of *S. dasyclados* and *S. viminalis* GA_1, GA_4, GA_8, GA_9, GA_{19}, GA_{20} and GA_{29} were all detected but levels of GA_{19} and GA_{20} were particularly high in both species in the vegetative shoot (Junttila et al. 1988).

7.3.3 Resource Availability and Recycling: Carbon (C) and Nitrogen (N)

A few weeks prior to any visible new growth in the spring, there is a mobilisation of resource along the willow stem. The latest formed vessels overwinter in mature state and are utilised first when growth recommences (Sennerby-Forsse 1986, 1995). Activation of the cambium follows, beginning at the apex and taking more than a week to spread to the base. Phloem differentiation starts at least two weeks before flowering and bud break, whilst xylem differentiation occurs around the same time as bud activation (Sennerby-Forsse 1986).

Spring xylem flow comprises a flux of concentrated sucrose solution derived from the mobilisation of starch reserves in xylem ray cells in the roots and stool. Sucrose is cleaved near sink tissues to form fructose and glucose by sucrose synthase and invertase. In poplar, sucrose and fructose represent 70 and 90 % respectively of the total soluble sugars available to the cambium during active growth (Deslauriers et al. 2009). As xylem differentiation progresses, the soluble sugars decrease, indicating that secondary growth rapidly becomes a considerable C sink. From August onwards, increasing cold hardiness is associated with starch breakdown and an increase in the levels of raffinose and other potential cryoprotectants (Druart et al. 2007). Parallel changes in flux occur in leaves; in quaking aspen, sucrose concentrations increase during leaf expansion whilst hexose sugars peak at the point of mid-expansion and then rapidly decrease (Jeong et al. 2004).

A xylem-to-phloem transfer facilitated by ray cells enables transport of N as amino acids in the phloem to developing leaves (Cooke et al. 2003). Leaf photosynthetic capacity shows a strong positive correlation with leaf N content and vertical leaf N gradient in the canopy is positively correlated with shoot biomass (Weih and Rönnberg-Wastljung 2007). Resorption of N from leaves and its remobilisation is an important factor in maintaining growth since the more effective this is the less N is lost from the plant. Pests and diseases, or drought, can result in premature senescence and reduce photosynthetic capacity too early in the season, whilst increased N availability can delay growth cessation and senescence and limit the time for efficient resorption from leaves. Seasonal N cycling is linked to phenology and is well characterised in poplars (Cooke et al. 2003). Glutamine and asparagine are the major forms in which N is transported in poplar (Cooke and Weih 2005). Over winter, N is stored principally in vegetative storage proteins (VSPs), particularly below the bark (bark storage proteins; BSPs). BSPs show a characteristic pattern of autumn accumulation and spring disappearance within the bark, wood and roots. Reactivation of cambial activity in spring coincides with degradation of BSPs and a rise in amino acid levels (Druart et al. 2007).

7.4 Modelling Growth in Willow

Models can increase the selection of improved phenotypes in breeding by identifying key elements on the basis of their relative importance within the system. Model-aided crop improvement is a succession of explorations achieved by dissecting the processes contributing to productivity into their elements. The number of processes describing these elements depends on the desired level of complexity and experimental evidence available. For a descriptive simulation of plant growth, the key components of any model would comprise elements of development (phases of phenology), resource capture (leaves, roots) and storage (biomass, reproductive organs).

A large number of models exist for simulating growth and productivity of trees, forest or SRC, which can be categorised into research and management tools and vary greatly in complexity. Models have been specifically developed for SRC willow (Eckersten and Slapokas 1990) and poplar (Deckmyn et al. 2004), of which the poplar model SECRETS is the most comprehensive and complex, describing a total of 22 processes and quantitative relationships (Deckmyn et al. 2004). The SECRETS model was adopted for a similar biological and management system and is thus closest to meeting the requirements needed to describe the trials in BSBEC-BioMASS. However, it was not considered adequate for our purposes because the carbon (C) allocation routine originates from a global, upscaled approach, which distinguishes between green C (leaf, fine roots, etc.) and structural C. This was considered unlikely to match our specific requirements for a detailed process description of canopy dynamics, of the interaction between above- and below-ground compartments and of the effect of reserves, within the sink-source control operating in the willow perennial cycle. To achieve the latter we implemented a sink-source balance described in LINGRA (Schapendonk et al. 1998) which has been successfully applied in a generic way to energy grasses (Richter et al. 2010b; Triana et al. 2011). In the following, we outline the conceptual basis of our process-based model and show the power of the model sensitivity analysis to rank model parameters (Richter et al. 2010a). Ultimately our vision is to link genetic information to the model parameterization to explain the gene to phenotype (G $\rightarrow$ P) relations (see Sect. 7.6.3).

Our newly developed model for Light Use and Carbon Allocation for Salix Species (LUCASS) simulates growth and development of SRC willow grown in monoculture (M. Cerasuolo, Rothamsted Research, UK, unpublished results). LUCASS considers the processes of phenological and morphological development of the plant (including senescence and dormancy), light interception, photosynthesis, and respiration, as well as biomass formation. The model defines the plant organs (buds, leaves, branches, stems, stool and roots) as sinks, to which carbohydrates (CHO) from a common source pool (photosynthates, mobilised reserves) are allocated. Partitioning follows the principles of balancing demand (sink strength) and supply (CHO sources). These processes are dependent on temperature and day length as well as on light and water availability. Boundaries for crop growth are defined by the soil hydrology and water and energy balance model.

The specificity of our modelling approach was first demonstrated for the simulation of light interception (Cerasuolo et al. 2013) which accounted for the horizontal and vertical structure of the canopy of SRC willow. The model represented the varietal differences for light interception observed for the four genotypes in the BSBEC field trial in terms of the distribution of leaf inclination and its effect on light extinction, but most prominently, in terms of the clumping index and vertical leaf area distribution. Compared to explicit 3D descriptions of the canopy structure, this model provides a parsimonious but effective means of identifying genotype-specific traits to improve varieties.

The sensitivity analysis applied to LUCASS identified other important varietal traits that characterise light interception, like the leaf shape and leaf extension rate. The parameters of sink formation; the onset of stem formation (phenological control related to daylength) and the stem extension rate, proved to be of predominant importance for yield, across a range of environments. The fraction allocated to aboveground biomass is unsurprisingly the most important parameter, and genotypes derived from harsh environments would invest more into belowground reserves to survive periods of growth inhibition. Experimentally, however, this ratio is challenging to establish, as biomass extraction methods are often incomplete (e.g. ignoring fine and subsoil roots) and low sampling frequencies do not allow relative allocation rates to be derived. The experimental setup, such as used in the BSBEC trial, also does not allow for the quantification of root exudates, which is crucial for estimating the total carbon allocation balance. Total assimilation rates based on accumulated biomass will therefore be underestimated, and it will be important to support the calibration of photosynthesis parameters with instantaneous measurements.

Different process-based models utilise a variable mixture of mechanistic and empirical understanding that is reflected in the variable number of elements and mathematical relationships. In developing LUCASS as a first modular step to developing a full process-based SRC willow model we have identified some important knowledge gaps and assumptions that need be filled by new experimentation. For the integration of genetic information (Sect. 7.5), the largest challenge will be to identify and dissect key processes, i.e. to become more mechanistic. For this, new evidence needs to be generated to bridge the gap between genomic (QTL, gene) and phenotypic (physiological, pheno-morphological) information and to calibrate new, dissected process parameters (see Sect. 7.6.3).

7.5 Genetic Determinants of Growth Processes Relevant to Willow

To determine the genetic basis of growth processes affecting yield in willow key resources for both linkage and association genetics have been developed and exploited (Karp et al. 2011). One of the largest mapping populations, K8 (n = 947), was planted at Long Ashton (Somerset, West UK) in 2000, at RRes (Hertfordshire, East UK) in 2002 and the RRes farm at Woburn (well-drained, nutrient-poor soils) in 2009. An additional eleven populations (n = ~500) were also developed at RRes. Several smaller families exist in Sweden and North America. The Swedish and UK groups also jointly formed an association mapping population (n = 380) which is planted at contrasting sites in both countries (Karp et al. 2011).

Using these resources QTLs have been mapped for a large number of traits in willow including rust resistance (Hanley 2003; Tsarouhas et al. 2003; Hanley et al. 2011), insect resistance (Rönnberg-Wastljung et al. 2006), shoot height, stem diameter, and stem number (Tsarouhas et al. 2002), frost tolerance, phenology (Tsarouhas et al. 2003, 2004), water-use efficiency and drought tolerance (Rönnberg-Wastljung et al. 2005; Weih et al. 2006). Willow QTLs are currently being interrogated with respect to candidate genes from Arabidopsis and poplar. It is not feasible to comprehensively cover the range of genes and gene networks that could have relevance here. Instead, an overview is given in Table 7.1.

Table 7.1 Overview of the main genetic controls of growth processes in Arabidopsis and corresponding knowledge in poplar/willow. Only some key genes/pathways are described. For more comprehensive coverage please see suggested publications in poplar and reviews

Growth process	Some key genes in Arabidopsis	Simplistic overview in Arabidopsis	Key genes/knowledge in poplar (and/or willow, if known)	References
Light/Photoperiod perception circadian clock	• *PHYTOCHROME* (*PHY*) • *PHYTOCHROME INTERACTING FACTORs* (*PIFs*) • *LATE ELONGATED HYPOCOTYL1* (*LHY1*) • *TIMING OF CAB EXPRESSION1* (*TOC1*) • *FLAVIN-BINDING, KELCH REPEAT F-BOX1* (*FKF1*) • *GIGANTEA* (*GI*)	Photoperiod is perceived in leaves by phytochromes (PHY) which entrain components of the circadian clock. Five *PHY* genes are known. They interact with phytochrome interacting factors (PIFs). Several genes (e.g. *LHY1* and *TOC1*) act as regulatory clock components, interacting with other pathways. E.g.. in blue light (daytime) a circadian clock-controlled FKF1-GI complex degrades transcriptional repressors of *CO* and regulates the timing of *CO* expression	Two *PHY* genes are known in poplar: *PHYA* is involved in dormancy and *PHYB* in bud flush. *PHYB2* co-locates with bud set QTLs on two linkage groups. Single nucleotide polymorphisms (SNPs) in *PHYB2* showed clinal variation, suggesting it may be involved in adaptive response to photoperiodic conditions. Poplar regulatory clock components include *PtLHY1*, *PtLHY2* and *PtTOC1*	Frewen et al. (2000), Ingvarsson et al. (2006), Ibanez et al. (2010), Olsen (2010), Kunihiro et al. (2011), Sawa and Kay (2011, 2012)

(continued)

Table 7.1 continued

Growth process	Some key genes in Arabidopsis	Simplistic overview in Arabidopsis	Key genes/knowledge in poplar (and/or willow, if known)	References
Vegetative to floral transition (at least five pathways are known: a key one is shown here)	• *FLOWERING LOCUS C (FLC)* • *CONSTANS (CO)* • *FLOWERING LOCUS T (FT)* • *bZIP TRANSCRIPTIONAL FACTOR* • *FRIGIDA (FRI)*	Time measurement in the photoperiodic flowering pathway is regulated by daytime expression of *CO*. *CO* upregulates *FT* expression whilst *FLC*, a strong repressor of flowering, directly represses *FT*. Under LD *CO* expression coincides with light, high *CO* in leaves upregulates *FT*, the FT florigen goes to the shoot apex and complexes with the bZIP transcriptional factor to activate expression of floral-meristem identity genes (below). In SD, light/*CO*/*FT* coincidence is lost, *FT* is repressed and floral meristems do not form. In vernalization-requiring accessions (e.g. perennial *A. lyrata*) *FRI* upregulates *FLC* whilst prolonged cold (vernalization) overrides *FRI* enabling flowering when temperatures warm	No structural orthologue of *FLC* is known in poplar but there is an *FT* duplication (*FT1* and *FT2*, on chromosomes VIII and X). Control via the CO/FT regulon model has been demonstrated in SD-induced growth cessation and bud set. However, over-expression of *CO* gave no evidence to support a role of *CO* in regulating bud set or bud flush. A pulse of *FT1* expression in winter appears to initiate the transition from vegetative to reproductive meristems. *FT2* forms molecular networks with different genes in response to various stress factors to control vegetative growth	Bohlenius et al. (2006), Hsu et al. (2011, 2012), Sawa and Kay (2011), Pin and Nilsson (2012), Kemi et al. (2013)

(continued)

Table 7.1 continued

Growth process	Some key genes in Arabidopsis	Simplistic overview in Arabidopsis	Key genes/knowledge in poplar (and/or willow, if known)	References
Meristem identity: floral primordial formation	• *LEAFY* (*LFY1*) • *APETALA1* (*AP1*)	All pathways involved in vegetative to reproductive transitions converge on meristem identity genes, of which *LFY* is the most important. *LFY* occurs at low levels in leaf primordia but once above a threshold identity switches to floral primordia. *AP1* encodes a transcription factor with a MADS-domain and determines sepal and petal development	Studies of a *LF* orthologue (*PtLFY*;) identified in poplar and *AP1* orthologue (*SAP1*-1. from *S. discolour* suggest similar controls will operate in willow to switch meristem identity from leaf to floral primordial)	Fernando and Zhang (2006), Srikanth and Schmid (2011), Hsu et al. (2012), Siriwardana and Lamb (2012)
Cell division	• *CYCLINs* (*CYCs*) • *CYCLIN-DEPENDENT KINASEs* (*CDK*)	The cyclin family of proteins regulate the cell cycle by interacting with cyclin (dependent kinases CDKs). Arabidopsis contains 10 *CYCD* genes	22 *CYCD* genes were identified in the poplar (Populus trichocarpa) genome; six *CYCD* subgroups are conserved across higher plants	Menges et al. (2007), Dong et al. (2011)
Cell wall synthesis and secondary growth	• *CELLULOSE SYNTHASE* (*CesA*) • *GLYCOSYLTRANSF ERASES GTs* (*IRXs*) • *VASCULAR-RELATED NAC DOMAIN* (*VND*)	*CesA* encode cellulose synthase catalytic subunits. Lignin biosynthesis is a branch of the phenylpropanoid pathway and includes several gene such ass *PAL, C4H, 4CL, CCR, HCT, C3H, CCoAOMT, CCR, F5H, COMT, CAD*. GTs (*IRX* genes) are involved in xylan (hemicellulose) biosynthesis	Many of the cell wall biosynthesis genes have been identified in poplar, particularly in the lignin pathway. In addition, a number of transcription factors regulate differentiation of vessels. proteins (VND6, VND7, SND1 and NST1) are key switches regulating a cascade of downstream transcription factors leading to secondary wall biosynthesis	Sarkar et al. (2009), Demura and Ye (2010), Douglas et al. (2011), Carroll et al. (2012), Li et al. (2012), Sanchez-Rodriguez et al. (2012), Schuetz et al. (2013)

(continued)

Table 7.1 continued

Growth process	Some key genes in Arabidopsis	Simplistic overview in Arabidopsis	Key genes/knowledge in poplar (and/or willow, if known)	References
Shoot apical meristems (SAMs) and root apical meristem (RAMs)	• *KNOTTED1* (*KN-1*) like (*KNOX1*) • *SHOOTMERISTEMLESS* (*STM*) • *BREVIPEDICELLUS* (*BP*) • *CUPSHAPED COTYLEDON 2* (*CUC2*) • *PINHEAD* (*PNH*) • *NO APICAL MERISTEM* (*NAM*) • *WUSCHEL* (*WUS*) • *CLAVATA3* (*CVL3*) • *CLAVATA-LIKE19* (*CVLE19*) • *WUS-LIKE* (*WOX5*) • *SCARECROW* (*SCR*)	*KNOX1* homeobox transcription factors *STM* and *BP* play key roles in formation of the embryogenic SAM. *CUC2*, *NAM* (identified in Petunia) all affect SAMs but through different mechanisms. The identity and maintenance of stem cells in the SAM central zone is regulated by *WUS* which interacts with *CLV3* to regulate the size of the organizing centre and the stem cell niches. Mutations *stm*, *cvl1*, *cvl3*, *pnh* and *wus* affect lateral and primary SAMs, indicating common regulation. A *WUS*-like gene (*WOX5*) and a *CVL3* homologue (*CLE19*) regulate the root apical meristem (RAM), whilst *SCR* is involved in specifying tissue identity in RAMs	In poplar the role of *WUS* and *STM* in both SAM and axillary meristems has been confirmed but functional orthologues of *WUS* and *STM* (*PttWUS* and *PttCLV3*) do not appear to play a role in the cambium. Instead *KNOX* genes show high expression and a poplar orthologue of *STM* (*ARBORKNOX1;ARK1*) regulates cambial functions and cell differentiation during secondary growth, including regulation of cell wall biosynthesis. Similarly, a poplar *BP* ortholog (*ARK2*) is expressed in both the cambium and in lignifying cells	Schrader et al. (2004), Bao et al. (2009), Demura and Ye (2010), Du and Groover (2010)

(continued)

Table 7.1 continued

Growth process	Some key genes in Arabidopsis	Simplistic overview in Arabidopsis	Key genes/knowledge in poplar (and/or willow, if known)	References
Shoot and root growth, apical dominance architecture	• *INDOLE ACETIC ACID* (*IAA*) • *ABSCISIC ACID* (*ABI*) • *GIBBERELLINS* (*GA*)) • DELLAs • *ISOPENTYL TRANSFERASE* (*IPTs*) • *CYTOKININ OXIDASE/DEHYDROGENASE* (*CKXs*) • *ARABIDOPSIS HISTIDINE KINASE* (*AHKs*) • *ARABIDOPSIS RESPONSE REGULATORS* (*ARRs*) • *CAROTENOID CLEAVAGE DIOXYGENASE* (*CCD7*)	Key phytohormone gene networks relate to their biosynthesis, degradation and signalling. The rate limiting step of cytokinin biosynthesis is catalysed by *IPTs*, degradation is by *CKXs* whilst receptors include the *AHK* family. Auxin synthesis involves *IAA* whilst signalling includes ARRs *GA* biosynthesis involves *GA* oxidases and signalling includes receptors such as GIDI. DELLA proteins also play key roles in the gibberellin signalling pathway. Strigolactones are derived from carotenoids, e.g. via *CCD7*. The ABI biosynthetic pathway originates in the chloroplast with hydroxylation of β-carotene to zeaxanthin. ABI interact with many genes related to stress, e.g. dehydrins	In poplar over expression of *PcGA2ox* resulted in reduced apical dominance, short trees and wide crowns. Involvement of ABI and auxin related genes has been demonstrated in relation to drought stress and root and shoot growth and auxins, cytokinins and strigolactones in branching	Koornneef et al. (1998), Hartweck (2008), Nieminen et al. (2008), Domagalska and Leyser (2011), Mueller and Leyser (2011)

Although "omic" studies are only just beginning in willow, the derivation of the whole genome sequence of *P. trichocarpa* (Tuskan 2006) has resulted in data sets and resources that are highly beneficial (Yang et al. 2009). Transfer of information was facilitated by direct alignment of the K8 map to the poplar genome and the high degree of macrosynteny found between them (Hanley et al. 2006). More recently, full genome sequencing of willow has been carried out and draft sequences are close to publication. Transcriptomic and metabolomic studies have also been conducted, but not yet published. Such approaches can be combined with QTL analyses, for example to identify e-QTLs or mQTL, respectively (Sect. 7.6), that can be informative in analysis of regulatory pathways. In terms of gene validation, the disadvantage of willow is that it currently lacks a robust transformation technology. However, advancements have been made in understanding the role played by key genes through their study in transgenic Arabidopsis and poplar. Moreover, functional variation of willow alleles can sometimes be assessed by transforming them into *Arabidopsis* (Ward et al. 2013) or poplar lines.

7.6 An Integrated Understanding of Growth and Biomass Yield in Willow

Given the complexity of growth process in willow (Sect. 7.2), the plethora of exogenous and endogenous cues and gene networks involved (Sects. 7.3 and 7.5), and the large lists of possible genes (e.g. Table 7.1), how is it possible to identify key targets for breeding or genes to select? There has been a technological revolution in the way that genomes and their expression can be interrogated in a global "systems" way and gene networks affecting traits can be searched for. However, these need to be combined with hypotheses based on biology and on knowledge of the variation used by the plants in competing for growth and meeting environmental challenges. For the use of models describing the interaction between plants and the environment (Sect. 7.4), new avenues will have to be developed for integrating experimental evidence with genetic information in the parameterization of predictive tools.

7.6.1 New Technology-led Approaches

Transcriptomics, metabolomics and/or proteomics can contribute to a systems-based understanding of a critical developmental transition and/or to compare different developmental stages, tissues or genotypes. Such approaches has been used to investigate many of the developmental stages of Fig. 7.1 in poplar, including senescence (Andersson et al. 2004), secondary wood formation (Schrader et al. 2004), dormancy and terminal bud formation (Ruttink et al. 2007).

From such studies a more holistic and informed picture can be developed of how different genes come into play, initially in sensing the environmental cues, then resulting in changes in meristem identity, development and resource allocation. For example Ruttink et al. (2007) showed that SD- induced bud formation was associated with genes involved in light signal transduction and the circadian clock (e.g. *PHYB, TOC1* and *CO/FT*, *GA*s and *ABA* Table 7.1). As bud structure became visible, changes in expression associated with *WUS*, *CLV* and *KNOX1* pathways became apparent, as well as down regulation of genes involved in the cell cycle, (e.g. *CYCA1* and *CDKB*). A few weeks after SD induction genes in the glyoxylate pathway were up-regulated and photosynthetic pathways down-regulated, and carbohydrate metabolism was increasingly modified towards accumulation of storage products and vcryoprotectants.

Due to the excessive numbers of genes whose expression is altered during these transitions, these approaches, alone, do not easily lead to the identification of a single critical gene to target in breeding. A transcriptomic study of short day-induced apical (terminal) bud formation in white spruce (*Picea glauca*), identified 4,460 differentially expressed sequences (El Kayal et al. 2011), for example. This was reduced to 108 genes differentially expressed only in developing buds but a large effort is still needed to identify which of these are important to change with respect to achieving crop improvement.

A step forward lies in combining "omic" approaches with genetics, for example, by transcriptome analysis of mutants, or genotypes differing in QTL loci/alleles, or transgenic plants. Expression profiles can also be treated as phenotypes. In "genetical genomics" thousands of gene expression levels can be assayed and the expression phenotypes are subjected to standard QTL analysis to derive expression QTLs (eQTLs). The eQTL location coincides with that of the regulated gene in *cis*-regulation, while *trans*-acting eQTLs identify regulatory elements elsewhere in the genome. eQTLs may be evenly spread or appear in "hotspots", depending on the genetic architecture of the gene interactions. Master regulons that underlie *trans*-eQTL hotspots ("hubs") are of particular interest as they tend to be at the center of gene expression networks (network eQTLs). Similar approaches can be used to derive mQTL (from metabolite profiling).

Bioinformatics tools have been developed to help mine the information collected in the large data sets that are accumulating from these kinds of "omics" studies. However, knowing how to spot something of significance (i.e. defining the question) for crop improvement remains an issue, especially for complex traits for which there is little prior knowledge.

7.6.2 New Biology-led Approaches

In willow, as we have shown, the biology of growth is well characterised although knowledge of the determinants of growth has mostly been inferred from model plants. To target key biology, we have pursued a systematic approach in which we framed a number of initial hypotheses for crop improvement and built empirical models based on this available knowledge. In BSBEC-BioMASS the hypotheses were based on: extending the growing season, improving architecture and increasing above-ground allocation. Data from phenotyping the BSBEC trial and mapping populations in the field is being used to refine these hypotheses and develop process-based models to identify critical processes, developmental stages and component traits. Simultaneously we have been building and deploying genetic and genomic approaches to map QTL and identify key genes associated with these targets, using some of the technological approaches outlined above.

Phenotypic data collected for the growth stages outlined in Fig. 7.1 have demonstrated that variation exists for the majority of traits examined in willow. This includes: bud flush, biomass yield (stem heights, diameters and numbers) growth cessation and senescence, canopy and leaf traits and above and below-ground allocation, as well compositional traits (e.g. lignin, cellulose and hemicellulose content). However, examination of the way the traits co-locate with respect to yield QTL and the use of process-based modelling and sensitivity analysis (Sect. 7.4) have shown that some of the traits which are variable and have a high heritability are not the most important ones to target in breeding for improved yield.

Efforts to map bud flush (as a way of extending the growing season) have been successful. A consistent QTL was detected in several mapping populations in the same location and a possible causal candidate identified, whose function makes sense with respect to the networks listed in Table 7.1. However, integration of the results with other data from our studies and the results of sensitivity analyses suggests that it is the events prior to bud flush (e.g. the chilling requirement and resource mobilisation), rather than bud flush *per se*, that may be more important for yield improvement, at least in UK environments. The sensitivity analysis also showed that allocation is a key trait affecting yield, again suggesting that resource mobilisation events prior to new spring growth and/or dormancy may be more important than bud flush or bud set *per se*. Importantly BSBEC-BioMASS results indicate that there is not a set relationship between above and below-ground allocations and that selecting for more above-ground will not necessarily be at the expense of biomass belowground. Experiments in which plants differing in the bud flush QTL are sampled in the weeks prior to bud flush are now being performed at RRes and specific experiments have been designed to investigate resource mobilisation throughout the perennial cycle.

Willows show variation in leaf shape, leaf area and leaf area index (LAI), but plants with quite different canopy architecture can attain similar high yields. Our process-modelling, (Sect. 7.4) revealed that key parameters are not leaf area or LAI *per se* but the vertical distribution profile of leaf area and clumping index. Efficient light penetration through the canopy can be achieved for willows with large LAI if leaves also show a high degree of clumping (Cerasuolo et al. 2013). These results provide guidelines with which to assess canopy structures associated with different stem numbers.

BSBEC-BioMASS also assessed willow biomass as a feedstock for sugar release in saccharification tests for biofuel production. Genetic variation in saccharification potential was assessed and QTL mapped (Brereton et al. 2010). This has led to a number of potential targets that are now being pursued. However, the most interesting data came when correlations were sought with biomass composition and with growth traits, and none were found. However, when a RW phenotype was induced in eight genotypes grown in pots, glucose release strongly correlated with the glucose release obtained for mature field-grown trees. No such correlation was found when RW was not induced. This suggests that genotypic differences in RW response may be a primary determinant of the variation observed in sugar release from willow biomass, and could be selected for (Brereton et al. 2012). Experiments are now underway to identify the genetic basis of the differences in RW response.

7.6.3 New Model-led Approaches

The advantages of modular model structure, and the separation of code and parameter space, of process-based models, enable the down-scaling (dissection) and parameterisation of key processes at higher granularity. Hammer et al. (2006) stated, new models can help in "navigating the biological complexity in breeding improved crop plants", developing predictive tools for the genotypic controls in phenotypic ($G \rightarrow P$) models. In their paper, they refer to the combined analyses of QTLs and an eco-physiological model to address the genetic variability of leaf growth among 100 recombinant inbred lines in response to meristem temperature, leaf water potential, and vapour pressure deficit (Reymond et al. 2003). Our SRC model has similar relationships between leaf extension rate and environmental variables, and it would be interesting to evaluate these for the K8 mapping population in the context of QTL analysis. Messina et al. (2011) conceptualise an iterative framework to build realistic $G \rightarrow P$ models based on extensive experimentation to establish the relationship between genomics and phenomics. To follow this, it may be necessary to expand physiological models implementing modules that represent 3D structural components as functional-structural plant models (FSPM; Vos et al. 2010), although a pseudo-3D model (Cerasuolo et al. 2013) might suffice. The dynamics of resprouting and die-back show similarity to the tillering and tiller suppression observed in cereals as a function of red:far red ratio, where FSPM proved useful (Vos et al. 2010). Certainly, the sink-source regulation implemented as an up-scaled morphogenesis in LUCASS can be down-scaled using the concepts established in an FSPM like GreenLab (Guo et al. 2006), or other such approaches (Dingkuhn et al. 2007). Parameters determining onset of sink formation and sink size proved to be of crucial importance in the sensitivity analysis of LUCASS; however, dynamics of stem number were modelled empirically, and not as an evidence-based process.

7.7 Concluding Remarks

Further improvement of willow as a biomass crop requires an understanding of the biology of growth and the key target traits that will result in gains in yield and/or improved environmental performance. We have found that integrating knowledge from phenotyping of specific genotypes in field trials with genetic mapping and crop modelling in an iterative way has enabled the identification of key component traits that contribute to useful variation in the field and the developmental stages that are most critical to this. QTL can then be mapped and advanced "omic" technologies applied, as appropriate, to identify markers for breeding. Process-based models can also be used to help predict yield gains and performance in different environments. Despite the fact that willows cannot be transformed, this approach has led us to the identification of likely casual genes and the first results for bud flush, yield (increase in height and stem diameters) are currently in preparation as publications.

Acknowledgments The authors would like to thank the UK Biotechnological and Biological Sciences Research Council (BBSRC) and Ceres Inc. for funding support of the "BBSRC Sustainable Bioenergy Centre (BSBEC): Perennial Bioenergy Crops Programme" (BB/G016216/1: BSBEC-BioMASS) and BBSRC for funding the RRes "Cropping Carbon" Institute Strategic Programme Grant. We would also like to thank Jennifer Cunniff, Marianna Cerasuolo, Cristina Gritsch, Tim Barraclough and March Castle (RRes) and Sarah Purdy, Lawrence Jones and Anne Maddison (IBERS) for their dedicated work as part of the BSBEC-BioMASS programme and William Macalpine, Rachel Rossiter and Peter Fruen (RRes) for general scientific support. Rothamsted Research is an Institute supported by the BBSRC.

References

Alvim R, Hewett EW, Saunders PF (1976) Seasonal variation in hormone content of willow.1. Changes in abscisic acid content and cytokinin activity in xylem sap. Plant Physiol 57:474–476

Alvim R, Saunders PF, Barros RS (1979) Abscisic acid and the photoperiodic induction of dormancy in *Salix viminalis* L. Plant Physiol 63:774–777

Andersson A, Keskitalo J, Sjodin A, Bhalerao R, Sterky F, Wissel K, Tandre K, Aspeborg H, Moyle R, Ohmiya Y, Brunner A, Gustafsson P, Karlsson J, Lundeberg J, Nilsson O, Sandberg G, Strauss S, Sundberg B, Uhlen M, Jansson S, Nilsson P (2004) A transcriptional timetable of autumn senescence. Genome Biol 5(R24):21–13

Arihan O, Guvenc A (2011) Studies of the anatomical structure of stems of willow (*Salix* L.) species (*Salicacea*) growing in Ankara Province, Turkey. Turk J Bot 35:1–17

Bao YH, Dharmawardhana P, Mockler TC, Strauss SH (2009) Genome scale transcriptome analysis of shoot organogenesis in *Populus*. BMC Plant Biol 9(132)

Bohlenius H, Huang T, Charbonnel-Campaa L, Brunner AM, Jansson S, Strauss SH, Nilsson O (2006) CO/FT regulatory module controls timing of flowering and seasonal growth cessation in trees. Science 312:1040–1043

Bowen MR, Hoad GV (1968) Inhibitor content of phloem and xylem sap obtained from willow (*Salix viminalis* L.) entering dormancy. Planta 81:64–70

Brereton NJ, Pitre FE, Hanley SJ, Ray M, Karp A, Murphy RJ (2010) Mapping of enzymatic saccharification in short rotation coppice willow and its independence from biomass yield. Bioenergy Res 3:251–261

Brereton NJB, Pitre FE, Ray MJ, Karp A, Murphy RJ (2011) Investigation of tension wood formation and 2,6-dichlorbenzonitrile application in short rotation coppice willow composition and enzymatic saccharification. Biotechnol Biofuels 4(83)

Brereton NJ, Ray MJ, Shield I, Martin P, Karp A, Murphy RJ (2012) Reaction wood—a key cause of variation in cell wall recalcitrance in willow. Biotechnol Biofuels 5:83

Cannell MGR, Milne R, Sheppard LJ, Unsworth MH (1987) Radiation interception and productivity of willow. J Appl Ecol 24:261–278

Carroll A, Mansoori N, Li S, Lei L, Vernhettes S, Visser RGF, Somerville C, Gu Y, Trindade LM (2012) Complexes with mixed primary and secondary cellulose synthases are functional in Arabidopsis plants. Plant Physiol 160:726–737

Cerasuolo M, Richter GM, Cunniff J, Purdy S, Shield I, Karp A (2013) A pseudo-3D model to optimise the target traits of light interception in short-rotation coppice willow. Agric Meteorol 173:127–138

Cooke JEK, Weih M (2005) Nitrogen storage and seasonal nitrogen cycling in *Populus*: bridging molecular physiology and ecophysiology. New Phytol 167:19–30

Cooke JEK, Brown KA, Wu R, Davis JM (2003) Gene expression associated with N-induced shifts in resource allocation in poplar. Plant, Cell Environ 26:757–770

Crow P, Houston TJ (2004) The influence of soil and coppice cycle on the rooting habit of short rotation poplar and willow coppice. Biomass Bioenergy 26:497–505

Davies JK, Jensen E, Juntilla O, Rivier L, Crozier A (1985) Identification of endogenous gibberellins from *Salix pentandra*. Plant Physiol 78:473–476

Deckmyn G, Laureysens I, Garcia J, Muys B, Ceulemans R (2004) Poplar growth and yield in short rotation coppice: model simulations using the process model SECRETS. Biomass Bioenergy 26:221–227

Demura T, Ye Z-H (2010) Regulation of plant biomass production. Curr Opin Plant Biol 13:299–304

Deslauriers A, Giovannelli A, Rossi S, Castro G, Fragnelli G, Traversi L (2009) Intra-annual cambial activity and carbon availability in stem of poplar. Tree Physiol 29:1223–1235

Dingkuhn M, Luquet D, Clement-Vidal A, Tambour L, Kim HK, Song YH (2007) Is plant growth driven by sink regulation? In: Spiertz JHJ, Struik PC, Van Laar HH (eds) Scale and complexity in plant systems research: gene-plant-crop relations. Springer, New York, pp 157–170

Domagalska MA, Leyser O (2011) Signal integration in the control of shoot branching. Nat Rev: Mol Cell Biol 12:211–221

Dong Q, Zhao Y, Jiang HJ, He H, Zhu S, Cheng B, Xiang Y (2011) Genome-wide identification and characterization of the cyclin gene family in *Populus trichocarpa*. Plant Cell Tiss Organ Cult 107:55–67

Douglas CJ, Ehlting J, Harding SA (2011) Phenylpropanoid and phenolic metabolism in *Populus*: gene family structure and comparative and functional genomics. In: Joshi CP (ed) Genetics genomics and breeding of poplar. CRC press, Boca Raton, pp 304–326

Druart N, Johansson A, Baba K, Schrader J, Sjodin A, Bhalerao RR, Resman L, Trygg J, Moritz T, Bhalerao RP (2007) Environmental and hormonal regulation of the activity-dormancy cycle in the cambial meristem involves stage-specific modulation of transcriptional and metabolic networks. Plant J 50:557–573

Du J, Groover A (2010) Transcriptional regulation of secondary growth and wood formation. J Integr Plant Biol 52:17–27

Eckersten H, Slapokas T (1990) Modeling nitrogen turnover and production in an irrigated short-rotation forest. Agric Meteorol 50:99–123

El Kayal W, Allen CCG, Ju CJT, Adams E, King-Jones S, Zaharia LI, Abrams SR, Cooke JEK (2011) Molecular events of apical bud formation in white spruce, *Picea glauca*. Plant, Cell Environ 34:480–500

Ericsson T (1984) Root biomass distribution in willow stands grown on bog. In: Perttu K (ed) Ecology and management of forest biomass. Swed Univ Agric Sci Rep 15:335–348

Fernando DD, Zhang SL (2006) Constitutive expression of the SAP1 gene from willow (*Salix discolor*) causes early flowering in *Arabidopsis thaliana*. Devel Genes Evol 216:19–28

Frewen BE, Chen THH, Howe GT, Davis J, Rohde A, Boerjan W, Bradshaw HD (2000) Quantitative trait loci and candidate gene mapping of bud set and bud flush in Populus. Genetics 154:837–845

Grime JP, Hodgson JG, Hunt R (1988) Comparative plant ecology. A functional approach to common British species. Unwin Hyman, London

Guo Y, Ma YT, Zhan ZG, Li BG, Dingkuhn M, Luquet D, De Reffye P (2006) Parameter optimization and field validation of the functional-structural model GREENLAB for maize. Ann Bot 97:217–230

Hammer G, Cooper M, Tardieu F, Welch S, Walsh B, van Eeuwijk F, Chapman S, Podlich D (2006) Models for navigating biological complexity in breeding improved crop plants. Trends Plant Sci 11:587–593

Hanley SJ (2003) Genetic mapping of important agronomic traits in biomass willow. PhD thesis, University of Bristol, UK

Hanley SJ, Mallott MD, Karp A (2006) Alignment of a *Salix* linkage map to the *Populus* genomic sequence reveals macrosynteny between willow and poplar genomes. Tree Genet Genom 3:35–48

Hanley SJ, Pei MH, Powers SJ, Ruiz C, Mallott MD, Barker JHA, Karp A (2011) Genetic mapping of rust resistance loci in biomass willow. Tree Genet Genom 7:597–608

Hartweck LM (2008) Gibberellin signaling. Planta 229:1–13

Hsu C-Y, Adams JP, Kim H, No K, Ma C, Strauss SH, Drnevich J, Vandervelde L, Ellis JD, Rice BM, Wickett N, Gunter LE, Tuskan GA, Brunner AM, Page GP, Barakat A, Carlson JE, dePamphilis CW, Luthe DS, Yuceer C (2011) *FLOWERING LOCUS T* duplication coordinates reproductive and vegetative growth in perennial poplar. Proc Natl Acad Sci USA 108:10756–10761

Hsu C-Y, Adams JP, No K, Liang H, Meilan R, Pechanova O, Barakat A, Carlson JE, Page GP, Yuceer C (2012) Overexpression of *constans* homologs *CO*1 and *CO*2 fails to alter normal reproductive onset and fall bud set in woody perennial poplar. Plos ONE 7(e45448)

Ibanez C, Kozarewa I, Johansson M, Ogren E, Rohde A, Eriksson ME (2010) Circadian clock components regulate entry and affect exit of seasonal dormancy as well as winter hardiness in *Populus* trees. Plant Physiol 153:1823–1833

Ingvarsson PK, Garcia MV, Hall D, Luquez V, Jansson S (2006) Clinal variation in phyB2, a candidate gene for day-length-induced growth cessation and bud set, across a latitudinal gradient in European aspen (*Populus tremula*). Genetics 172:1845–1853

Jeong ML, Jiang HY, Chen HS, Tsai CJ, Harding SA (2004) Metabolic profiling of the sink-to-source transition in developing leaves of quaking aspen. Plant Physiol 136:3364–3375

Jimenez S, Reighard GL, Bielenberg DG (2010) Gene expression of DAM5 and DAM6 is suppressed by chilling temperatures and inversely correlated with bud break rate. Plant Mol Biol 73:157–167

Junttila O, Kaurin A (1990) Environmental control of cold acclimation in *Salix pentandra*. Scand J For Res 5:195–204

Junttila O, Abe H, Pharis RP (1988) Endogenous gibberellins in elongating shoots of clones of *Salix dasyclados* and *Salix viminalis*. Plant Physiol 87:781–784

Karp A, Hanley SJ, Trybush SO, Macalpine W, Pei M, Shield I (2011) Genetic improvement of willow for bioenergy and biofuels. J Integr Biol 53:151–165

Kemi U, Niittyvuopio A, Toivainen T, Pasanen A, Quilot-Turion B, Holm K, Lagercrantz U, Savolainen O, Kuittinen H (2013) Role of vernalization and of duplicated *FLOWERING LOCUS C* in the perennial *Arabidopsis lyrata*. New Phytol 197:323–335

Koornneef M, Leon-Kloosterziel KM, Schwartz SH, Zeevaart JAD (1998) The genetic and molecular dissection of abscisic acid biosynthesis and signal transduction in Arabidopsis. Plant Physiol Biochem 36:83–89

Kunihiro A, Yamashino T, Nakamichi N, Niwa Y, Nakanishi H, Mizuno T (2011) Phytochrome-Interacting Factor 4 and 5 (PIF4 and PIF5) activate the homeobox ATHB2 and auxin-inducible IAA29 genes in the coincidence mechanism underlying photoperiodic control of plant growth of *Arabidopsis thaliana*. Plant Cell Physiol 52:1315–1329

Lennartsson M, Ogren E (2004) Clonal variation in temperature requirements for budburst and dehardening in *Salix* species used for biomass production. Scand J For Res 19:295–302

Li S, Lei L, Somerville CR, Gu Y (2012) Cellulose synthase interactive protein 1 (CSI1) links microtubules and cellulose synthase complexes. Proc Natl Acad Sci USA 109:185–190

Liu Y, Liu S (2010) Formation and anatomical characteristics of tension wood in stem of poplar I-107 seedlings (*Populus euramericana* cv. "74/76") induced by artifical inclination. Scientia Silvae Sinicae 46:133–140

MacDonald AJS (1989) Nitrate availability and shoot area development in small willow (*Salix viminalis*). Plant, Cell Environ 12:417–424

Menges M, Pavesi G, Morandini P, Bogre L, Murray JAH (2007) Genomic organization and evolutionary conservation of plant D-Type cyclins(1 C W). Plant Physiol 145:1558–1576

Messina CD, Podlich D, Dong Z, Samples M, Cooper M (2011) Yield-trait performance landscapes: from theory to application in breeding maize for drought tolerance. J Exp Bot 62:855–868

Mueller D, Leyser O (2011) Auxin, cytokinin and control of shoot branching. Ann Bot 107:1203–1212

Nieminen K, Immanen J, Laxell M, Kauppinen L, Tarkowski P, Dolezal K, Tahtiharju S, Elo A, Decourteix M, Ljung K, Bhalerao R, Keinonen K, Albert VA, Helariutta Y (2008) Cytokinin signaling regulates cambial development in poplar. Proc Natl Acad Sci USA 105:20032–20037

Olsen JE (2010) Light and temperature sensing and signalling in induction of bud dormancy in woody plants. Plant Mol Biol 73:37–47

Pin PA, Nilsson O (2012) The multifaceted roles of FLOWERING LOCUS T in plant development. Plant, Cell Environ 35:1742–1755

Reymond M, Muller B, Leonardi A, Charcosset A, Tardieu F (2003) Combining quantitative trait loci analysis and an ecophysiological model to analyze the genetic variability of the responses of maize leaf growth to temperature and water deficit. Plant Physiol 131:664–675

Richter GM, Acutis M, Trevisiol P, Latiri K, Confalonieri R (2010a) Sensitivity analysis for a complex crop model applied to durum wheat in the Mediterranean. Eur J Agron 32:141–150

Richter GM, Dailey GA, Riche AB (2010b) Identifying key process parameters in a sink-source interaction model for energy crops. In: Wery J S-TI, Perrin A ed. Proceeding of the XI Congress of the Eur Soc of Agronomy (Agro 2010). Montpellier, p 97–98.

Rohde A, Prinsen E, De Rycke R, Engler G, Van Montagu M, Boerjan W (2002) PtABI3 impinges on the growth and differentiation of embryonic leaves during bud set in poplar. Plant Cell 14:1885–1901

Rönnberg-Wastljung AC, Glynn C, Weih M (2005) QTL analyses of drought tolerance and growth for a *Salix dasyclados* x *Salix viminalis* hybrid in contrasting water regimes. Theor Appl Genet 110:537–549

Rönnberg-Wastljung AC, Ahman I, Glynn C, Widenfalk O (2006) Quantitative trait loci for resistance to herbivores in willow: field experiments with varying soils and climates. Entom Exp Appl 118:163–174

Ruttink T, Arend M, Morreel K, Storme V, Rombauts S, Fromm J, Bhalerao RP, Boerjan W, Rohde A (2007) A molecular timetable for apical bud formation and dormancy induction in poplar. Plant Cell 19:2370–2390

Rytter RM (1999) Fine-root production and turnover in a willow plantation estimated by different calculation methods. Scand J Res 14:526–537

Rytter RM (2001) Biomass production and allocation, including fine-root turnover, and annual N uptake in lysimeter-grown basket willows. For Ecol Man 140:177–192

Rytter RM, Hansson AC (1996) Seasonal amount, growth and depth distribution of fine roots in an irrigated and fertilized *Salix viminalis* L-plantation. Biomass Bioenergy 11:129–137

Rytter RM, Rytter L (1998) Growth, decay, and turnover rates of fine roots of basket willows. Can J For Res-Rev Can Rech For 28:893–902

Sanchez-Rodriguez C, Bauer S, Hematy K, Saxe F, Ibanez AB, Vodermaier V, Konlechner C, Sampathkumar A, Rueggeberg M, Aichinger E, Neumetzler L, Burgert I, Somerville C, Hauser M-T, Persson S (2012) CHITINASE-LIKE1/POM-POM1 and its homolog CTL2 are glucan-interacting proteins important for cellulose biosynthesis in Arabidopsis. Plant Cell 24:589–607

Sarkar P, Bosneaga E, Auer M (2009) Plant cell walls throughout evolution: towards a molecular understanding of their design principles. J Exp Bot 60:3615–3635

Sawa M, Kay SA (2011) GIGANTEA directly activates Flowering Locus T in *Arabidopsis thaliana*. Proc Natl Acad Sci USA 108:11698–11703

Sawa M, Kay SA (2012) Circadian clocks and photoperiodism. Fukuoka igaku zasshi (= Hukuoka acta medica) 103:29–34

Schapendonk AHCM, Stol W, van Kraalingen DWG, Bouman BAM (1998) LINGRA, a sink/source model to simulate grassland productivity in Europe. Eur J Agron 9:87–100

Schrader J, Nilsson J, Mellerowicz E, Berglund A, Nilsson P, Hertzberg M, Sandberg G (2004) A high-resolution transcript profile across the wood-forming meristem of poplar identifies potential regulators of cambial stem cell identity. Plant Cell 16:2278–2292

Schuetz M, Smith R, Ellis B (2013) Xylem tissue specification, patterning, and differentiation mechanisms. J Exp Bot 64:11–31

Sennerby-Forsse L (1986) Seasonal variation in the ultrastructure of the cambium in young stems of willow (*Salix viminalis*) in relation to phenology. Physiol Plant 67:529–537

Sennerby-Forsse L (1995) Growth processes. Biomass Bioenergy 9:35–43

Sennerby-Forsse L, Zsuffa L (1995) Bud structure and resprouting in coppiced stools of *Salix viminalis* L, *S. eriocephala* Michx, and *S. amygdaloides* Anders. Trees-Struct Func 9:224–234

Sennerby-Forsse L, Berggren B, Brunkener L, Fjell I (1984) Growth behaviour and anatomy of meristems in Salix. In: Perttu K (ed) Ecology and management of forest biomass. Swed Univ Agric Sci Rep 15:481–501

Siriwardana NS, Lamb RS (2012) The poetry of reproduction: the role of LEAFY in *Arabidopsis thaliana* flower formation. Int J Devel Biol 56:207–221

Srikanth A, Schmid M (2011) Regulation of flowering time: all roads lead to Rome. Cell Mol Life Sci 68:2013–2037

Stott KG (1961) The rooting pattern of the basket willow—*Salix triandra* var. Black maul—at Long Asthon. In: annual report for long Ashton research station for 1961. Long Ashton Research Station, Long Ashton, pp 187–190

Triana F, Ragaglini G, Bonari E, Cerasuolo M, Richter GM (2011) Modelling the water balance of different grass species used for bioenergy. Biomass Energy Crops IV 112:163–170

Tsarouhas V, Gullberg U, Lagercrantz U (2002) An AFLP and RFLP linkage map and quantitative trait locus (QTL) analysis of growth traits in *Salix*. Theor Appl Genet 105:277–288

Tsarouhas V, Gullberg U, Lagercrantz U (2003) Mapping of quantitative trait loci controlling timing of bud flush in *Salix*. Hereditas 138:172–178

Tsarouhas V, Gullberg U, Lagercrantz U (2004) Mapping of quantitative trait loci (QTLs) affecting autumn freezing resistance and phenology in *Salix*. Theor Appl Genet 108:1335–1342

Tuskan GA (2006) The genome of black cottonwood. Populus tricharpa Science 313:1596–1604

Verwijst T (1996a) Cyclic and progressive changes in short-rotation willow coppice systems. Biomass Bioenergy 11:161–165

Verwijst T (1996b) Stool mortality and development of a competitive hierarchy in a *Salix viminalis* coppice system. Biomass Bioenergy 10:245–250

Vos J, Evers JB, Buck-Sorlin GH, Andrieu B, Chelle M, de Visser PHB (2010) Functional-structural plant modelling: a new versatile tool in crop science. J Exp Bot 61:2101–2115

Ward SP, Salmon J, Hanley SJ, Karp A, Leyser O (2013) Using Arabidopsis to study shoot branching in biomass willow. Plant Physiol 162:800–811

Weih M (2009) Genetic and environmental variation in spring and autumn phenology of biomass willows (*Salix* spp.): effects on shoot growth and nitrogen economy. Tree Physiol 29:1479–1490

Weih M, Rönnberg-Wastljung AC (2007) Shoot biomass growth is related to the vertical leaf nitrogen gradient in *Salix* canopies. Tree Physiol 27:1551–1559

Weih M, Rönnberg-Wastljung AC, Glynn C (2006) Genetic basis of phenotypic correlations among growth traits in hybrid willow (*Salix dasyclados* x *S.viminalis*) grown under two water regimes. New Phytol 170:467–477

Yang XH, Kalluri UC, DiFazio SP, Wullschleger SD, Tschaplinski TJ, Cheng ZMM, Tuskan GA (2009) Poplar genomics: state of the science. Crit Rev Plant Sci 28:285–308

Chapter 8
Camelina: An Emerging Oilseed Platform for Advanced Biofuels and Bio-Based Materials

Umidjon Iskandarov, Hae Jin Kim and Edgar B. Cahoon

Abstract *Camelina* (*Camelina sativa* (L.) Crantz) is a Brassicaceae oilseed crop with valuable agronomic and biotechnological attributes that make it an attractive renewable feedstock for biofuels and bio-based materials. *Camelina* seeds contain 30–40 % oil and can achieve oil yields per hectare that surpass established oilseed crops such as soybean. *Camelina* is also productive under conditions of limited rainfall and low soil fertility. As a short season, frost tolerant oilseed, *Camelina* is amenable to double cropping systems and fallow year production. Simple, non-labor intensive *Agrobacterium*-based transformation methods have recently been described for *Camelina* that can be used in combination with breeding to rapidly improve seed quality and agronomic traits to advance *Camelina* as a production platform for biofuels and industrial feedstocks in geographical regions such as the North American Great Plains that currently have little oilseed production for edible vegetable oils.

Keywords *Camelina* • Oilseed crop • Biofuels • Bio-based materials • Biodiesel • Fatty acids

---

U. Iskandarov (✉) · H. J. Kim (✉) · E. B. Cahoon
Center for Plant Science Innovation and Department of Biochemistry, University of Nebraska-Lincoln, E318 Beadle Center, 1901 Vine Street, Lincoln, NE 68588, USA
e-mail: uiskandarov2@unl.edu

H. J. Kim
e-mail: haejin.kim@unl.edu

E. B. Cahoon
e-mail: ecahoon2@unl.edu

M. C. McCann et al. (eds.), *Plants and BioEnergy*, Advances in Plant Biology 4,
DOI: 10.1007/978-1-4614-9329-7_8,

8.1 Introduction

Camelina sativa or *Camelina*, known also as false flax or gold of pleasure, is an annual oilseed species of the Brassicaceae family. *Camelina* has received growing interest as biofuel crop for the production of vegetable oils for biodiesel and aviation fuel because of its productivity in geographic regions that are not currently used for large-scale oilseed production and the possibility of growing *Camelina* in farming systems (e.g., double cropping) that do not compete with crops for food production. *Camelina* currently has a number of seed quality and agronomic issues that limit its wider use for biofuel and possibly other industrial feedstock production. These limitations can now be addressed by advances in biotechnology and the increasing availability of genomic resources to facilitate breeding.

Camelina is native to Eastern Europe and Central Asia (Putnam et al. 1993). It has been cultivated in Europe since the Bronze age, as early as 1500–400 BC (Bouby 1998; Zubr 2010). It was introduced to North America from Europe most likely as a weed along with flax and is well adapted to Southern Canada, Northern Great Plains and Pacific Northwest of the U.S. *Camelina* grows 0.3–1 m high, stems are smooth and branched with arrow shaped leaves that are 5–8 cm long (Fig. 8.1). Flowers are small and consist of four pale or greenish yellow petals and give rise to pear-shaped pods (5–10 mm) containing as many as 16 seeds. Seeds are small

Fig. 8.1 *Camelina sativa* (*Camelina*). **a** Flowering *Camelina* plant. **b** Mature pods. **c** Comparison of *Arabidopsis* (*Arabidopsis thaliana*), *Camelina*, and rapeseed (*Brassica napus*) seed sizes

yellow–brown or brown, oblong, rough with rigid surface (Putnam et al. 1993). Seeds contain 30–40 % oil on dry weight basis of which 64 % is polyunsaturated, including 30–45 % of the omega-3 fatty acid α-linolenic acid (Vollmann et al. 1996).

Historically *Camelina* has been valued for the vegetable oil extracted from its seeds (Moloney et al. 1998; Jaskiewicz and Matyka 2003; Flachowsky et al. 2011). The ancient Romans used *Camelina* oil as massage oil, lamp fuel, as well as cooking oil. In modern times, *Camelina* oil has gained a niche market, particularly in Europe, for its nutraceutical value because of its high content of α-linolenic acid (Hurtaud and Peyraud 2007; Ehrensing and Guy 2008). *Camelina* meal obtained after oil extraction, contains 10–11 % fiber, 10–14 % oil, and about 40 % protein making it a valuable product for use (Korsrud et al. 1978; Onyilagha et al. 2012) in cattle, chicken and hog feed (Ehrensing and Guy 2008). However, due to the presence of glucosinolates (19–23 μmol/g) in *Camelina* meal regulations require limited daily use to avoid negative impacts on livestock productivity although the amount is not higher than that of the widely used canola meal (Matthäus and Angelini 2005; Moser 2010). U.S. Food and Drug Administration regulations currently restrict the use of *Camelina* meal to $\leq$10 % of the total diet of beef cattle, broiler chickens, and laying hens and $\leq$2 % of the diet of growing swine for United States livestock production.

8.2 Agronomic Properties

Camelina has recently gained renewed interest as a low input biofuel feedstock because of its minimal agronomic input requirements for productivity on marginal lands with limited fertility and water resources. *Camelina* requires low to moderate amounts of fertilizers and is productive with nitrogen and phosphorus levels of as low as 90–100 and 67 kg/ha, respectively (McVay and Lamb 2007; Ehrensing and Guy 2008; Sipalova et al. 2011; Solis et al. 2013). Field trials with cambic chernozem soil in Timisoara, Romania showed that only addition of the fertilizers nitrogen (100 kg/ha) and phosphorus (60 kg/ha) increased seed yield from 932 to 1,813 kg/ha and oil up to 584 kg/ha increasing the oil yield by 25 % when plants were sowed at 25 cm row distance which was found to be optimal (Imbrea et al. 2011). Similar doses of nitrogen fertilizer were recommended for *Camelina* growth to achieve optimal yield based on the results of the laboratory experiments (Sipalova et al. 2011).

Despite being a cool season crop, *Camelina* is well suited for dryland cropping systems in which soil moisture and rainfall can be maximized by planting in early months of spring depending on location. For example, as high as 1,912 kg/ha of grain and 506 kg/ha oil yield were achieved in dryland conditions near Havre, Montana USA, while grain yield of 3,250 kg/ha was obtained in Austria if water was not limiting throughout vegetation period (Vollmann et al. 1996; McVay and Lamb 2007; Ehrensing and Guy 2008). The 2 year trials in arid zones of Southwestern USA showed that *Camelina* can be a low water use crop. *Camelina* grown from January to May in Maricopa, Arizona yielded in over 1,500 kg/ha seed with maximum seasonal water use of just 47–49 cm, furthermore the seed yield loss was not significant unless the soil water depletion before irrigation reached 70 % and higher (Hunsaker et al. 2011).

Camelina has traditionally been grown without pesticides and has allelopathic effects on weed species (Ehrensing and Guy 2008; Gesch and Cermak 2011). Herbicide research on *Camelina* is currently ongoing. Being a minor weed, *Camelina* is usually not a problem in other crops and does not have seed dormancy. Research showed that by planting *Camelina* seeds in winter or early spring, herbicide use can be avoided since *Camelina* seeds can germinate at low temperatures and seedlings are frost tolerant (Robinson 1987) and suppress many weed species, except perennial weeds, especially if seeded at high density (Ehrensing and Guy 2008; Gesch and Cermak 2011). *Camelina* was shown to be a viable winter crop for the Northern Corn Belt of the U.S., where seed yields as high as 1,317 kg/ha and oil yields as high as 420 g/kg were obtained from fall sown crop promoting good weed suppression (Gesch and Cermak 2011). Thus sowing does not require pre-emergence weed control, greatly reducing both production costs and environmental damage. Susceptibility of *Camelina* to herbicides inhibiting acetolactate synthase (ALS), that are commonly used for wheat cultivation in the Pacific Northwest of the U.S., has limited cultivation of *Camelina* as an oilseed crop in the area (Hanson et al. 2004; Pavlista et al. 2011).

Unlike other Brassicaceae crops such as rapeseed, *Camelina* does not require insecticide application. *Camelina* is resistant to crucifer insect pests due to large concentrations of the insect deterrent quercetin glycosides in its tissues (Onyilagha et al. 2012; Naranjo and Stefanek 2012). *Camelina* is also highly resistant to blackleg disease caused by the fungus *Leptosphaeria maculans*, which is a major pathognen of many Brassicaceae crops, such as canola. *Camelina* is susceptible to downy *Peronospora Camelinae* (downy mildew), *Alternaria brassicae* and the saprotrophic fungus *Rhizoctonia*, but none of these pathogens has been reported to cause major yield losses in *Camelina* (Robinson 1987; Salisbury et al. 1995; Ehrensing and Guy 2008).

Because it has a relatively short growth cycle (80–100 days) and frost tolerance, *Camelina* can be used in double cropping systems during cool periods of the year (Gesch and Archer 2012). Field trials showed that short season cultivars of soybean could be double cropped after winter *Camelina* in the upper U.S. Midwest (Gesch and Archer 2012). The net return from the *Camelina*-soybean double crop was higher than that of mono-cropped soybean in the period analyzed (Gesch and Archer 2012). *Camelina* can also be grown following wheat, barley, peas and lentils, but should not be planted following Brassicaceae crops to avoid increased risks by pests and diseases (Fleenor 2011). In addition, *Camelina* has been considered as a tertiary crop for rotations in northeastern Colorado where wheat is rotated with crops such as corn (Brandess 2012). In this region, *Camelina* could be planted in October after the first crop in year one and harvested before July of year two. This could then be followed by land recuperation period of July to mid-September when winter wheat could be planted and harvested in July of year three. This is followed by a fallow period until April or May of the year four when the rotation restarts (Brandess 2012). This rotation would allow farmers to grow three crops versus two in three years (Brandess 2012). Since it is best adapted to cooler climates, *Camelina* could also be grown in winter in areas with mild winters (Gesch and Cermak 2011).

Although the oil content of seeds is lower than that of *Brassica napus* on dry weight basis, the oil yield per hectare can reach as high as that of *B. napus* if good agronomic management is practiced (Putnam et al. 1993; Imbrea et al. 2011). *Camelina* has a low seeding rate (as low as 3–5 kg/ha) compared to other agronomic crops, including canola to establish dense stands (McVay and Lamb 2007; Pilgeram et al. 2007). In addition, existing equipment that is used for harvesting and processing of other crops can be adapted to *Camelina* (Brandess 2012). These agricultural attributes of *Camelina* give it compelling properties and make it a favorable oilseed crop to be grown in agronomically demanding lands with low-inputs.

8.3 Genetic Improvement: Variety Selection, Breeding, Genomic Resources, Biotechnology

Limiting the full potential of *Camelina* as a biofuel oilseed crop is the need to improve a number of agronomic, yield, and oil quality traits. In contrast to other Brassicaceae oilseeds such as canola and rapeseed, *Camelina* has not undergone extensive breeding and only a relatively small number of cultivars are available for commercial production. From an agronomic production standpoint, improvement in traits such as heat tolerance, downy mildew resistance, water and nitrogen use efficiencies, and herbicide resistance are desirable. For biodiesel and industrial uses such as bio-based lubricants, enhancement in seed oil content from the current 30–40 % of seed weight to levels of 40–50 % of seed weight, as is currently found in elite rapeseed germplasm, is a major target for *Camelina* crop improvement. An additional target for biodiesel and bio-lubricants is the reduction of the high polyunsaturated fatty acid content of the seed oil [35–39 % linolenic acid (18:3) and 20–25 % linoleic acid (18:2)] and replacement with high content of the more oxidatively stable monounsaturated fatty acid oleic acid (18:1). Furthermore, reductions in seed glucosinolate levels would allow for increased use of *Camelina* meal for livestock production.

Varietal selection and screening of germplasm following mutagenesis are among the approaches used to date for *Camelina* crop improvement. Several *Camelina* varieties have been selected for higher oil content and improved fatty acid composition. For example, the cultivar Blaine Creek is richer in ω-3 fatty acids, while Suneson has 2–3 % higher oil content, and is rich in α-linolenic acid (Ehrensing and Guy 2008). Lines with resistance to acetolactate synthase (ALS)-targeting herbicides imazethapyr and sulfosulfuron and altered seed fatty acid composition, including increased oleic acid content, have also been identified screening of mutagenized populations (Vollamnn et al. 1997; Buchsenschutz-Nothdurft et al. 1998; Kang et al. 2011; Walsh et al. 2012).

Recent advances in *Camelina* genomics are also providing avenues for *Camelina* improvement through marker-assisted breeding. Molecular genetic maps have been assembled for *Camelina* using random amplified polymorphic DNA (RAPD) markers and amplified fragment length polymorphisms (Vollmann et al. 2005; Gehringer et al. 2006). These maps have been used to localize QTLs for agronomic characteristics such as seed yield, oil content, 1,000-seed weight, and plant height (Gehringer et al. 2006). More recent AFLP fingerprinting data using 53 accessions from different origins showed a high genetic diversity in the species which could offer opportunities for breeding (Ghamkhar et al. 2010). The chromosome number of *Camelina* is 2n = 40, which was confirmed by linkage map using 157 AFLP markers and 3 *Brassica* SSR markers (Gehringer et al. 2006). Genetic mapping based on AFLP, SSR and ILP makers indicated that *Camelina* is a hexaploid (Hutcheon et al. 2010). This was supported by isolation of three copies of *FATTY ACID DESATURASE 2* (*FAD2*) and *FATTY ACID ELONGASE 1* (*FAE1*) genes, both of which are single copy in *Arabidopsis* (Gehringer et al. 2006; Galasso et al. 2011). Like other important crops, polyploidy of *Camelina* will likely complicate efforts to develop molecular markers and assemble whole genome sequence. Advanced technologies for molecular genetics and genomics, including RNA-seq and next-generation genome and transcriptome sequencing, will likely provide unprecedented opportunities to accelerate improvement of agronomic and seed quality traits for *Camelina* (Varshney et al. 2009; Edwards et al. 2012).

Complementing the impact of breeding on crop improvement, *Camelina* is highly amenable to biotechnological enhancement through the use of *Agrobacterium tumefaciens*-mediated transformation. *Camelina* can be easily transformed using protocols similar those routinely used for *Arabidopsis thaliana* transformation. These methods include vacuum infiltration of flowers with a solution of *Agrobacterium* harboring a binary vector that contains the desired transgene (Lu and Kang 2008), or by simple floral dip with the *Agrobacterium* solution (Liu et al. 2012). Plants with *Agrobacterium*-infiltrated or -dipped flowers are grown to maturity. Seeds obtained from these plants are then screened to identify those containing the transgene. For this process, genes for resistance to antibiotics (e.g., kamamycin and hygromycin) or herbicides (e.g., glufosinate) can be used as selection markers for obtaining transgenic plants by screening of seeds on media containing the selective agent or by spraying seedlings with the selective agent in the case of herbicides (Lu and Kang 2008; Liu et al. 2012). Fluorescent protein selection markers such as DsRed under control of a seed-specific or constitutive promoter can also be used to identify transgenic seeds based on fluorescence of seeds with equipment as low tech as a green LED flashlight and red camera filter (Lu and Kang 2008). A wide variety of seed specific promoter/3'UTR cassettes can easily be inserted into binary vectors to express several candidate genes to modify seed oil traits. With these transformation methods and metabolic engineering toolbox, transgenic seeds can be obtained in as little as 6–8 weeks following *Agrobacterium* infiltration or dipping of flowers. Unlike transformation protocols for most crops, *Camelina* transformation can be done with minimal labor input and without the need for specialized technical skills. As such, *Agrobacterium*-based transformation offers a relatively simple and rapid, cost-effective approach for improvement of agronomic and seed quality traits.

Recent biotechnological efforts to improve the agronomic properties of *Camelina* have included transgenic expression of *Arabidopsis* purple acid phosphatase 2 that resulted in increased seed size and yield and faster growing plants relative to non-transformants. In addition, a recent report described the enhancement of the oleic acid content, a desirable biofuel trait, in *Camelina* seeds by anti-sense suppression using an inverted portion of the *Camelina CsFAD*2-1 gene under the control of the seed-specific promoter for the phaseolin gene (Kang et al. 2011). The resulting transgenic seeds contained 38–51 % oleic acid compared to 13–18 % oleic acid in seeds from non-transformed plants (Kang et al. 2011). In this study, seeds from a mutant of the *CsFAD*2-2 locus obtained from random mutagenesis that contained a premature stop at the Trp288 codon had ~27 % oleic acid (Kang et al. 2011). An interpretation of this result is that two or more of the three known *FAD*2 loci in *Camelina* contribute to the desaturation of oleic acid in seeds. Because of the high identity of these genes, it is possible to use antisense or RNA interference suppression with sequence from only one of these genes to get a mid- to high-oleic acid trait in *Camelina* seeds. It can be envisioned that additional enhancements in oleic acid content can be achieved through transgenic suppression of the *FAE*1 genes that are responsible for the elongation of oleic acid (18:1) to gondoic acid (20:1). Moreover, genes from other species could be transferred to *Camelina* to obtain additional biofuel-type traits, such as short- and medium-chain fatty acid-specific FatB-type thioesterases to achieve an oil functionality mimicking Jet A1 fuel. For industrial uses of *Camelina* oil, the castor bean (*Ricinus communis*) fatty acid hydroxylase has been successfully transferred to *Camelina* to produce ricinoleic acid and other hydroxy fatty acid in the seed oil of transformants (Lu and Kang 2008). As *Camelina* crop improvement progresses, it is likely that these efforts may involve a combination of varietal selection, mutagenic breeding, and biotechnological approaches. Indeed, the simple transformation protocols for *Camelina* hold considerable promise for rapid genetic improvement of this crop.

8.4 Current and Future Prospects for Commercial Production

Despite its considerable potential as a low input oilseed for biofuel production, *Camelina* has yet to see extensive commercial production. In the United States, large-scale production of *Camelina* has largely been restricted to the state of Montana where 8,400 ha (20,800 acres) were grown in 2009 (Anonymous 2011a). Spurring the recent interest in *Camelina* has been successful tests that have used *Camelina* oil as an ingredient of aviation fuel for commercial airliners and military jets. To stimulate increased commercial planting of *Camelina*, the U.S. Department of Agriculture Farm Service Agency announced in July 2011 a program targeted for the states of California, Washington, and Montana under the Biomass Crop Assistance Program to provide 5 year contracts for the production of up to 20,200 ha (50,000 acres) of *Camelina* for aviation fuel or other biomass conversion (Anonymous 2011b).

Table 8.1 Fatty acid composition of seed oils of *Camelina*, soybean, sunflower, and rapeseed

Oil source	Fatty acid composition (% of total fatty acids)							
	16:0	18:0	18:1	18:2	18:3	20:0	20:1	22:1
Camelina	7.5	4.0	14.9	21.4	31.7	5.1	11.8	3.3
Soybean	11.6	4.6	23.9	52.0	6.8	0.3	0.2	0.4
Sunflower	6.1	6.4	24.8	61.7	0.1	0.4	0.2	0.3
Rapeseed	3.3	1.7	18.1	14.1	7.7	2.8	8.2	44.2

More widespread production of *Camelina* in the North American Great Plains and the Pacific Northwest of the United States will undoubtedly require ready markets for *Camelina* oil and meal with cost-competitive pricing as well as the development of more extensive infrastructure for crushing of *Camelina* seeds and conversion of *Camelina* oil to biodiesel, aviation, or other fuel (Table 8.1).

Acknowledgments Research in the Cahoon lab for *Camelina* genetic improvement is supported by grants from the Center for Advanced Biofuel Systems (CABS), an Energy Frontier Research Center funded by the U.S. Department of Energy, Office of Science, Office of Basic Energy Sciences under Award Number DE-SC0001295, U.S. Department of Agriculture–Agriculture and Food Research Initiative 2009-05988, and NSF Plant Genome IOS 0701919.

References

Anonymous (2011a) 2010 *Camelina* Crop United States Department of Agriculture National Agricultural Statistics Service, Washington, D.C. http://www.nass.usda.gov/Statistics_by_State/Montana/Publications/Press_Releases_Crops/camelina.pdf

Anonymous (2011b) Fact sheet: biomass crop assistance program—project area number 8 *Camelina* growers in California. Montana and Washington United States Department of Agriculture Farm Service Agency, Washington, D.C. http://www.fsa.usda.gov/Internet/FSA_File/bcap_8_fact_sheet.pdf

Bouby L (1998) Two early finds of gold-of-pleasure (*Camelina* sp.) in middle Neolithic and Chalcolithic sites in western France. Antiquity 72:391–398

Brandess A (2012) Modeling the profitability of *Camelina* sativa as a biofuel feedstock in eastern Colorado. Colorado State University

Buchsenschutz-Nothdurft A, Schuster A, Friedt W (1998) Breeding for modified fatty acid composition via experimental mutagenesis in *Camelina sativa* (L.) Crtz. Ind Crops Prod 7:291–295

Edwards D, Batley J, Snowdon RJ (2012) Accessing complex crop genomes with next-generation sequencing. Theor Appl Genet (in press)

Ehrensing DT, Guy SO (2008) *Camelina*. Oregon State University, Corvalis. http://extension.oregonstate.edu/catalog/pdf/em/em8953-e.pdf

Flachowsky G, Langbein T, Böhme H, Schneider A, Aulrich K (2011) Effect of false flax expeller combined with short-term vitamin E supplementation in pig feeding on the fatty acid pattern, vitamin E concentration and oxidative stability of various tissues. J Anim Physiol Anim Nutr 78:187–195

Fleenor RA (2011) Plant guide for *Camelina* (*Camelina sativa*) USDA-natural resources conservation service, Spokane. http://plants.usda.gov/plantguide/pdf/pg_casa2.pdf

Galasso I, Manca A, Braglia L, Martinelli T, Morello L, Breviario D (2011) h-TBP: an approach based on intron-length polymorphism for the rapid isolation and characterization of the multiple members of the β-tubulin gene family in *Camelina sativa* (L.) Crantz. Mol Breed 28:635–645

Gehringer A, Friedt W, Lühs W, Snowdon RJ (2006) Genetic mapping of agronomic traits in false flax (*Camelina sativa* subsp. sativa). Genome 49:1555–1563

Gesch RW, Cermak SC (2011) Sowing date and tillage effects on fall-seeded *Camelina* in the Northern Corn Belt. Agron J 103:980–987

Gesch RW, Archer DW (2012) Double-cropping with winter *Camelina* in the northern Corn Belt to produce fuel and food. Ind Crops Prod (in press)

Ghamkhar K, Croser J, Aryamanesh N, Campbell M, Kon'kova N, Francis C (2010) *Camelina* (*Camelina sativa* (L.) Crantz) as an alternative oilseed: molecular and ecogeographic analyses. Genome 53:558–567

Hanson B, Park K, Mallory-Smith C, Thill D (2004) Resistance of *Camelina microcarpa* to acetolactate synthase inhibiting herbicides. Weed Res 44:187–194

Hunsaker D, French A, Clarke T, El-Shikha D (2011) Water use, crop coefficients, and irrigation management criteria for *Camelina* production in arid regions. Irrig Sci 29:27–43

Hutcheon C, Ditt RF, Beilstein M, Comai L, Schroeder J, Goldstein E, Shewmaker CK, Nguyen T, De Rocher J, Kiser J (2010) Polyploid genome of *Camelina sativa* revealed by isolation of fatty acid synthesis genes. BMC Plant Biol 10:233

Hurtaud C, Peyraud J (2007) Effects of feeding *Camelina* (seeds or meal) on milk fatty acid composition and butter spreadability. J Dairy Sci 90:5134–5145

Imbrea F, Jurcoane S, Halmajan H, Duda M, Botos L (2011) *Camelina sativa*: a new source of vegetal oils. Rom Biotechnol Lett 16

Jaskiewicz T, Matyka S (2003) Application of *Camelina sativa*, its seeds, extrudate and oil cake in diets for broiler chickens and the effect on rearing indices and carcass quality. Ann Anim Sci Suppl 2:181–184

Kang J, Snapp AR, Lu C (2011) Identification of three genes encoding microsomal oleate desaturases (FAD2) from the oilseed crop *Camelina sativa*. Plant Physiol Biochem 49:223–229

Korsrud GO, Keith MO, Bell JM (1978) Comparison of nutritional-value of crambe and *Camelina* seed meals with egg and casein. Can J Anim Sci 58:493–499

Liu X, Brost J, Hutcheon C, Guilfoil R, Wilson AK, Leung S, Shewmaker CK, Rooke S, Nguyen T, Kiser J, De Rocher J (2012) Transformation of the oilseed crop *Camelina sativa* by *Agrobacterium*-mediated floral dip and simple large-scale screening of transformants. In Vitro Cell Devel Biol-Plant 48:462–468

Lu C, Kang J (2008) Generation of transgenic plants of a potential oilseed crop *Camelina sativa* by *Agrobacterium*-mediated transformation. Plant Cell Rep 27:273–278

Matthäus B, Angelini LG (2005) Anti-nutritive constituents in oilseed crops from Italy. Ind Crops Prod 21:89–99

McVay K, Lamb P (2007) *Camelina production in Montana*. Montana State University Extension, Bozeman. http://msuextension.org/publications/AgandNaturalResources/MT200701AG.pdf

Moloney A, Woods V, Crowley J (1998) A note on the nutritive value of *Camelina* meal for beef cattle. Ir J Agric Food Res 243–247

Moser BR (2010) *Camelina* (*Camelina sativa* L.) oil as a biofuels feedstock: golden opportunity or false hope? Lipid Technol 22:270–273

Naranjo SE, Stefanek MA (2012) Feeding behavior of a potential insect pest, Lygus hesperus, on four new industrial crops for the arid southwestern USA. Ind Crops Prod 37:358–361

Onyilagha JC, Gruber MY, Hallett RH, Holowachuk J, Buckner A, Soroka JJ (2012) Constitutive flavonoids deter flea beetle insect feeding in *Camelina sativa* L. Biochem Syst Ecol 42:128–133

Pavlista A, Isbell T, Baltensperger D, Hergert G (2011) Planting date and development of spring-seeded irrigated canola, brown mustard and *Camelina*. Ind Crops Prod 33:451–456

Pilgeram AL, Sands DC, Boss D, Dale N, Wichman D, Lamb P, Lu C, Barrows R, Kirkpatrick M, Thompson B (2007) *Camelina sativa*, a Montana *omega*-3 and fuel crop. Issues New Crops New Uses 129–131

Putnam D, Budin J, Field L, Breene W (1993) *Camelina*: a promising low-input oilseed. In: Janick J, Simon J (eds) New crops. Wiley, New York, pp 314–322

Robinson RG (1987) *Camelina*: a useful research crop and a potential oilseed crop. University of Minnesota

Salisbury P, Ballinger D, Wratten N, Plummer K, Howlett B (1995) Blackleg disease on oilseed *Brassica* in Australia: a review. Anim Prod Sci 35:665–672

Sipalova M, Losak T, Hlusek J, Vollmann J, Hudec J, Filipcik R, Macek M, Kracmar S (2011) Fatty acid composition of *Camelina sativa* as affected by combined nitrogen and sulphur fertilisation. Afr J Agric Res 6:3919–3923

Solis A, Vidal I, Paulino L, Johnson BL, Berti MT (2013) *Camelina* seed yield response to nitrogen, sulfur, and phosphorus fertilizer in South Central Chile. Ind Crops Prod 44:132–138

Varshney RK, Nayak SN, May GD, Jackson SA (2009) Next-generation sequencing technologies and their implications for crop genetics and breeding. Trends Biotechnol 27:522–530

Vollmann J, Damboeck A, Eckl A, Schrems H, Ruckenbauer P (1996) Improvement of *Camelina* sativa, an underexploited oilseed, vol 1. ASHS Press, Alexandria

Vollmann J, Damboeck A, Baumgartner S, Ruckenbauer P (1997) Selection of induced mutants with improved linolenic acid content in *Camelina*. Lipid/Fett 99:357–361

Vollmann J, Grausgruber H, Stift G, Dryzhyruk V, Lelley T (2005) Genetic diversity in *Camelina* germplasm as revealed by seed quality characteristics and RAPD polymorphism. Plant Breed 124:446–453

Walsh DT, Babiker EM, Burke IC, Hulbert SH (2012) *Camelina* mutants resistant to acetolactate synthase inhibitor herbicides. Mol Breed 30:1053–1063

Zubr J (2010) Carbohydrates, vitamins and minerals of *Camelina sativa* seed. Nutr Food Sci 40:523–531

Chapter 9
Perspectives in Brazil of the Contribution of Palm Trees to Biodiesel Production

Janaina M. Meyer and Antonio Salatino

Abstract The awareness of the depletion and contamination derived from fossil fuels and the resultant environmental crisis has led to the recognition of the necessity of research and development aiming the production of biofuels. An escalation of oil prices has been the outcome of political instabilities in areas of the world which traditionally have been massive petroleum providers, allied to oil deficits and economic crises. The present paper discusses the potentialities of palm species native to Brazil as feedstock of biodiesel.

9.1 Introduction

The awareness of the depletion and contamination derived from fossil fuels and the resultant environmental crisis has led to the recognition of the necessity of research and development aiming the production of biofuels. An escalation of oil prices has been the outcome of political instabilities in areas of the world which traditionally have been massive petroleum providers, allied to oil deficits and economic crises. This has contributed to a world claim for alternatives to fossil oils, among which products derived from plants has been the aim of great expectation, particularly in developing countries (Zhuang et al. 2011). Thus, new strategies for power generation have become urgent all over the world. Some alternatives based on plant products have been implemented, such as the production of ethanol from corn and sugar-cane, as well as biodiesel from seed oils. Biofuels not only may reduce the dependency on imported mineral oil, but also have the advantage of giving off emissions with lower load of damaging environmental pollutants.

J. M. Meyer · A. Salatino (✉)
Department of Botany, Institute of Biosciences, University of São Paulo,
Rua do Matão 277, São Paulo, SP 05508-090, Brazil
e-mail: asalatin@ib.usp.br

M. C. McCann et al. (eds.), *Plants and BioEnergy*, Advances in Plant Biology 4,
DOI: 10.1007/978-1-4614-9329-7_9, © Springer Science+Business Media New York 2014

However, Brazil and most developing countries face problems of inefficiencies regarding transport systems and energy distribution. Especially in countries with continental dimensions such as Brazil, these constraints tend to give rise to wide isolated areas, where many people remain with no possibilities to advance toward biomass utilization and production of liquid fuels, gas and electricity (REN21 2005). These difficulties and the success of the sugar-cane ethanol program encouraged the Brazilian government to create in 2003 the National Program for Production and Use of Biodiesel.

Biodiesel is a fuel prepared from vegetable oil or animal fat (triglycerides), by means of a process of transesterification with alcohols, mainly methanol or ethanol, using a catalyst (sodium or potassium hydroxide). The product obtained is a mixture of methyl or ethyl esters of fatty acids, plus glycerol as a by-product (Buckeridge and Salatino 2010). Finding commercial uses for the latter substance is important to achieve reductions in the final cost of biodiesel. Glycerol may be used in chemistry and livestock industries. It has recently been proposed the fermentative conversion of glycerol in ethanol by *Escherichia coli* (Yasdani and Gonzales 2008). A study by the US Environmental Protection Agency (EPA 2002) concluded that the addition of biodiesel to mineral diesel (ecodiesel) reduces pollutant emission from the engine exhaust. This effect is due to the higher oxygen content in biodiesel, which reduces the amount of unburned hydrocarbons, carbon monoxide and particulate material in the engine exhausts. A reduction also occurs in the amounts of sulfur and aromatic pollutants in the emissions. The amount of nitrogen oxides (NO_X) tend to be higher in ecodiesel emissions, but this problem may be circumvented by tuning the engines properly (Sze et al. 2007). Xue et al. (2011) reported similar amounts of NO_X in biodiesel and mineral diesel emissions.

The production of biodiesel worldwide has grown considerably. It was negligible in 2000 and has grown incessantly since then, with a strong ramping up beginning in 2005 (Fig. 9.1). A combination of factors has accounted for the increase in biodiesel production. First, the stead threat of sudden increases in mineral oil prices may be pointed out. Second, the independence from imported fossil energy sources may bring about some economic security. Also important has been the development of new technologies and introduction of alternative feedstocks for biodiesel production. A report from Pike Research (2011) estimates that revenues derived from biodiesel production will be three times as much in 2020 (US$ 71 billion) as they were in 2010 (US$ 18.4 billion).

Given the environmental and economic perspectives, it is expected for the next decades a continuous rise in the production of biodiesel in emerging countries, mainly China and Brazil. Forecasts assume that Brazilian biodiesel production will surpass the European production by 2015. By 2020, it is estimated that 10 % of all road diesel in BRICs countries (Brazil, Russia, India and China) will be biodiesel (Emerging Market Online 2012).

Brazilian government claims that the 2003 program for biodiesel production was conceived with focus on regional development and social inclusion. Official data assumes that thousands of jobs in familiar agriculture have recently been the result of implementation of the program for cultivation of oleaginous plants for biodiesel

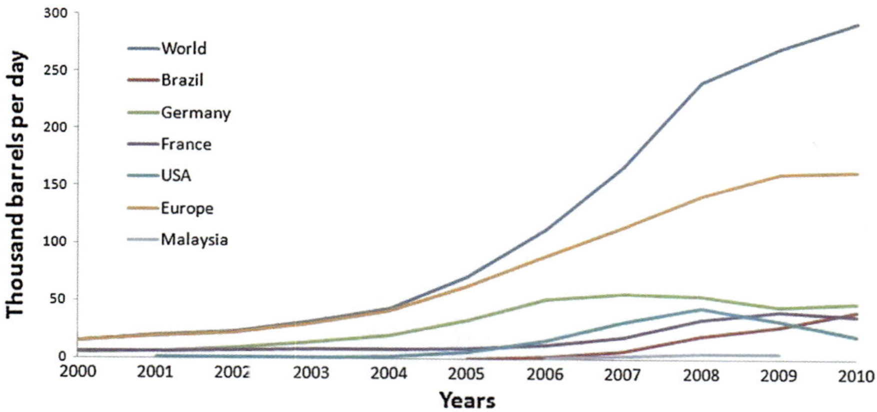

Fig. 9.1 Evolution of the world biodiesel production. *Sources* International energy agency, Agência Nacional de Petróleo (Brazil) and United States Department of Agriculture

production. Cultivation of plants as biodiesel feedstock in Brazilian regions most seriously affected by economic and social problems (mainly in the north and northeast) has contributed to bolster a modest but detectable lessening of regional disparities inside the country. Companies buying raw material from familiar farmers for biodiesel production enjoy tax privileges and government financial support, including a Social Fuel Stamp. They must, however, assume the commitment to purchase the total production at pre-set prices, ensuring thus a financial safety for the low-scale farmer.

The production of biodiesel in Brazil has considerably risen in the last few years. From a virtually null production up to 2006, the production rapidly grew to an amount comparable to the output of France and Germany (world leaders) in 2010 (Fig. 9.1). From a current mandatory blend of 3 % (B3), expected to rise to B5 in 2013, the hopeful expectation for 2015 is B20 (European Biofuels 2012).

Biodiesel cost optimization in Brazil requires that the distance between feedstock plantations and final consumers have to lie inside a 200 km radius, in order to reduce costs derived from transportation (Dias 2007). A means to help achieving this aim is the bioprospection of oleaginous potential sources as alternatives to the species currently exploited as feedstock, most of them cultivated in huge monocultures considerably harmful to the environment. The extent of the Brazilian territory and the wide diversity of its flora and habitats provide immense opportunities to find alternative sources of biodiesel feedstock.

9.2 Current Biodiesel Feedstock Sources in Brazil

Worldwide, biodiesel is produced mainly from vegetable oils (chiefly palm oil, soybean and canola) and secondarily from animal fat and microalgae. In Brazil, biodiesel is also produced chiefly from seed oils. Major sources are soybean

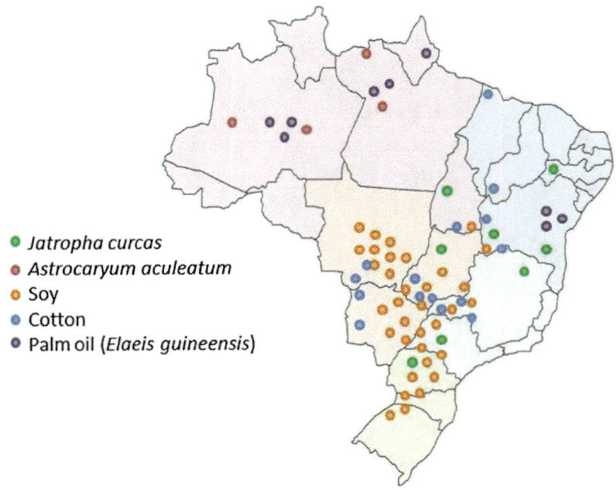

Fig. 9.2 Locations of plantation of oleaginous plants in Brazil for provision of biodiesel feedstock. *Sources* AGROPALMA, ABIOVE and ABRAPA

(*Glicyne max* L., Leguminosae), cottonseed (*Gossypium hirsutum* L., Malvaceae) and palm oil (*Elaeis guineensis* Jacq., Arecaceae). Minor contributors are sunflower oil (*Helianthus annuus* L., Asteraceae) and the fruit of the tucumã palm tree (*Astrocaryum aculeatum* Meyer, Arecaceae). Except for the latter source, the four other species are not native to Brazil, despite the megadiversity of the country (the widest in the world).

Soybean is a major biodiesel feedstock in many countries and Brazil is its second largest producer. According to the Brazilian Institute of Geography and Statistics, the cultivated area of soy plants in Brazil reaches up to 24.5 million ha (IBGE 2012), or 2,9 % of the territory. The regions mostly used for soybean cultivation are the south, southeast and central-west (Fig. 9.2). The contribution to biodiesel production of soybean oil in Brazil is 70–80 %. The second important feedstock in Brazil is cottonseed oil, although contributing far below soybean oil: 3–5 % (ANP 2012). Cottonseed plants are cultivated in distinct parts of the country, such as southeast, northeast and central-west (Fig. 9.2).

The dependence on one or two major biodiesel feedstock is neither reliable nor feasible, especially taking into consideration the expectation of the Brazilian government of a continuous and rapid increase of biodiesel blending (B20 in 2020). For this reason, incentives have been invested in cultivation of several other species, according to traditional and climatic characteristics of the Brazilian regions. By encouraging the production of feedstock at many points of all regions of the country, the program aims also to optimize costs of its transportation to biodiesel plants. Palm oil tree is a species native from Africa. It has been one of the major crops in tropical countries in the New and Old Worlds for production of edible oil. The species has been assumed in Brazil as an important alternative for production of oil in regions not suitable for cultivation of soybean and where it has been cultivated and consumed traditionally: the northern and northeastern regions (Fig. 9.2). The tucumã palm tree (*Astrocaryum aculeatum*, Arecaceae) is native in the Amazon. It has been cultivated in several localities of the Brazilian Northern regions as biodiesel feedstock (Fig. 9.2), but achieving so far very little contribution to the national production. *Jatropha curcas* oil (Euphorbiaceae) has been rated as one of the most promising feedstock for biodiesel production in the tropical world. It was assumed as highly productive, drought resistant and requiring little care for cultivation. Official funds were invested in several tropical countries for cultivation of jatropha plants aiming the biodiesel production. For several reasons (including ecological and economic unexpected problems) the former optimistic forecasts were frustrated (Axelsson and Franzén 2010). Incentives from the Brazilian government in the last decade has encouraged programs for cultivation of jatropha plants in several areas of Brazil (Fig. 9.2), and also the set-up of a biodiesel plant exclusively for production of jatropha biodiesel. However, the contribution in Brazil from jatropha oil for biodiesel production has been negligible.

Given the so far limited resources being exploited, mostly from non-native species, in addition to the immense potential possibilities offered by the diversity of the Brazilian flora, bioprospection of alternative sources of biodiesel feedstocks is crucial for technological development and improvement of the social-economic condition of people from Brazilian less-favored areas.

9.3 Potentialities of Oils from Palm Trees as Biodiesel Feedstock

The family of the palm trees (Arecaceae) is one of the largest among the monocotyledons, comprising about 1,500 species (Henderson et al. 1995). Palm trees are one of the main physiognomic characteristics of tropical forests from both New and Old Worlds. There are approximately 200 species of palms in the Brazilian flora (Souza and Lorenzi 2005). Palm species are distributed in all Brazilian ecosystems. The Amazon and Atlantic forests are among the world ecosystems with wider diversity of palm species (Cintra et al. 2005). Examples of palm trees from Brazilian rain forests are *Euterpe oleracea* (the açaí palm) and *Astrocaryum sciophilum*. *Euterpe edulis*, *Mauritia flexuosa*, *Orbygnia phalerata* and *Syagrus oleracea* are palm species from the "cerrados" (savanna ecosystems from the central-west and southeast Brazil). Examples of palm species from the "caatinga" (semiarid ecosystem from northeast Brazil) are *Copernicia prunifera* ("carnaúba"), *Syagrus coronata* and *Syagrus oleracea*. In the Brazilian "restingas" (coastal ecosystems with sandy and salty soils) occur *Astrocaryum aculeatissum*, *Attalea humilis*, *Bactris vulgaris* and *Syagrus romanzoffiana*, among other palm species.

Many species of palm trees have enormous oleaginous potential. Not only the palm oil tree, but also other palm species have long been important sources, such as coconut (*Cocos nucifera*) and babaçu (*Orbygnia* spp.). The dry endosperm of coconut may attain 60 % of fat. The content of oil in the kernel of the babaçu fruit may reach up to 70 %. In addition, palm species may be prodigal fruit producers, such as coconut, açaí and babaçu. A single raceme of babaçu contains normally around 250 fruits. These characteristics, allied to their capacity to grow and reproduce well in tropical environments, demanding only little care, turn the palm species interesting feedstock sources for biodiesel production in less-favored regions of the Americas, Africa and Asia. They have been regarded as suitable crops to be cultivated in degraded environments. As commented above, two palm species (palm oil tree and tucumã) have been recommended for cultivation in the Brazilian Amazon, aiming biodiesel production.

In the biodiesel context, another point worth considering is the distribution of fatty acids of lipids from palm species. They seem to contain low contents of polyunsaturated fatty acids. In both coconut fat and babaçu oil the main constituent is lauric acid (dodecanoic acid—saturated C_{12}), with very low or negligible contents of linoleic and linolenic acids (C_{18} di- and triunsaturated acids, respectively). Low content of polyunsaturated acids is a desirable characteristic in biodiesel feedstock. The increase in the number of double bonds of fatty acids favors peroxidation processes during biodiesel burn, promoting the accumulation of carbon deposits inside the engine cylinders (Anand et al. 2010). In addition, polyunsaturated acids favor the emission of NO_x (McCormick et al. 2001).

It is likely that lipids from many Arecaceae other than coconut and babaçu have low content of polyunsaturated fatty acids. The distribution of fatty acids has been shown to bear taxonomic meaning, i.e. closely related species tend to have similar distribution of fatty acids (Santos and Salatino 1998; Mayworm and Salatino 2002). Hence many palm species are expected to produce seed oils with low contents of polyunsaturated fatty acids. Data on Table 9.1 strengthen this hypothesis. Among the 14 listed species, only 2 (*Euterpe edulis* and *Oenocarpus bacaba*) have oils with high content of diunsaturated acid (linoleic). The oils from the other species may be combined in two groups: (1) oils with predominance of palmitic (saturated) and oleic (monounsaturated) acids—*Acrocomia aculeata*, *Astrocaryum vulgare*, *Bactris gasipaes*, *Elaeis guineensis*, *Euterpe oleracea*, *E. precatoria*, *Mauritia flexuosa*, *M. vinifera*, *Oenocarpus bacaba* and *O. bataua*; (2) oils with predominance of lauric and myristic acids, both with saturated and medium length carbon chains—*Acrocomia aculeata*, *Astrocaryum murumuru*, *Cocos nucifera* and *Orbygnia phalerata*. Both species groups have low or negligible contents of the polyunsaturated acids linoleic and linolenic (Table 9.1). It is interesting to note in Table 9.1 that oils from palm species seem to have practically no fatty acids with carbon chains longer than C_{18}. Few species on the table have contents of oil lower than 20 %, and the oil yield of several species are considerably higher than the yield of oleaginous species currently exploited (e.g. soybean: 20 %). Taking also into account the common high production of fruits, favorable perspectives of productivity per hectare may be expected from the exploitation of palm species as biodiesel sources.

Despite the many advantages and high potential of palm trees as biodiesel feedstock, care should be taken during its exploitation. For example, it is necessary to take into consideration the area to be deforested in order to avoid threats to the preservation of plant and animal species. The palm oil boom production in Indonesia has provided vital income to small-scale farmers, but certainly too large extents of natural tropical forests have given over to palm oil production.

Table 9.1 Oil yield and distribution of fatty acids of lipids of *Elaeis guineensis* (palm oil tree) and Brazilian native species of palms (Arecaceae)

Species	Oil yield (%)	Fatty acids (%)											
		Caprilic (8:0)	Capric (10:0)	Lauric (12:0)	Myristic (14:0)	Palmitic (16:0)	Palmitoleic (16:1)	Stearic (18:0)	Oleic (18:1)	Vaccenic (18:1 *cis*11)	Linoleic (18:2)	Linolenic (18:3)	Araquidic (20:0)
Acrocomia aculeata (Jack) Lood. ex Martius[1]	53.0		5.0	50.9	13.1	7.6		3.0	17.9		2.5		
Astrocaryum murumuru[2]	31.0	2.7	2.0	51.6	25.8	6.0		2.9	5.7		3.0	0.1	0.1
Astrocaryum vulgare Mart.[3]	18.0		0.8			22.9		2.9	67.6		1.2		
Bactris gasipaes Kunth[4]	8.0				0.5	41.3	5.9	3.3	43.1		1.7	4.1	2.3
Cocos nucifera L.[5]	27.7	5.8	4.8	49.1	21.8	8.4		2.8	6.1		1.2		
Elaeis guineensis Jacq[6]	62.0				0.3	31.1		8.8	49.6		9.2	0.8	
Euterpe edulis Martius[7]	39.7				3.4	23.1	0.8	6.3	15.6	15.6	36.4	1.5	
Euterpe oleracea Martius[8]	31.0			0.07	0.1	26.2	4.9	1.8	52.0	3.4	7.8	0.6	0.1
Euterpe precatoria Martius[9]	29.6				0.3	14.8	0.4	0.3	80.6		3.1		

(continued)

Table 9.1 continued

Species	Oil yield (%)	Fatty acids (%)											
		Caprilic (8:0)	Capric (10:0)	Lauric (12:0)	Myristic (14:0)	Palmitic (16:0)	Palmitoleic (16:1)	Stearic (18:0)	Oleic (18:1)	Vaccenic (18:1 *cis*11)	Linoleic (18:2)	Linolenic (18:3)	Araquidic (20:0)
Mauritia flexuosa L.[10]	11.2				0.1	17.3		2.0	73.3		2.4	2.2	
Mauritia vinifera L.[10]	66.0			0.03	0.1	16.8	0.3	1.8	76.1		4.9	1.0	
Oenocarpus bacaba Mart.[11]	23.6					22.5	0.7	4.7	34.41		34.9	0.9	
Oenocarpus bataua (Mart.) Burret.[11]	28.6					15.8	1.2	2.3	73.5		5.4	0.7	
Orbignya phalerata Mart.[12]	66.0	6.0	5.0	44.0	17.0	8.0		4.5	14.0		2.0		

[1]Belén-Camacho et al. (2005); [2]Mambrim and Barrera-Arellano (1997); [3]Schirmann et al. (2011); [4]Clement et al. (1998); [5]Bhatnagar et al. (2009); [6]Monde et al. (2009); [7]Panza et al. (2009); [8]Bora and Rocha (2004); [9]Escriche et al. (2009); [10]Silva et al. (2009); [11]Unpublished data; [12]Machado et al. (2006)

Tree burning and soil degradation, sometimes in carbon-rich peatlands, have given off enormous amount of global-warming gases (Gilbert 2012). For this reason, the choice of adequate localities for palm plantation in rainforest is important, for example, avoiding marsh or wetland environments. Many internet sites have blamed palm oil plantations in Indonesia for having caused the death of hundreds of orangutans, a primate species already assumed as under serious risk of extinction. A possibility of cultivation in tropical forest with reduced or no risk of ecological harm is the establishment of plantations in already degraded Amazonian zones, such as has been done with black pepper (*Piper nigrum*) (Kato et al. 2001).

A seemingly interesting way of exploitation of native products is the harvest of fruits in extractive regimen in preserved areas, created for this specific aim (Mayworm et al. 2011). Extensive extractive areas have been established in several locations of the Brazilian Amazon (Dubois 1996). The adequate conservation of extractive areas depends on the sustainable exploitation of a high diversity of species. Given the several favorable aspects of frequency of occurrence of palm species in the Amazon, the commonly high productivity of fruits and the adequate chemical profiles, the attainment of biodiesel feedstock in extractive areas is seemingly promising from the economic, social and environmental viewpoints.

9.4 Concluding Remarks

It is important to widen the exploitation of biodiesel feedstock toward a wider spectrum of plant species. Little has been done in Brazil aiming the exploitation of native species. Plants of Arecaceae are particularly interesting in this regard. The introduction of products from native palm species may help minimizing the food versus fuel dispute in biodiesel production (Lam et al. 2009): most biodiesel feedstocks produced in Brazil are important edible products. The exploitation of native palm species in Brazil may ameliorate the social-economic condition of many people and help in the preservation of species and ecosystems, as far as precautions are taken and adequate and environmental friendly measures are put into effect.

Acknowledgments The authors thank provision of funds by CNPq (Conselho Nacional do Desenvolvimento Científico e Tecnológico) and FAPESP (Fundação de Amparo à Pesquisa do Estado de São Paulo).

References

Agência Nacional do Petróleo (ANP) (2012) Available at htpp://www.anp.gov.br. Accessed on Oct 2012

Anand A, Ranjan A, Mehta PS (2010) Estimating the viscosity of vegetable oil and biodiesel fuels. Energy Fuels 24:664–672

Axelsson L, Franzén M (2010) Performance of jatropha biodiesel production and its environmental and socio-economic impacts—a case study. Chalmers, Department of Energy and Environment, Göteborg, 79 (available at http://www.focali.se/filer/FRT%20201006%20-%20Jatropha.pdf. Accessed on Jan 2013

Belén-Camacho DR, López I, García D, González M, Moreno-Álvarez MJ, Medina C (2005) Physicochemical evaluation of seed and seed oil of corozo (*Acrocomia aculeata* Jacq.). Grasas Aceites 56:311–316

Bhatnagar AS, Kumar PKP, Hemavathy J, Krishnae AGG (2009) Fatty acid composition, oxidative stability, and radical scavenging activity of vegetable oil blends with coconut oil. J Am Oil Chem Soc 86:991–999

Bora PS, Rocha RVM (2004) Macaiba palm: fatty and amino acids composition of fruits. Ciencia y Tecnología Alimentaria 4:158–162

Buckeridge MS, Salatino A (2010) Feedstock biochemistry applied to biofuels in Brazil. In: Coser TR, Davis MJ (eds) Biofuels: reasonable steps towards a renewable energy source. Fulbright Comission Brazil, Brasília, 1, pp 85–89

Cintra R, Ximenes AC, Gondim FR, Kropf MS (2005) Forest spacial heterogeneity and palm richness, abundance and community composition in Terra Firme forest, Central Amazon. Braz J Bot 28: 75–84

Clement CR, Aguiar JPL, Arkcoll DB (1998) Composição química do mesocarpo e do óleo de três populações de pupunha (*Bactris gasipaes*) do Rio Solimões, Amazonas, Brasil. Rev Bras Fru 20:115–118

Dias GLS (2007) Um desafio novo: o biodiesel. Estudos Avançados 21:179–183

Dubois JCL (1996) Uses of wood and non-wood forest products by Amazonian dwellers. Unasylva 186:8–15

Emerging Markets Online (2013) Biodiesel 2020: A global market survey, 2nd ed. Available at http://www.emerging-markets.com. Accessed on Jan 2013

Environmental Protection Agency (EPA) (2002) Available at: http://epa.gov/air/caaac/mstrs/korotney.pdf. Accessed on Oct 2012

Escriche I, Restrepo J, Serra JA, Herrera LF (1999) Composition and nutritive value of Amazonian palm fruits. Food Nutr Bull 20:361–365

European Biofuels (2013) Technology platforms. Biofuels in South America and Central America. Available at http://www.biofuelstp.eu/s_america.html. Accessed on Jan 2013

Gilbert N (2012) Palm-oil boom raises conservation concerns. Nature 487 (available at http://www.nature.com/news/palm-oil-boom-raises-conservation-concerns-1.10936. Accessed on Jan 2013

Henderson A, Galeano G, Bernal R (1995) Field guide to the palms of the Americas. Princeton University Press, Princeton

Kato AK, Albuquerque FC, Duarte MLR, Carvalho JEU, Hamada M (2001) Cultivo da Pimenta-do-Reino sob Sistema Sombreado. Comunicado técnico, Embrapa, http://www.infoteca.cnptia.embrapa.br/bitstream/doc/403378/1/ComTec55.pdf. Accessed on Jan 2013

Lam MK, Tan KT, Lee KT, Mohamed AR (2009) Malaysian palm oil: Surviving the food versus fuel dispute for a sustainable future. Renew Sustain Energy Rev 13:1456–1464

Machado GC, Chaves JBP, Antoniassi R (2006) Composição em ácidos graxos e caracterização física química de óleos hidrogenados de coco babaçu. Ceres 53:463–470

Mambrim MCT, Barrera-Arellano D (1997) Characterization of palm tree fruit oils from Brazilian Amazonia region. Grasas Aceites 48:154–158

Mayworm MS, Salatino A (2002) Distribution of seed fatty acids and the taxonomy of Vochysiaceae. Biochem Syst Ecol 30:961–972

Mayworm MS, Buckeridge MS, Marquez UML, Salatino A (2011) Nutritional reserves of Vochysiaceae and potential economic uses. An Acad Bras Cienc 83:523–531

McCormick RL, Grabowski MS, Alleman TL, Herring AM, Tyson KS (2001) Impact of biodiesel source material and chemical structure on emissions of criteria pollutants from a heavy-duty engine. Environ Sci Tech 35:1742–1747

Monde AA, Carbonneau M-A, Eymard-Duvernay S, Tiahou G, Badiou S, Adon B, Cristol J-P (2009) Comparative study of fatty acid composition, vitamin E and carotenoid contents of palm oils from four varieties of oil palm from Côte d' Ivoire. J Sci Food Agric 89:2535–2540

Panza V, Pighin D, Láinez V, Pollero RJ, Maldonado S (2009) Storage lipids and proteins of *Euterpe edulis* seeds. Biocell 33:99–106

Pike Research (2011) Biofuels markets and technologies—Global ethanol, biodiesel, and feedstock markets: technology assessments, policy issues, key supply chain stakeholders and worldwide supply and demand forecasts. p 144. http://www.pikeresearch.com. Accessed on Oct 2012

REN21 (2005) renewable energy policy network for the 21st Century. Renewables global status report. http://www.ren21.net/Resources/Publications/REN21Publications.aspx. Accessed on Oct 2012

Santos DYAC, Salatino A (1998) Fatty acids and seed oils of species of Diplusodon Pohl (Lythraceae). Biochem Syst Ecol 26:109–115

Schirmann GS, Rebollar PM, Miller PRM (2011) Açaí na dieta brasilera—tradições regionais e valor nutricional. VII Congresso Brasileiro de Sistemas Agroflorestais. 19–22

Silva SM, Sampaio KA, Taham T, Rocco SA, Ceriani R, Meirelles AJA (2009) Characterization of oil extracted from buriti fruit (*Mauritia flexuosa*) grown in the Brazilian Amazon Region. J Am Oil Chem Soc 86:611–616

Souza VC, Lorenzi H (2005) Botânica Sistemática. Instituto Plantarum, São Paulo

Sze C, Whinihan J, Olson B, Schenk C, Sobotowski RA (2007) Impact of test cycle and biodiesel concentration on emissions. SAE technical paper 2007-01-4040

Xue J, Grift TE, Hansen AC (2011) Effect of biodiesel on engine performances and emissions. Renew Sust Energ Rev 15:1098–1116

Yazdani SS, Gonzalez R (2008) Engineering *Escherichia coli* for the efficient conversion of glycerol to ethanol and co-products. Metabol Eng 10:340–351

Zhuang D, Jiang D, Liu L, Huang Y (2011) Assessment of bioenergy potential on marginal land in China. Renew Sust Energ Rev 15:1050–1056

Chapter 10
Xylan Biosynthesis in Plants, Simply Complex

Ahmed Faik, Nan Jiang and Michael A. Held

Abstract Xylans are major non-cellulosic polysaccharides in grasses and trees and represent the third most abundant biopolymer on earth (after cellulose and chitin). Xylans have important impacts on biofuel production because they are contributors to plant biomass recalcitrance, yet plants deficient in xylans synthesis grow abnormally. Therefore, deciphering the biochemical mechanisms of xylan biosynthesis will undoubtedly contribute to identifying ways to improve biofuel yields from plant biomass. *Arabidopsis irregular xylem* (*irx*) mutants have shown that genes from GT43 and GT47 CAZy families encode proteins associated with xylan biosynthesis. These genes are duplicated, have overlapping expression patterns, and thus exhibit partial functional redundancy. However, genes from one pair are incapable of complementing mutations in the other, suggesting that their encoded proteins may function cooperatively in xylan synthase complexes (XSCs), and recent work in wheat supports the existence of such XSCs. More recent genetic studies in *Arabidopsis* suggest that xylan backbone elongation/synthesis can be uncoupled from side chains additions to a certain extent. However, what we still don't know is how xylan backbone synthesis is initiated and then elongated by XSCs? And how many different XSCs does a plant employ to make xylans? In this chapter we will discuss what we know about xylan backbone initiation, elongation, and uncoupled substitution of the backbone. We will also discuss recent advances in the regulatory mechanisms of xylan synthesis.

Keywords Xylan synthesis • Hemicellulose • Arabidopsis • Wheat • Grasses • Cell wall • Irregular xylem • GT43 • GT47 • GT8 • Glycosyltransferases

---

A. Faik (✉) · N. Jiang
Environmental and Plant Biology Department, Ohio University, Athens, OH 45701, USA
e-mail: faik@ohio.edu

M. A. Held
Department of Chemistry and Biochemistry, Ohio University, Athens, OH 45701, USA

M. C. McCann et al. (eds.), *Plants and BioEnergy*, Advances in Plant Biology 4,
DOI: 10.1007/978-1-4614-9329-7_10, © Springer Science+Business Media New York 2014

10.1 Introduction

Plant lignocellulosic biomass consists mostly of plant cell walls and holds great promise as a raw material for renewable biofuels. It is expected that manipulation of the biosynthetic pathways of plant cell wall polysaccharides will lead to improvements in plant biomass yield, digestibility, and energy recovery (conversion rates). Xylans are the major non-cellulosic polysaccharides in plant lignocellulosic biomass from grasses and trees and represent the third most abundant polymer on earth (after cellulose and chitin). Xylan polymers are found mainly in the primary cell walls of grasses and the secondary cell walls of grasses and dicots. In lignocellulosic biomass, xylans interact directly with lignin and cellulose and significantly contribute to its recalcitrance to enzymatic digestion (saccharification). Overcoming this recalcitrance requires the inclusion of one or more steps of biomass pretreatment (Faik 2013). Xylans are polymers made mostly of pentoses (5-carbon sugars), and this adds another level of complexity for their efficient fermentation, as most fermenting microorganisms currently used in the biofuel industry prefer hexoses (6-carbon sugars). For plants, xylans in primary cell walls affect many aspects of plant growth and survival. Any deficiency in xylan synthesis directly impacts plant growth and survival. It seems that plants xylans have evolved many structural variations to fit numerous functions in the wall, which would allow plant growth under various environmental conditions. Thus, the designation "one polymer for all seasons" applies to xylans. The structures of xylans vary depending upon species, tissues, or even cells within the same tissue (Ebringerova et al. 2005). For instance, the structure of xylan in cell walls of the starchy endosperm is different from that of the aleurone or vascular walls. These structural variations suggest that plants have mechanistic and/or regulatory strategies to control xylan fine structure. Therefore, deciphering the biochemical mechanisms of xylan synthesis will undoubtedly contribute toward improving the quantity and quality of plant biomass for biofuels production. Xylan biosynthesis is far from understood and the process turns out to be surprisingly difficult to grasp. The primary thrust of this chapter is to critically analyze recent advances in the field of xylan synthesis and structure leading to our current understanding of the biochemical mechanisms and regulation of xylan synthesis in plant, and how these advances might improve biofuels production. To start, we will review some of the most recent work related to this topic in dicots and grasses and attempt to answer questions such as: How many genes does a plant need to make xylans? And how can we reconcile the genetic data with biochemical data that suggest the involvement of cooperative, multi-protein complexes? Next, we will discuss the sophisticated, multi-level regulatory mechanisms that control xylan synthesis and secretion. Finally, we will discuss how our current knowledge of xylan synthesis may be used to improve plant biomass for biofuels, as well as aspects of xylan synthesis that are lagging behind and deserve more attention.

10.2 Biochemical Mechanisms of Xylan Synthesis: So Many Genes, So Little Known

Despite the increasing attention paid to the field of xylan biosynthesis, there is limited progress on the biochemical side of the process. It is clear that plants have developed a sophisticated biosynthetic mechanism with a complex regulation system to control the fine structural details of xylan biosynthesis. This system likely includes a large number of genes and adapted synthesis and secretion pathways. In fact, we have learned (and are still learning) from genetics in *Arabidopsis* that many genes are involved in glucuronoxylan (GX) biosynthesis in secondary cell walls (in xylem and fiber cells). These genes are members of a large family, called *IRREGULAR XYLEM* (*IRX*) genes, that also includes genes associated with cellulose synthesis (Turner and Somerville 1997; Brown et al. 2005). Of particular interest are three glycosyltransferase (GT) families classified in the CAZy database as GT8, GT43, and GT47. Interestingly, *IRX* genes from GT43 and GT47 families exist as duplicates, namely *IRX*10/*IRX*10-*L*, *IRX*14/*IRX*14-*L*, *FRA*8(*IRX*7)/*F*8*H*(*IRX*7-*L*), *IRX*9/*IRX*9-*L*, and *IRX*15/*IRX*15-*L*, and exhibit partial functional redundancy and overlapping expression patterns. This gene redundancy is an indication of the importance of xylan synthesis to plant development and survival. On the other hand, *IRX* genes from GT8 family don't show such redundancy, which may suggest more specific roles limited to a particular tissue types or developmental stages. It is important that our readers keep in mind that the current genetic advances in xylan biosynthesis are limited to *irx* phenotypes in *Arabidopsis*. These phenotypes occur in a specialized tissue and any analysis of genetic data should be put in this particular context. Generalizations regarding the mechanisms of xylan synthesis may not yet be possible with these data alone. For example, the description of GX synthesis in secondary cell wall should be understood under the physiological and cell biological context of that particular tissue. This resembles the fabled description of an elephant by blind men, each one describing the elephant through limited touch perception. With this in mind, the *Arabidopsis irx* mutants have revealed the involvement of many genes in xylan biosynthesis, but questions about how these gene products interact to make xylan polymers in primary and secondary cell walls remain. Nevertheless, the *irx* mutants indicate that *IRX* genes impact GX synthesis during secondary cell wall deposition to varying degrees. These impacts are described in Table 10.1, which will be discussed as we progress through the text.

Biochemical advances in grasses has revealed new aspects of glucurono(arabino)xylan (GAX) biosynthesis, such as the presence of multi-enzyme complexes and a cooperative biosynthetic mechanism that genetics alone could not demonstrate. We recognize that these aspects may well be specific to growing tissues in grasses. Nonetheless, this provides us the opportunity in this chapter to compare and contrast data from *Arabidopsis* (on secondary cell wall) with the data from grasses (on primary cell wall), and develop a bigger picture of xylan synthesis in plants. Therefore, we will separate our discussion of the biosynthetic process of xylan biosynthesis into four sub-sections: II-1 xylan backbone initiation/termination, II-2 xylan backbone elongation, II-3 xylan backbone additional decoration through an uncoupled mechanism, and II-4 xylan secretion/delivery. These sections will be used to help integrate genetic and biochemical advances into a more general mechanism that may be applied to both dicots and grasses, as well as to different plant tissues.

10.2.1 Xylan Backbone Inititiation/Termination in Arabidopsis: Does Sequence 1 Have a Say?

It has been shown in dicots that GX polymers of the secondary cell wall have a tetrasaccharide (called sequence 1), 4-β-D-Xyl*p*-(1,4)-β-D-Xyl*p*-(1,3)-α-L-Rha*p*-(1,2)-α-D-Gal*p*A-(1,4)-D-Xyl*p*, located at the reducing end of their chains (Table 10.1) (Johansson and Samuelson 1977; Andersson and Samuelson 1983; Pena et al. 2007). It was hypothesized that sequence 1 is important for the initiation or termination of GX backbone elongation, based on the observation that this sequence is absent in GX-deficient *Arabidopsis* mutants *fra*8(*irx*7), *irx*8, and *parvus* (Pena et al. 2007; Lee et al. 2007). However the fact that *fra*8(*irx*7) and *irx*8 mutants can still produce some GX (Lee et al. 2007), suggests that it is not absolutely required for priming xylan backbone synthesis. Other *irx* mutants having shorter xylan backbone chains (such as *irx*9, *irx*10, and *irx*14), or double mutants having very little GX (such as *irx*10/*irx*10*L* or *irx*14/*irx*14-*L*, Table 10.1), still show the presence of this oligosaccharide in their GX, which may suggest that somehow sequence 1 prematurely terminates xylan backbone chain elongation (Brown et al. 2009; Wu et al. 2009, 2010). In support of this hypothesis, *irx*7 and *irx*8 mutants (both have GXs, but lack sequence 1) have xylan backbone chains with increased length and heterodispersity (Brown et al. 2007; Persson et al. 2007; Lee et al. 2009) (Table 10.1). However, it is still unclear what mechanism allows a plant cell to sense the length of xylan backbone and terminate the elongation. One possibility is that the addition of sequence 1 occurs in one of the late Golgi compartments, while GX backbone synthesis starts in early Golgi compartments. Thus, the distance traveled between the two compartments could represent the time for xylan backbone elongation. Defective synthetic machinery may have a slower synthesis rate to allow the production of normal sized GX chains. It also appears that GlcA substitution does not dictate the addition of sequence 1, as the Xyl to GlcA substitution ratio is unaltered in all *irx* mutants (regardless the length of xylan backbone, Table 10.1).

Taking all of these observations into consideration, we hypothesize that sequence 1 may have an additional/alternative role in xylan synthesis in secondary cell walls. Specifically, sequence 1 might be involved in driving the secretion of the newly synthesized GX to the cell surface. Secondary cell wall deposition occurs at a specific developmental stage (e.g., after cell elongation is ceased), and may require a rapid, signal-mediated secretion mechanism to deliver the massive

Table 10.1 Comparison of the effects of mutations in *IRX7*/*IRX7-L*, *IRX9*/*IRX9-L*, *IRX*10/*IRX*10-*L*, *IRX*14/*IRX*14-*L*, and *IRX*15/*IRX*15-*L* pairs along with *IRX*8 and *PARVUS* genes on GX biosynthesis and phenotype of Arabidopsis plants

Mutants (accession #)	irx phenotype	Xylan chain length	GX content compared to WT	Xyl substitution	X6-dependent XylT activity	Sequence 1 at reducing end	References
*fra*8(*irx*7) (*At*2*g*28110)	Strong	Increased and hetero-disperse	−59 %	Unaffected	Unaffected	Absent	Pena et al. (2007), Brown et al. (2009), Lee et al. (2009)
*f*8 *h*(*irx*7-*L*) (*At*5*g*22940)	Mild	Not known	−20 %	Unaffected	Unaffected	Unknown	Brown et al. (2009), Lee et al. (2009)
*irx*7/*irx*7-*L*	Strong	Reduced	−85 %	Not known	Unaffected	Absent	Wu et al. (2010), Persson et al. (2007)
*irx*10 (*At*1*g*27440)	Mild	Reduced	−10 %	Unaffected	Unaffected	Present	
*irx*10-*L* (*At*5*g*61840)	No	Unaffected	100–110 %	Unaffected	Unaffected	Present	Brown et al. (2009)
*irx*10 *irx*10-*L*	Strong	Not known	Absent	–	Decreased	Present	Wu et al. (2010)
*irx*9 (*At*2*g*37090)	Strong	Reduced	−50 %	Unaffected	Decreased	Present	Pena et al. (2007)
*irx*9-*L* (*At*1*g*27600)	Mild	Not known	−20 %	Unaffected	Not known	Present	
*irx*9 *irx*9-*L*	Strong	Reduced	−85 %	Unaffected	Decreased	Present	
*irx*14 (*At*4*g*36890)	Mild	Reduced	−50 %	Unaffected	Decreased	Present	Wu et al. (2010)
*irx*14-*L* (*At*5*g*67230)	No	Not known	−20 %	Unaffected	Not known	Present	Wu et al. (2010)
*irx*14 *irx*14-*L*	Strong	Not known	Absent	–	Decreased	Present	Wu et al. (2010)
*irx*8 (*At*5*g*54690)	Strong	Increased and hetero-disperse	−50 %	Unaffected	Unaffected	Absent	Pena et al. (2007)
Parvus (*At*1*g*19300)	Strong	Not known	−50 %	Unaffected	Unaffected	Absent	Lee et al. (2007)

(continued)

Table 10.1 continued

Mutants (accession #)	irx phenotype	Xylan chain length	GX content compared to WT	Xyl substitution	X6-dependent XylT activity	Sequence 1 at reducing end	References
*irx*15 (*At*3*g*50220)	No	Unaffected	−2–5 %	Unaffected	Unaffected	Present	Brown et al. (2011), Jensen et al. (2011)
*irx*15-*L* (*At*5*g*67210)	No	Unaffected	Unaffected	Unaffected	Unaffected	Present	Brown et al. (2011), Jensen et al. (2011)
*irx*15/*irx*15-*L*	Mild, uneven xylan distribution)	Reduced	−35 %	Unaffected	Slightly decreased	Present	Brown et al. (2011), Jensen et al. (2011)

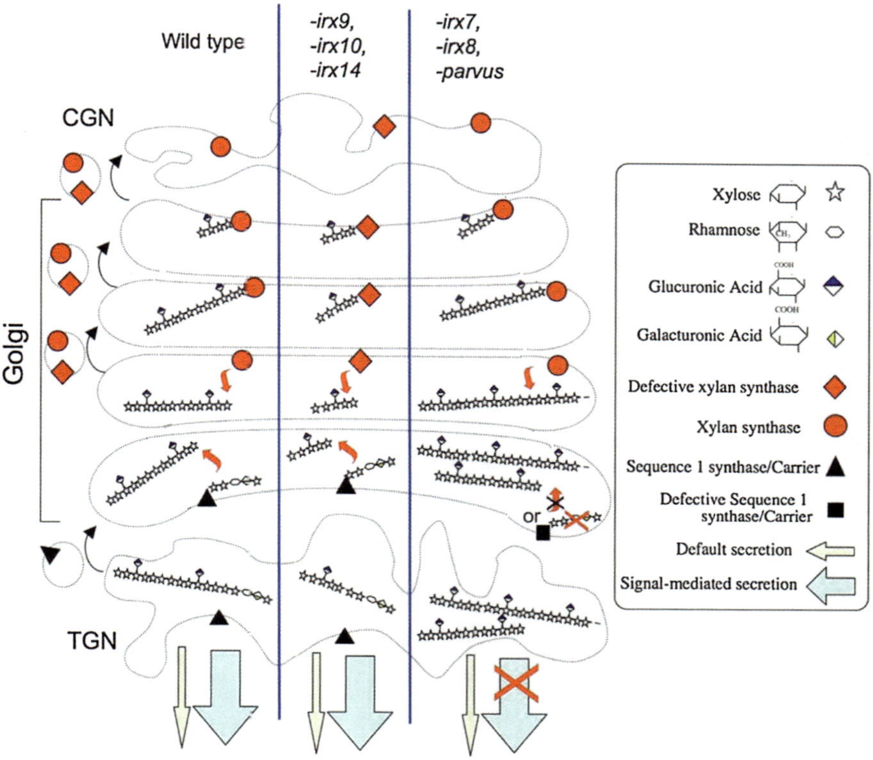

Fig. 10.1 Model for sequence 1 as a secretion signal during GX synthesis in secondary cell wall. GX secretion would require a fast signal-mediated pathway from *trans* Golgi network (*TGN*) to cell surface. Xylan backbone synthesis is initiated in *cis* Golgi network (*CGN*) or *cis*-Golgi stacks and completed before reaching *trans* Golgi stacks or TGN, where sequence 1 terminate xylan backbone elongation. Mutations in *IRX9/IRX9-L*, *IRX10/IRX10-L*, and *IRX14/IRX14-L* genes would result in a defective xylan synthase complex with a slow elongation process, and early termination by the transfer of sequence 1 onto GX with shorter chains. Mutations in IRX7/IRX7-L, IRX8, and PARVUS genes would affect either the synthesis of sequence 1 or its transfer onto GX, which would result in slower secretion to the secondary cell wall and accumulation of GX with slightly longer backbone chains in the Golgi

amounts of GX polymer needed for building the secondary cell wall. The absence of such a secretory mechanism would result in the accumulation of GX in the Golgi of mature plant cells that usually lacking high vesicular trafficking to accommodate this need. In addition to this fast-delivery pathway for GX, a default secretion pathway with a much slower rate may exist for other secretory cargo including some GX (Fig. 10.1). In this model, sequence 1 would be synthesized in the *trans*-Golgi or *trans* Golgi network (TGN) by a sequence 1 synthase complex (indicated as ▲ in Fig. 10.1) that can also act as a carrier. This sequence 1 synthase complex would transfer sequence 1 to the newly synthesized GX, which would facilitate its secretion, possibly with the help of a protein or lipid receptor that recognizes sequence 1 as a secretion signal. In this model, xylan backbone synthesis would be initiated early in the Golgi (e.g., *cis* Golgi) by xylan synthase complexes (indicated as red ● in Fig. 10.1) and completely decorated before reaching the late Golgi compartments (i.e., *trans*-Golgi stacks or TGN). The whole "GX-sequence 1" polymer can then be recognized by its putative receptor, packaged in vesicles (most likely of clatherin type), and then delivered to the cell surface (Fig. 10.1). This model is in agreement with the current genetic data. For example mutations that yield defective GX synthases (*irx*9, *irx*10, and *irx*14, indicated as red ◆ in Fig. 10.1) would produce GX with shorter xylan chain lengths (at much slower rate) compared to wild type. These shorter GX chains would still be terminated by sequence 1 in the *trans*-Golgi or TGN and delivered to the cell surface (Fig. 10.1). In the case of *irx*7, *irx*8 and *parvus* mutants, defects in the transfer reaction of sequence 1 onto the newly synthesized GX or in sequence 1 synthase complex itself (indicated as ■ in Fig. 10.1) would result in slower secretion of GX to the cell wall and possibly the accumulation of large amounts of GX in the Golgi. Another possibility is that the absence of sequence 1 may delay the release of newly synthesized GX from xylan synthase complexes. This would result in the accumulation of GX having longer xylan backbone chains in *trans*-Golgi or TGN and could explain the longer xylan backbone chains observed in GX from *irx*7 and *irx*8 mutants (Fig. 10.1). We can imagine that the accumulation of GX in *trans*-Golgi or TGN might trigger a reduction in GX biosynthesis by a negative feedback mechanism (yet to be discovered). Signal-mediated secretion would appear to be specific to GX in secondary cell wall of vascular tissues, as cellulose-deficient *Arabidopsis* mutants (such as *irx*3) do not show a decrease in GX content (Brown et al. 2011). The same situation was observed in *irx*4, an *Arabidopsis* mutant defective in lignin biosynthesis (mutation in a *cinnamoyl CoA reductase* gene), which exhibits collapsed xylem and altered plant development, but retains normal levels of GX (Jones et al. 2001).

Another possible function of sequence 1 could be for the integration of GX into the secondary cell wall. For example, sequence 1 may be needed to attach GX at specific sites in the cell wall (i.e., to lignin or pectins), which would also drive GX secretion. Several types of covalent bonds that connect xylans to lignin (and other cell wall polymers) within the secondary cell wall have been documented (Joseleau et al. 1992). These covalent bonds include glycosidic linkages between free C-1 positions (reducing end) of Xyl residues and *p*-Coumaric acid of lignin; benzyl ester linkages between the side chains of GlcA and/or Ara residues and lignin (via ferulic or coumaric acids); and ether linkages between lignin and either Ara or Xyl residues of side chains. Most of these linkages are susceptible to alkaline treatment, but can be identified by enzymatic treatment of xylans (Joseleau et al. 1992).

So far, there is no experimental evidence of the presence of sequence 1 in monocots (Fincher 2009). However, work by Zeng et al. (2010) showed that a purified wheat xylan synthase complex (XSC) could produce in vitro a GAX-like polymer that releases two main oligosaccharides (called peak I and peak II) after digestion with endoxylanase III. Structural analysis of peak I suggested that it was an oligosaccharide with a degree of polymerization (DP) of 8 made exclusively of Xyl and Ara residues in a ratio of 3:1, respectively. Peak II, on the other hand was not fully characterized, but contained mainly Xyl and GlcA (with very little Ara). This uncharacterized oligosaccharide could be a variation of sequence 1 needed for initiation/termination/secretion of GAX biosynthesis in monocot primary cell walls. The fact that grass homologs of *Arabidopsis IRX7*, *IRX8*, and *PARVUS* genes (required for synthesis of sequence 1) are not highly expressed in tissues that produce large amount of AX in grasses strongly suggests that xylans in primary cell walls don't need sequence 1 for chain elongation termination and secretion. Similarly, low expression of *Arabidopsis IRX7*, *IRX8*, and *PARVUS* genes was observed in transcriptional profiling studies carried out in psyllium (*Plantago ovata Forsk*) seed mucilage, a tissue that is rich in xylans and primary cell walls (Jensen et al. 2011). This is in agreement with the fact that psyllium mucilage xylans lack the terminal sequence 1 oligosaccharide (Jensen et al. 2011). Our hypothesis is that a GlcA-rich oligosaccharide, with a structure yet to be determined, may act as a terminator in xylan biosynthesis in grasses. In fact, intact wheat microsomes generated only short GAX polymers in vitro, having a DP ranging from 50 to 80 (Zeng et al. 2008). It would be interesting to test the effect of sequence 1 on GAX synthesis in secondary cell walls of grasses. Current data suggest that xylan chain termination mechanisms might be different between grasses and dicots. For example, grass homologs of *Arabidopsis IRX7*, *IRX8*, and *PARVUS* genes, which are required for synthesis of sequence 1, are not highly expressed in tissues that produce large amount of AX in grasses.

It is becoming evident that dissecting the steps of GX biosynthesis within *Arabidopsis* Golgi apparatus will be critical toward a full understanding of the biosynthetic process. It will be important to determine where sequence 1 is synthesized and transferred onto GX, and what the exact roles of IRX7, IRX8, and PARVUS are in the process. Developing a biochemical assay for GX synthesis in secondary cell walls in *Arabidopsis* will be a major breakthrough for testing the biochemical function of these GTs. More importantly, it will be important to assess the role of sequence 1 in the secretion of GX to the cell surface, and identify possible GTs that catalyze the transfer of sequence 1 onto GX. Although these types of experiments are difficult to implement, any information gained from them would help tremendously advance the field of xylan biosynthesis. Plant cell wall polysaccharide secretion is still lagging behind and more efforts are needed to understand this area. Our proposed model for sequence 1 in GX secretion provides a framework to design experiments to answer specific questions in this field.

10.2.2 Xylan Backbone Elongation: Cooperative Mechanism and Core Xylan Synthase Complexes

Several observations provide direct evidence that xylan backbone synthesis and elongation is under the control of more than one protein. For example, genes from one *IRX* pair are incapable of complementing mutations in the other pairs (i.e., *ixr9/irx9-L* double mutant is not complemented by *IRX*10, *IRX*14, *F8H*, or *FRA*8), and mutations in *IRX* genes have additive effects on *irx* phenotype and plant growth (Table 10.1). Other observations provide indirect evidence that the IRX proteins function cooperatively in XSCs. For instance, the degree of substitution of the xylan backbone with GlcA residues is not affected by mutations in *IRX* genes regardless of the length of the xylan backbone chain (Table 10.1). This is only possible if xylan backbone elongation is coupled with the addition of GlcA side chains via the cooperative action of multi-protein complexes. Although our work in wheat supports the existence of XSCs and provides evidence of the involvement of a cooperative mechanism for elongating of GAX-like xylan backbone polymers (Zeng et al. 2008, 2010), there is no equivalent GAX synthesis assay in mature *Arabidopsis* plants, making it difficult to translate data from etiolated wheat seedlings (grass) to mature *Arabidopsis* plants (dicots). To gain insights into GAX biosynthesis in dicots, we sought to test if etiolated wild type *Arabidopsis* seedlings have GAX synthesis activity similar to that observed in etiolated wheat seedlings. To do so, Golgi-enriched microsomes were prepared from 5 to 7-day old etiolated *Arabidopsis* seedlings and tested for GAX synthesis activity as described in Zeng et al. (2010). This previous work showed a stimulatory effect of UDP-Xyl on the incorporation of [^{14}C]GlcA, from UDP-[^{14}C]GlcA, into ethanol-insoluble products using the wheat microsomes. Similarly, intact *Arabidopsis* microsomes showed substantial [^{14}C]GlcA incorporation in presence of UDP-Xyl, and no [^{14}C]GlcA incorporation in the absence of UDP-Xyl (Fig. 10.2). Monosaccharide composition analysis of the [^{14}C]-product generated by *Arabidopsis* microsomal membranes in presence of UDP-[^{14}C]GlcA, UDP-[^{14}C]Ara*p*, and UDP-[^{14}C]Xyl was then performed by acid hydrolysis (2 M TFA, 1 h at 120 °C) of the [^{14}C]-product and fractionation by High pH anion exchange chromatography (HPAEC) on a CarboPac PA20 column (Dionex). Our data show that ~70 % of the [^{14}C]-radiolabel co-elutes with Ara, Xyl, and GlcA in a ratio Xyl:Ara:GlcA of 9:0.5:1 (according to UDP-sugars specific [^{14}C]-radioactivity), suggesting less Ara incorporation (Fig. 10.2).

General conclusions from these preliminary data are (1) *Arabidopsis* microsomes, like in wheat, also require the presence of both UDP-Xyl and UDP-GlcA to produce a GAX-like polymer. Although the incorporation of [^{14}C]GlcA was ~10 times lower compared with wheat microsomes, GAX synthesis activity was clearly present. This was expected, as GAX is only a minor constituent in *Arabidopsis* primary cell walls compared to wheat; (2) GAX-like polymers produced by *Arabidopsis* microsomes have a ratio of Xyl:GlcA of 9:1, which is similar to the ratio of 8:1 observed in GX from the secondary cell walls of *Arabidopsis* (Pena et al. 2007); and (3) like wheat, *Arabidopsis* uses a cooperative mechanism to simultaneously incorporate GlcA and Xyl, which also suggests conserved mechanisms of xylan synthesis in primary cell walls. It will be important to investigate GAX content and structure in cell walls from etiolated *Arabidopsis* seedlings to confirm the presence of Ara residues in xylan. Although GX polymers from secondary cell walls have been shown to be devoid of Ara, earlier studies

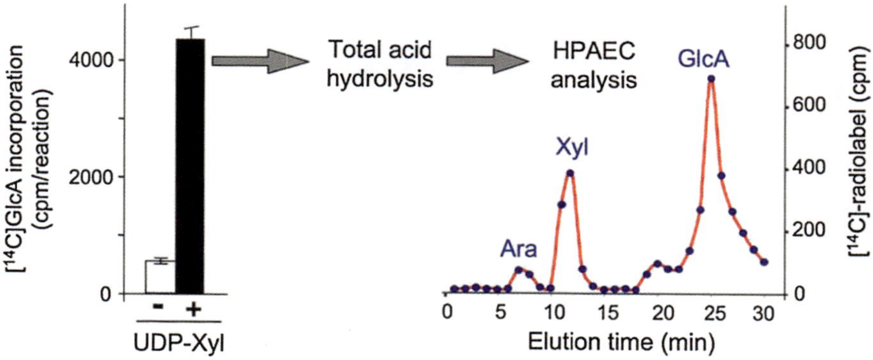

Fig. 10.2 Analysis of glucurono(arabino)xylan (*GAX*) synthase activity in etiolated wild type *Arabidopsis* seedlings grown on vertical plates. The activity was monitored via transfer of [^{14}C]GlcA from UDP-[^{14}C]GlcA into ethanol-insoluble products in presence of Golgi enriched micrsomal preparations (~0.2 mg proteins from 5 to 7-day old seedlings). [^{14}C]GlcA incorporation was measured in the presence (black bar) or absence (white bar) of UDP-Xyl as described in Zeng et al. (2010). For product analysis all sugars were [^{14}C]-radiolabeled and the [^{14}C]-products formed were acid hydrolyzed (2 M TFA, 120 °C, 1 h), and the monosaccharides released were fractionated by High pH anion Exchange Chromatography (HPAEC, Dionex). The elution of [^{14}C]-radiolabeled material was monitored by counts (cpm) using a scintillation counter. These analyses were conducted in triplicate

on primary cell walls in *Arabidopsis* suggested the presence of small amounts of GAX (~4 %) (Zablackis et al. 1995), which is comparable to most dicot plants (Darville et al. 1980; Scheller and Ulvskov 2010).

Genetic analyses in *Arabidopsis* suggest that xylan backbone elongation of GX synthesis is tightly linked to the addition of GlcA side chains. Unfortunately, there is no experimental data from *Arabidopsis irx* mutants to directly support the involvement of a cooperative mechanism and/or a multi-protein complex (such as XSC from wheat). Purified wheat XSCs seem to contain only orthologs to *Arabidopsis* "IRX-L" proteins (no "IRX" orthologs were present in the complex), namely, TaGT43-4 (putative ortholog of IRX14-L) and TaGT47-13 (putative ortholog of IRX10-L) (Zeng et al. 2010, Faik unpublished data). This finding raises several important questions related to xylan backbone elongation: (1) do all "IRX" and "IRX-L" proteins interact with each other to form core XSCs? and (2) how many of different XSCs does a plant need to make xylans? Our hypothesis is that, like in the wheat core XSC, *Arabidopsis* IRX7/IRX7-L, IRX9/IRX9-L, IRX10/IRX10-L, and IRX14/IRX14-L pairs form several core XSCs by pairing partners from GT43 and GT47 families. For example, four putative core XSCs are possible just from the combinations of IRX10/IRX10-L and IRX14/IRX14-L pairs ([IRX10-IRX14]; [IRX10-IRX14-L]; [IRX10-L-IRX14]; and [IRX10-L-IRX14-L] complexes), and another four putative core XSCs could be formed from the IRX9/IRX9-L and IRX7/IRX7-L pairs, generating a total of eight possible core XSCs. Experimental data will be needed to substantiate the number of core XSCs in *Arabidopsis*.

The question then becomes, why have so many core XSCs? And how are they functionally different? One possible answer is that some of these core XSCs are specialized for certain tissues or at certain developmental time points (e.g., for making xylans in primary vs. secondary cell wall). Since a wheat core XSC was purified from etiolated seedlings rich in primary cell wall, it is reasonable to suggest that some of these XSCs are involved in xylan synthesis in primary cell walls (i.e., GAX and AX), while others have major roles in producing xylans for secondary cell walls (i.e., GX), reminiscent of the unique rosette terminal complexes employed for primary and secondary cell wall biosynthesis. Nonetheless, the reason for a plant to have specialized core XSCs for each type of cell wall is puzzling.

10.2.3 Substitution of the Xylan Backbone: Uncoupled Versus Coupled Addition of Side Chains

Recently, work by Anders et al. (2012) showed that silencing two α-(1,3)-AraT genes (*XAT*1 and *XAT*2, members of the GT61 family) did not affect the number of substituted Xyl residues in AXs from the endosperm of wheat. It is known that Ara residues in endosperm AXs are either α-(1,2)- or α-(1,3)-linked to the xylan backbone, and that some Xyl residues can be di-arabinosylated with both linkages. Thus, there are technically four types of α-AraTs that may be required for arabinosylation of AXs in the wheat endosperm: α-(1,3)-AraTs and α-(1,2)-AraTs that would add Ara onto unsubstituted Xyl residues of the backbone, and α-(1,2)-AraTs or α-(1,3)-AraTs that would transfer Ara onto mono-substituted α-(1,3)-Ara or α-(1,2)-Ara-linked Xyl residues, respectively to form di-substituted Xyl residues.

The findings by Anders et al. (2012) have important implications regarding AX synthesis in grasses. First, it suggests that despite the absence of α-(1,3)-linked Ara, xylan backbone elongation is still linked to the incorporation of Ara (mirroring the situation of GlcA and Xyl incorporation in GX). Second, it also suggests that α-(1,2)-AraT activity, but not α-(1,3)-AraT activity, may work cooperatively with xylan synthase to synthesize AX polymer. Interestingly, the proteomics studies described by Zeng et al. (2010) did not identify any members of the GT61 family associated with fractions enriched in wheat GAX synthase activity, and this observation was confirmed by further proteomics analysis of the affinity-purified GAX synthase complex (Faik and Jiang unpublished data). Thus, we are tempted to conclude that the AraT detected with the purified wheat GAX synthesis activity is an α-(1,2)-AraT activity. This α-(1,2)-AraT may belong to a GT family other than GT61. Members of the GT47 family could be candidates for this type of α-(1,2)-AraT activity. It will be necessary to provide experimental data to support this hypothesis either by directly testing the enzyme activity of the candidates or through over-expressing them in *Arabidopsis* and analyzing xylan structures from transgenic plants. Together, these observations still support a cooperative mechanism for AX biosynthesis in wheat, but subsequent substitutions with Ara, acetyl groups, or GlcA could be uncoupled from xylan backbone elongation.

The situation is more complicated for GX synthesis in secondary cell walls. A recent work by Mortimer et al. (2010) showed that two *Arabidopsis* genes, *GUX*1 and *GUX*2 (both member of the GT8 family), were responsible for the addition of GlcA side chains onto GX in secondary cell wall. Interestingly, *gux*1 *gux*2 double mutant plants have xylan contents and xylan chain lengths similar to wild type, yet have almost no detectable GlcA residues. Additionally, these mutant plants have no detectable growth phenotype. Although this finding is difficult to reconcile with a cooperative mechanism of the XSC complex, it has some important implications toward understanding the GX biosynthetic mechanism. First, the data show that somehow xylan backbone elongation can be uncoupled from GlcA substitution during GX synthesis. A possible explanation is that GX from *gux*1 *gux*2 double mutant plants may still have low amounts of GlcA not detectable by the carbohydrate gel electrophoresis (PACE) technique. Although the PACE method is sensitive (detection of <1picomol of released sugars) (Goubet et al. 2002), it also relies on derivatization of the reducing ends of sugars with fluorophores, which could be a limiting step if the efficiency is low. In fact, MALDI-TOF MS analysis of oligosaccharides from GX prepared from *gux*1 *gux*2 double mutant plants still shows a small peak at *m/z* 963.50 corresponding to MeGlcA$(Xyl)_4$ (Mortimer et al. 2010). A second possible explanation is that other enzymes may have taken the place of the missing GlcAT in the XSC allowing xylan backbone elongation to occur with low amounts of GlcA side chains incorporation. An acetyltransferase (AcetylT) that adds acetyl groups to the C-6 positions of Xyl residues could be that enzyme. Lee et al. (2011) identified four *Arabidopsis REDUCED WALL ACETYLATION* (*RWA*) genes (called *RWA*1, *RWA*2, *RWA*3, and *RWA*4) required for xylan acetylation in the secondary cell wall. When the four *RWA* genes were mutated, a significant reduction in secondary cell wall thickening was observed. Although mutant plants showed a milder *irx* phenotype, there was no reduction in the amount of xylan Lee et al. (2011). The authors did not analyze the xylan chain length from *rwa*1/2/3/4 mutant plants. Although the mechanisms of xylan *O*-acetylation (or any plant cell wall polymer) are not known (Gille and Pauly 2012), studies in mammalian systems indicate that *O*-acetylation of sialic acid takes place in the lumen of the Golgi using acetyl-CoA as a donor substrate to form acetylated intermediates (Higa et al. 1989). The existence of a GX polymer with low GlcA content and higher acetylation has been documented in many species. For example, Goncalves et al. (2008) isolated a GX polymer from hybrid *Paulownia elongata*/*Paulownia fortunei* (a deciduous tree) that has much less GlcA substitution (Xyl:GlcA ratio of 20:1), but much higher acetylation (almost 50 % of Xyl are substituted with acetyl group), which may suggest that AcetylT can substitute for GlcAT in a XSC. From these observations, we propose a model for xylan backbone elongation whereby all xylan types are produced by core XSCs that elongate xylans by a coupled mechanism. These XSCs are modular, depending of the species and tissues, and can be substituted with alternate enzymes to overcome defects in xylan biosynthesis. The central

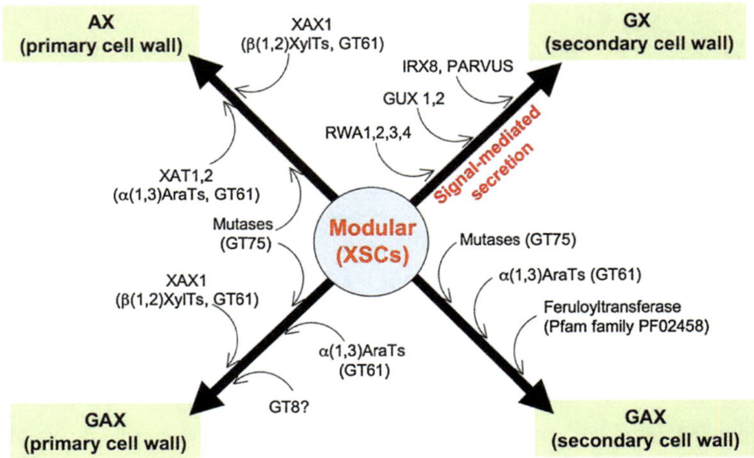

Fig. 10.3 Model for synthesis of various types of xylans (GAX, GX, and AX) in primary and secondary plant cell walls. Xylan synthase complexes (*XSCs*) are modular with a core complex formed by a pair of proteins from "IRX" and "IRX-L" sets (see text). These cores XSCs are responsible for the synthesis of the xylan backbone coupled with simultaneous incorporation of some GlcA and/or Ara side chains. Depending of the degree of substitution, additiona enzymes such as GUX1, 2 (GT8), XAT1, 2 and XAX1 (GT61), or RWA1,2,3, and 4, can be involved in adding GlcA, Ara, Xyl, or acetyl groups, respectively, to the backbone in an uncoupled manner. The mechanism of ferulic acid incorporation onto Ara or Xyl-Ara side chains by a family of putative feruloyltransferases (Pfam PF02458) is still unknown

core of these XSCs is composed of members of the "IRX" and "IRX-L" proteins. In this model (Fig. 10.3), additional enzymes/GTs can work independently from XSC (in an uncoupled manner), which would provide flexibility and adaptability of the biosynthetic machinery for the production of many xylan types.

10.2.4 Secretion/Deposition of Xylans to the Cell Wall

Plant cell wall polysaccharide secretion/deposition to the cell wall is still lagging behind other aspects of cell wall biology, and certainly more efforts are needed to advance this area of research. Our proposed model for sequence 1 in GX secretion during secondary cell wall deposition would give a framework to design experiments specifically to answer questions such as: Which proteins are involved in recognizing sequence 1 in endomembrane system? Or which GTs catalyze the transfer of this sequence 1 onto GX before secretion/deposition?

The recent characterization of two new *irx* mutants in *Arabidopsis* DUF579 genes (*irx*15 and *irx*15-*L*, Table 10.1) start to shed the light on the mechanisms of secondary cell wall deposition (Brown et al. 2011; Jensen et al. 2011). The double mutant *irx*15 *irx*15-*L* plants have a mild irx phenotype with a decrease of up to

50 % in GX content in stem secondary cell walls, with cellulose contents similar to many other GX-deficient mutants, see Table 10.1. However, what is most interesting about this double mutant is the uneven deposition of their secondary cell walls (compared to the smooth and even deposition of secondary cell walls in other *irx* mutants). Thus, it was proposed that IRX15 and IRX15-L proteins might have a role in the transport/delivery of polysaccharides to the cell wall (Brown et al. 2011). Although this proposed function needs experimental confirmation, there is indirect evidence that supports it. For example, one might expect the *IRX*15 and *IRX*15-*L* genes to be highly expressed in tissues that are actively synthesizing and secreting xylans and this is indeed true in the case of xylem/vessel tissues, and the mucilaginous layer (called husk) of Psylium seeds, which contains more than 60 % dry weight in heteroxylans (Fischer et al. 2004) and is a developing tissue actively synthesizing and secreting xylans. Also, structural analysis of a DUF579-containing protein (YPL225 W) from *S. cerevisiae* showed that it contains a coiled-coil domain necessary for protein–protein interactions, and can interact with components of the cytoskeleton (i.e., ARP23 complex) and the endomembranes (Costanzo et al. 2010; Brown et al. 2011). Therefore, it is possible that the DUF579-containing IRX15 and IRX15-L proteins may have the capacity to interact with cytoskeleton and participate in GX/polysaccharide trafficking. It is also possible that IRX15 and IRX15-L play a role during the biosynthesis of GX. For example, they could act as polysaccharides chaperones to maintain GX in a soluble form during xylan backbone elongation and delivery to the cell surface. Their putative role as polysaccharide chaperones could explain the shorter xylan chain lengths observed in the *irx*15 *irx*15-*L* double mutant plants, since the absence of these proteins would result in GX insolubility after a certain length. This would cause a reduction in the rate of synthesis by xylan synthase complexes, and aggregations of newly synthesized GX in the Golgi. The fact that X6-dependent XylT activity and the ratio Xyl to MeGlcA are unaffected in the *irx*15 *irx*15-*L* double mutant suggests that the xylan synthase complexes are functional in this mutant (Brown et al. 2011) (Table 10.1). Newly synthesized GX in the *irx*15 *irx*15-*L* double mutant would still be decorated with sequence 1 and delivered to cell wall, however, the absence of IRX15 and IRX15-L proteins may restrict vesicular trafficking to a certain area of the cell surface through smaller vesicles that are delivered in a patchy manner. Another possibility is that IRX15 and IRX15-L may contribute indirectly to GX solubility in the Golgi by playing a role in the *O*-acetylation of xylan backbone. It is known that most hemicelluloses, including GX, are *O*-acetylated (Kiefer et al. 1989; Carpita 1996; Scheller and Ulvskov 2010; Gille et al. 2011). In this scenario, IRX15 and IRX15-L could stabilize GX-specific acetyltransferases (i.e., reduced wall acetylation proteins: RWA1, RWA2, RWA3, and RWA4) on the newly synthesized GX. It would be interesting to know whether RWA proteins interact with IRX15 and/or IRX15-L. The *Arabidopsis At*1*g*33800 gene has been recently characterized and shown to encode a 4-*O*-methyltransferase specific to GX (GXMT) (Urbanowicz et al. 2012). This protein has also a DUF579 domain but clusters phylogenetically in a different group from IRX15 and IRX15-L sharing less than 30 % identity at the amino acid

level. Thus, determining the biochemical function of IRX15 and IRX15-L will be necessary before a clear conclusion regarding the roles of these proteins can be drawn.

10.3 Xylan Synthase Complexes: Primary Versus Secondary Cell Wall

According to studies from *Arabidopsis irx* mutants (summarized in Table 10.1), not all *IRX* genes have equal importance in GX synthesis in the secondary cell wall of mature *Arabidopsis* plants. For instance, while GX synthesis is not abolished in the *irx9/irx9-L* double mutant, both the *irx14/irx14-L* and the *irx10/irx10-L* double mutant make very little GX, indicating that *IRX10/IRX10-L*, and *IRX14/IRX14-L* pairs have more important roles in secondary cell wall GX biosynthesis than the *IRX9/IRX9-L* pair elongation (Wu et al. 2009, 2010). Furthermore, analysis of promoter-GUS constructs of *IRX* genes expressed in *Arabidopsis* indicate that *IRX7*, *IRX9*, *IRX14*, and *IRX14-L* are predominantly expressed in the central stele of root tissues which are rich in secondary cell wall, while *IRX9-L* and *IRX7-L* are expressed in the peripheral cell layers (tissues that are not making any secondary cell walls) as well as in central stele of root (Wu et al. 2009). Similarly, vascular tissues of *Arabidopsis* leaves showed restricted expression of *IRX7*, *IRX9*, *IRX14*, and *IRX14-L*, while *IRX9-L* and *IRX7-L* are expressed throughout leaf tissues (Wu et al. 2009). Additional promoter analyses using yellow fluorescence protein (YFP) fusions and confocal microscopy support the general conclusion that "IRX" genes are associated with the synthesis of secondary cell walls of xylem, while "IRX-L" genes are weakly expressed in xylem yet their expression seems to be more widespread (Wu et al. 2009). These observations give support to the hypothesis that xylan biosynthesis in primary and secondary cell walls in *Arabidopsis* may involve different sets of genes or XSCs, as was shown for cellulose synthesis (for review see Joshi and Mansfield 2007; Somerville 2006). Confirmation of this hypothesis may require purification of functional GAX synthase complexes from etiolated *Arabidopsis* seedlings (rich in primary cell walls). In fact, all the conclusions regarding *Arabidopsis* "IRX" and "IRX-L" genes involved in GX biosynthesis (Table 10.1) are based on experimental work in the stems of mature plants making secondary walls. Unfortunately, little work has been done on xylan synthesis and structure in etiolated *Arabidopsis* seedlings.

Etiolated seedlings are comprised mostly of rapidly expanding mesocotyl (in grasses) and hypocotyl cells (in dicots). These cells are rich in primary cell wall material and therefore offer an excellent experimental model to study the biochemistry, enzymology, and regulation of GAX synthesis and for testing the genes associated with the process. One would expect to see a phenotype in seedlings having mutations in genes important for GAX synthesis in primary cell walls. In an attempt to address this hypothesis and determine the role of "IRX" and "IRX-L" genes in

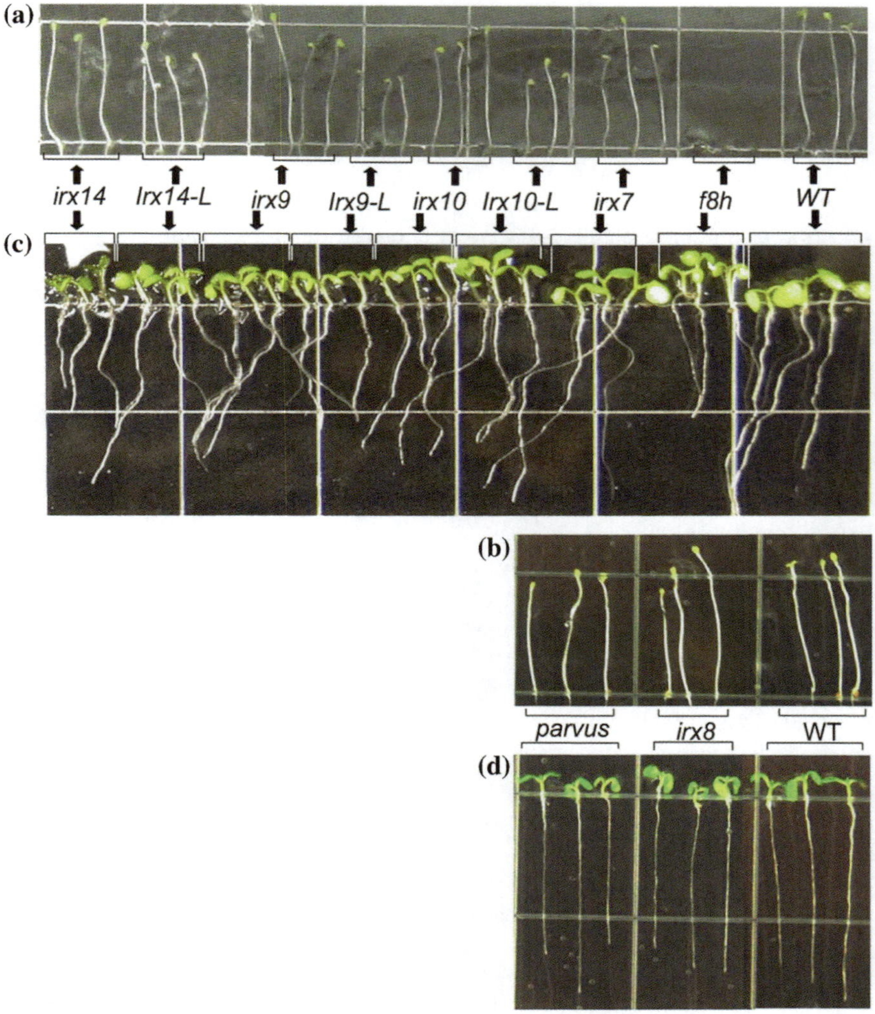

Fig. 10.4 Analysis of the growth of *Arabidopsis irx*7, *f8h*(*irx*7-*L*), *irx*9, *irx*9-*L*, *irx*10, *irx*10-*L*, *irx*14, *irx*14-*L*, *irx*8, and *parvus* mutants on *vertical plates* in *dark* (*panels* **a** and **b**) or *under light* (*panels* **c** and **d**). Seeds were surface sterilized and cold treated for two days before planting. Pictures were taken 5–7 days after germination

xylan synthesis in primary cell walls, we monitored the growth of *Arabidopsis irx* mutants (*irx*7, *irx*7-*L*, *irx*9, *irx*9-*L*, *irx*10, *irx*10-*L*, *irx*14, *irx*14-*L*, *irx*8, and *parvus*) on vertical plates under light and dark conditions. When *Arabidopsis irx* mutant seedlings were grown in the dark (etiolated), hypocotyl elongation seemed to be affected by mutations in *IRX*9-*L*, *IRX*10-*L*, or *IRX*14-*L* genes, as the seedlings from these mutants were 30–50 % shorter than wild type (Fig. 10.4a). On the other hand, mutations in *IRX*7, *IRX*9, *IRX*10, or *IRX*14 had only minor effects on hypocotyl elongation (~10 % shorter compared to wild type, Fig. 10.4a). Perhaps the most dramatic effect observed was the very delayed germination of seeds from *f8h*(*irx*7-*L*) mutants (Fig. 10.4a). Surprisingly, when the same *irx* mutants were grown under lighted condition, no significant differences in hypocotyl lengths were observed among all seedlings of *irx* mutants and wild type (Fig. 10.4b). However, a reduction in root length by ~50 % was observed for *irx*14 and *f8h*(*irx*7-*L*) compared to wild type (Fig. 10.4b). Seeds from *f8h*(*irx*7-*L*) mutant seem to have shorter germination delay under light conditions. It is worth noting that both *irx*8 and *parvus* mutants did not show any significant growth differences compared to wild type under light and dark conditions (Fig. 10.4c and d). This result is expected since both *IRX*8 and *PARVUS* genes are required for sequence 1 synthesis and the secretion of GX during secondary cell wall deposition. Sequence 1 is absent in other xylans such as GAX and AX. This preliminary finding supports the notion that sequence 1 may not be required for GAX synthesis in growing tissues with primary cell walls.

Taken together, these observations indicate that mutations in both sets of xylan synthesizing genes, "IRX" and "IRX-L" from GT43 and GT47 families, affect hypocotyl elongation of *Arabidopsis* seedlings, but "IRX-L" genes have a stronger effect, which is the reverse situation when *Arabidopsis* is grown on soil and making stems with secondary cell walls (Wu et al. 2009, 2010). Although this finding gives additional support to the conclusion that "IRX" and "IRX-L" genes have varying importance in xylan biosynthesis in primary and secondary cell wall, further analyses will be needed to determine the exact roles of each of these genes on plant development. We propose that "IRX" genes are more important in GX biosynthesis in secondary cell walls, while the "IRX-L" genes are more critical for xylans synthesis in tissues rich in primary cell walls. The general distribution trend of "IRX" and "IRX-L" gene expression within *Arabidopsis* body would look like two opposite gradients as depicted in Fig. 10.5. Figure 10.6 summarizes expression profiles of *IRX*7/*IRX*7-*L*, *IRX*9/*IRX*9-*L*, *IRX*10/*IRX*10-*L*, and *IRX*14/*IRX*14-*L* pairs in *Arabidopsis* roots using publicly available microarray data (http://bar.utoronto.ca/efp/cgi-bin/efpWeb.cgi; Brady et al. 2007; Winter et al. 2007). These expression data are in good agreement with our hypothesis and show that, in general, the expression of "IRX" genes are more associated with tissues making secondary cell walls, while "IRX-L" genes are more associated with growing/dividing tissues making primary cell walls. This differential role (and expression) of two very close genes highlights another level of regulation of xylan synthesis in dicots.

10.4 Regulation of Xylan Biosynthesis

In theory, cell wall polysaccharide biosynthesis can be regulated at several levels, including gene expression, the abundance and stability of mRNAs, the production and stability of the protein (translation and post-translational modifications), the effects of protein complex formation, and the secretion/delivery of newly synthesized polysaccharides to the cell surface and their integration into the cell wall. We are just starting

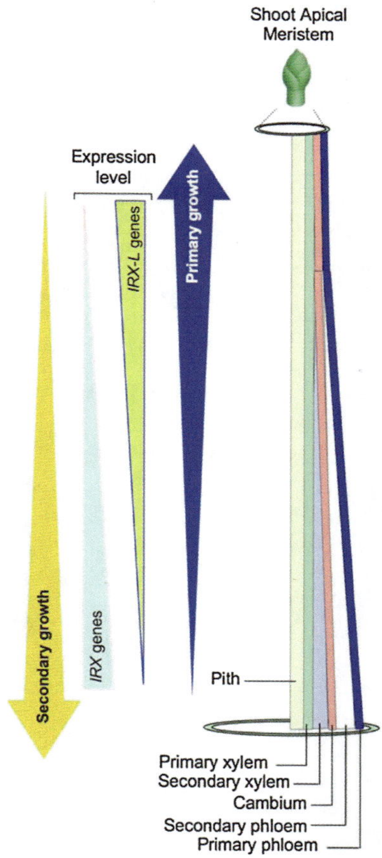

Fig. 10.5 Schematic presentation of the general distribution of "IRX" and "IRX-Like" genes in *Arabidopsis* stems. Primary and secondary growth is indicated by *arrows*. The expression of "IRX" genes increases with secondary growth, while the expression of "IRX-Like" genes increases in tissues with primary growth

to elucidate the roles of some transcription factors (TFs) and small RNA regulators of cell wall biosynthesis. We are also starting to identify some of the protein–protein interactions needed in assembling functional polysaccharide synthase complexes. However, we are still far from developing a larger picture of the regulatory landscape associated with the synthesis of any given cell wall polymer. Xylan biosynthesis is no exception, considering their importance in plant growth and survival.

10.4.1 Regulators of IRX Gene Expression: Xylan Versus Other Cell Wall Polymers

Currently there are about 20 known proteins that constitute the *Arabidopsis* TF regulatory network for cell wall biosynthesis. Elucidating the transcriptional regulatory mechanisms of cell wall synthesis is hampered by the fact that promoter sequences

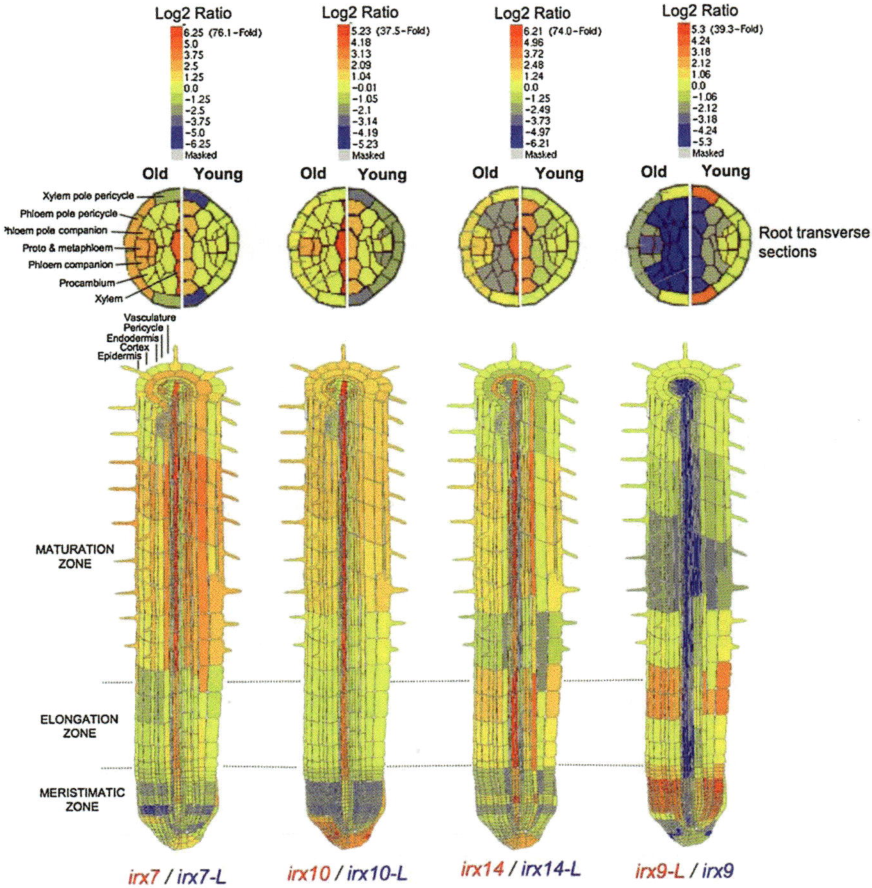

Fig. 10.6 Expression profiles of *IRX/IRX-L* pairs in *Arabidopsis* roots. The expression of each *IRX/IRX-L* pair was examined using the electronic fluorescenec pictograph (*eFP*) Browser in "Compare Mode" (http://bar.utoronto.ca/efp/cgi-bin/efpWeb.cgi) (Winter et al. 2007) of publicly available microarray data (Brady et al. 2007). *IRX/IRX-L* pairs were submitted as either primary or secondary genes and are colored *red* or *blue*, respectively below each root. Tissues colored in *red* indicate that the primary gene is more highly expressed, while tissues colored in *blue* indicate that the secondary gene is more expressed. The *yellow* indicates tissues that express the primary and secondary genes equally. Heat maps are included to show the scale of the log2 ration of the expression of gene 1 to the expression of gene 2 and color scales associated with these values are presented

are complex and several TFs cooperate to simultaneously regulate gene expression of cell wall synthesizing genes. For economical and technical reasons, the first studies on cell wall TFs targeted secondary cell wall formation. These works identified a subgroup of plant specific TFs: NAC (NAM/ATAF1/CUC2) and MYB families that control secondary cell wall thickening (Olsen et al. 2005; Demura and Fukuda 2007; Zhong and Ye 2007). These TFs appear to function as a developmental switch that induces the transition from primary to secondary cell wall deposition. VASCULAR-RELATED NAC-DOMAIN1-7 (VND1-7) are members of NAC TF family that are preferentially expressed in differentiating xylem cells in *Arabidopsis* (Kubo et al. 2005), and two of these proteins (VND6 and VND7) act as master switches for protoxylem and metaxylem development and are negatively regulated by VND-INTERACTING2 NAC protein (Yamaguchi et al. 2010). No TFs specific for inducing xylan synthesis have been identified. While five NAC SECONDARY WALL THICKENING PROMOTING FACTOR 1 (NST1) and NST2, and VND7 proteins appear to function as activators of cell wall genes involved in cellulose, xylan, and lignin biosynthesis, their direct targets appear to be other transcriptional activators (Zhong et al. 2008; Ko et al. 2009; McCarthy et al. 2009). Some TFs such as SND3, SND2, and MYB103 induce the expression of cellulose synthase genes (Sonbol et al. 2009), but the identification of equivalent TFs for any *IRX* genes is still elusive.

Progress is lagging behind in grasses. Currently, only two TFs (ZmMYB31 and ZmMYB42) are known to affect cell wall biosynthesis. These two maize proteins have been shown to function as repressors of lignin biosynthesis in *Arabidopsis*, but not maize (Fornalé et al. 2006; Sonbol et al. 2009). Ambavaram et al. (2011) showed that the expression of an *Arabidopsis* TF *SHINE* gene (a member of the AP2/ERF family) in rice caused coordinated down-regulation of lignin biosynthesis and up-regulation of cellulose and other cell wall biosynthetic genes. Interestingly, SHINE regulates cutin polymer biosynthesis in *Arabidopsis* (not cellulose or lignin) (Kannangara et al. 2007). This underscores the need for continued efforts to identify and characterize regulatory proteins for GTs associated with primary and secondary cell wall biosynthesis in various tissues and species. An attractive strategy for the identification of TFs specific to xylan biosynthesis in primary cell walls is to use the endosperm tissues in wheat as a model system. The cereal endosperm represents an excellent model for investigating plant cell wall biosynthesis and regulation since the walls of cereal endosperm have a simple composition, consisting mainly of AX, mixed linkage (1 → 3),(1 → 4)-β-D-glucan (MLG), and low amount of cellulose (Philippe et al. 2006a, b; Wilson et al. 2006). Thus, transcriptomics and/or proteomics of these tissues might allow for the identification of putative activators and repressors that are specific to AX and/or MLG synthesis. Several genetic studies have already linked AX biosynthesis and content in wheat grains to a few puantitative trait loci (QTLs) (Martinant et al. 1998; Charmet et al. 2009). More recently, Nguyen et al. (2011) conducted an extensive study on AX content in the grains of the Berkut x Krichauff doubled haploid (DH) population grown at two different environmental conditions using a linkage map of 528 genetic markers for QTL mapping. According to this study, QTLs associated with grain AX contents are located on chromosomes 1A, 2A, 7A, 3D, 4D, and 6B, but carrying only the two QTLs 2A, and 4D was sufficient to significantly increase AX content in the grain (Nguyen et al. 2011). Effects of these two QTLs were further confirmed and validated on other wheat varieties. Two important QTLs associated with AX accumulation showed epistatic interactions, which may suggest regulation at the genetic level, as one QTL may contain one or several genes coding for one or several proteins that prevent transcription of the gene(s) from the other QTL. This would be a major contribution toward understanding xylan biosynthesis in grasses.

10.4.2 Protein–Protein Interactions: Regulators of XSCs Assemby

Genetic analyses in *Arabidopsis* have shown that xylan synthesis can greatly impact the growth and reproduction of plants at various developmental stages. Thus, it is not surprising that plants would develop a regulatory step during the assembly and trafficking of core XSCs in the endomembranes. The importance of "IRX" and "IRX-L" protein interactions (core components of XSCs) can be deduced from a recent genetic study in *Arabidopsis* that showed that the truncated *Arabidopsis* IRX10 protein (lacking 22 amino acids from its N-terminal) cannot rescue *Arabidopsis irx*10/*irx*10-*L* double mutants (affected in xylan backbone elongation), and that the first 32 amino acids of IRX10 protein sequence were necessary for successful complementation (Wu et al. 2009). Furthermore, two tobacco proteins NpGUT1 (BAC20928.1) and NtGUT1 (BAD04923.1), homologous to *Arabidopsis* IRX10 and IRX10-L, exist in public databases and are both truncated at the N-terminus (lacking the first 71–74 amino acids) (Iwai et al. 2002). The truncated NpGUT1 could not complement the *Arabidopsis irx*10/*irx*10-*L* double mutant, but did complement the mutant when fused with the 32-amino acid N-terminus of IRX10 (Wu et al. 2009). Interestingly, *Arabidopsis* IRX10 and IRX10-L proteins and their putative orthologs from other plants, are predicted to have a cleavable signal peptide, suggesting that these proteins would be soluble in the Golgi lumen. Therefore, they would require interactions with a Golgi-localized protein for a proper Golgi targeting. We predict that the sequence between the cleavage site and the 70th amino acid position play an important role in protein–protein interactions within the XSC. Although this hypothesis is confirmed by genetic analysis, it will need to be confirmed by direct biochemical evidence of such protein interactions.

The general hypothesis is that XSCs assemble in the endoplasmic reticulum (ER), the location of the synthesis of all secretory proteins. One might expect that deficiencies in the interactions of XSC proteins would result in the accumulation of some of these proteins in the ER, which in turn may trigger proteasome-assisted degradation of these proteins. One also might expect that for proper ER export and Golgi accumulation, the assembled XSCs would need both ER exit and Golgi retention signals on at least one of the protein complex members. Nothing is known about xylan synthase trafficking, but the recent discovery of wheat XSC will allow us to explore these issues. Regulatory mechanisms that maintain an equilibrium between complex assembly and trafficking would be advantageous for situations that require a rapid physiological response. For example, controlling the production of only one key protein of the complex may be sufficient to trigger the physiological response. Furthermore, nothing is known about the composition of each XSC (i.e., number of each protein within the complex), or what induces a proper assembly of a functional XSC. Cell biological tools, such as bimolecular fluorescence complementation (BiFC) or surface plasmon resonance (SPR), will be of great help in elucidating the protein–protein interactions important for regulating XSC trafficking and for XSC complex assembly.

10.4.3 Modulation of Xylan Synthesis by Phytohormones: The Missing Link

Plants use phytohormones to translate environmental changes (external factors) into physiological responses to adapt to such changes. For example, coleoptiles and hypocotyls respond to light by inhibiting mesocotyl/hypocotyl cell elongation, and this light signal is perceived almost exclusively through phytochromes (Pjon and Furuya 1967; Takano et al. 2001). Interestingly, the inhibition of coleoptile elongation seems to be a direct result of the inhibition of cell wall polysaccharides synthesis (Ueda et al. 1994, 1995; Miyamato et al. 1997). Coleoptiles represent an excellent system to study the regulatory mechanisms of phytohormones on polysaccharide synthesis and growth, since coleoptile elongation is under the control of only three phytohormones: auxin (inducer of cell elongation), abscisic acid (ABA), and jasmonic acid (JA) (the latter two are both inhibitors of cell elongation) (Hoffmann-Benning and Kende 1992; Ueda et al. 1995; Giani et al. 1998). Gibberellins and brassinosteroids do not affect coleoptile development (Toyomasu et al. 1994; Sekimata et al. 2001). Thus, it will be important to identify the proteins involved in JA and ABA signaling pathways that lead to the regulation of GAX biosynthesis in grasses and dicots. The recent characterization of the rice (*Oryza sativa*) mutant *hebiba*, a mutant in JA biosynthesis, provided the first genetic evidence of crosstalk between light and JA signaling involved in photomorphogenesis in monocots (Riemann et al. 2003). In the dark, *hebiba* seedlings have a short coleoptile and long mesocotyl, which is opposite the phenotype of wild type plants (Riemann et al. 2003). Treatment of this mutant with exogenous methyl jasmonate (Me-JA) restored the wild type phenotype. Furthermore, the *Arabidopsis cev*1 mutant has a point mutation in the *cellulose synthase A*3 (*CesSA*3) gene, and showed constitutive production of JA (Ellis and Turner 2001). These findings further support the notion that changes in cell wall biosynthesis might be coupled to the JA-dependent signaling pathway used to induce plant defense responses. It is tempting to speculate whether JA can modulate (directly or indirectly) xylan synthesis in grasses, and if so, it would be of great interest to elucidate the mechanism(s) and the intermediate effectors involved in the process. Although most JA signal transduction pathways and intermediate effectors have been discovered in *Arabidopsis* (Thines et al. 2007), we still do not know how these mechanisms and target proteins directly modulate cell wall synthesis in grasses.

10.5 Concluding Remarks: Impacts of Xylan Biosynthesis on Biofuel Production

Despite progress in the genetics of xylan synthesis in *Arabidopsis*, our current knowledge of the biochemistry of the process is limited. The enzymology, active sites, functions of the predicted GTs, secretion of the newly synthesized polymer, coordination of gene expression, and assembly/organization of multi-enzyme complexes for xylan biosynthesis are not well understood. This chapter has discussed some of the critical aspects of xylan synthesis integrating the current biochemical and genetic data. Xylans are structurally complex and genetic studies have demonstrated the importance of these polymers for plant growth and survival. Manipulating their composition will impact the manner by which they interact with other polymers within the cell wall, which in turn will strongly influence energy recovery from plant biomass.

How might manipulating xylan biosynthesis impact biofuel production? We already know that biofuel yields from grass biomass depend on structural variations of the chemical composition of xylans. For example, the substitution rate by GlcA, Ara, or acetylation can either limit the accessibility of hydrolases to the xylan backbone, or limit fermentable yields by inhibiting yeast fermentation. Since cellulose microfibrils are thought to be tethered by crosslinking glycans such as xylans, manipulating xylan biosynthesis to produce polymers with reduced side chain contents should improve biofuel production by making cellulose more accessible. Mortimer et al. (2010) demonstrated that *Arabidopsis gux* mutants could produce xylans with very low GlcA branches without affecting plant growth. Although this progress shows potential for simplification of lignocellulosic biomass conversion in dicots, it still needs to be demonstrated in grasses. Similarly, many hemicelluloses (including xylans) in dicot cell walls are acetylated (Goncalves et al. 2008; Scheller and Ulvskov 2010; Lee et al. 2011). It has been shown that hemicellulose acetylation (including xylans) affects the downstream sugar conversion process and requires pretreatment of biomass for biofuel production (Helle et al. 2003; Carroll and Somerville 2009; Selig et al. 2009). Using techno-economic models (Klein-Marcuschamer et al. 2010), Gille et al. (2011) estimate that a 20 % reduction in *O*-acetylation would lead to a 10 % reduction in the costs of ethanol production.

In grasses, manipulating α-(1,3)-arabinosylation of xylans showed the potential to produce xylans with less Ara side chains but this did not seem to improve cell wall saccharification in the transgenic wheat plants (Anders et al. 2012). The only report of improved saccharification through AX manipulation was described in rice using the *xylosyl arabinosyl substitution of xylan* (*xax*1) mutant (Chiniquy et al. 2012). The XAX1 enzyme is a member of the GT 61 family and catalyzes the incorporation of β-(1,2)-linked Xyl onto Ara side chains to produce Xyl-Ara side chains. This β-(1,2)Xyl usually bears a ferulic acid in grass xylans (Hoije et al. 2006). Thus, xylans from this rice mutant have lower ferulic acid and Xyl contents. The lower ferulate content affects "xylan–xylan" and "xylan-lignin" cross-linking, which in turn improves saccharification in *xax*1 mutant plant (Chiniquy et al. 2012). Unfortunately, the *xax*1 mutant plants are dwarfed, which translates to a reduction in total biomass. Therefore, more research is needed to design strategies to manipulate GAX biosynthesis in grasses for improved digestibility without affecting plant growth and development.

Acknowledgments This material is based upon work partially supported by the National Science Foundation under Grant No. 1145887 to AF.

References

Ambavaram MMR, Krishnan A, Trijatmiko KR, Pereira A (2011) Coordinated activation of cellulose and repression of lignin biosynthesis pathways in rice. Plant Physiol 155:916–931

Andersson SI, Samuelson O (1983) Structure of the reducing end-groups in spruce xylan. Carbohydr Res 111:283–288

Anders N, Wilkinson MD, Lovegrove A, Freeman J, Tryfona T, Pellny TK, Weimar T, Mortimer JC, Stott K, Baker JM, Defoin-Platel M, Shewry PR, Dupree P, Mitchell RA (2012) Glycosyl transferases in family 61 mediate arabinofuranosyl transfer onto xylan in grasses. Proc Natl Acad Sci USA 109:989–993

Brady SM, Otlando DA, Lee J-Y, Wang JY, Koch J, Dinneny JR, Mace D, Ohler U, Benefey PN (2007) A high-resolution root spatiotemporal map reveals dominant expression patterns. Science 318:801–806

Brown DM, Zeef LAH, Ellis J, Goodacre R, Turner SR (2005) Identification of novel genes in *Arabidopsis* involved in secondary cell wall formation using expression profiling and reverse genetics. Plant Cell 17:2281–2295

Brown DM, Goubet F, Vicky WWA, Goodacre R, Stephens E, Dupree P, Turner SR (2007) Comparison of five xylan synthase mutants reveals new insight into the mechanisms of xylan synthesis. Plant J 52:1154–1168

Brown DM, Zhang Z, Stephens E, Dupree P, Turner SR (2009) Characterization of IRX10 and IRX10-like reveals an essential role in glucuronoxylan biosynthesis in *Arabidopsis*. Plant J 57:732–746

Brown D, Wightman R, Zhang Z, Gomez LD, Atanassov I, Bukowski J-P, Tryfona T, McQueen-Mason SJ, Dupree P, Turner SR (2011) *Arabidopsis* genes IRREGULAR XYLEM (IRX15) IRX15L encode DUF579-containing proteins that are essential for normal xylan deposition in the secondary cell wall. Plant J 66:401–413

Carpita NC (1996) Structure and biogenesis of the cell walls of grasses. Annu Rev Plant Physiol Plant Mol Biol 47:445–476

Carroll A, Somerville C (2009) Cellulosic biofuels. Annu Rev Plant Biol 60:165–182

Charmet G, Masood-Quraishi U, Ravel C, Romeuf I, Balfourier F, Perretant M, Joseph J, Rakszegi M, Guillon F, Sado P, Bedo Z, Saulnier L (2009) Genetics of dietary fiber in bread wheat. Euphytica 170:155–168

Chiniquy D, Sharma V, Schultink A, Baidoo E, Rautengarten C, Cheng K, Carroll A, Ulvskov P, Harholt J, Keasling J, Pauly M, Scheller H, Ronald P (2012) XAX1 from glycosyltransferase family 61 mediates xylosyltransfer to rice xylan. Proc Natl Acad Sci USA 109:17117–17122

Costanza M, Baryshnkova A, Bellay J et al (2010) The genetic landscape of a cell. Science 327:425–431

Darvill JE, McNeil M, Darvill AG, Albersheim P (1980) Structure of plant cell walls XI. Glucuronoarabinoxylan, a second hemicellulose in the primary cell walls of suspension-cultured sycamore cells. Plant Physiol 66:1135–1139

Demura T, Fukuda H (2007) Transcriptional regulation in wood formation. Trends Plant Sci 12:64–70

Ebringerova A, Hromadkova Z, Heinze T (2005) Hemicellulose. Adv Polym Sci 186:1–67

Ellis C, Turner JG (2001) The Arabidopsis mutant *cev1* has constitutively active jasmonate and ethylene signal pathways and enhanced resistance to pathogens. Plant Cell 13:1025–1033

Faik A (2013) "Plant cell wall structure-pretreatment" the critical relationship in biomass conversion to fermentable sugars. In: Gu T (ed) Green biomass pretreatment for biofuels production. Springer, Berlin, New York, pp 1–30

Fincher GB (2009) Revolutionary times in our understanding of cell wall biosynthesis and remodeling in the grasses. Plant Physiol 149:27–37

Fischer MH, Yu N, Gray GR, Ralph J, Anderson L, Marlett JA (2004) The gel-forming polysaccharide of psyllium husk (*Plantago ovata* Forski). Carbohydr Res 239:2009–2017

Fornalé S, Sonbol F-M, Maes T, Capellades M, Uigdomènech P, Rigau J, Caparrósruiz D (2006) Down-regulation of the maize and Arabidopsis thaliana caffeic acid *O*-methyl-transferase genes by two new maize R2R3-MYB transcription factors. Plant Mol Biol 62:809–823

Giani S, Qin X, Faono F, Breviario D (1998) In rice, oryzalin and abscisic acid differentially affect tubulin mRNA and protein level. Planta 205:334–341

Gille S, de Souza A, Xiong G, Benz M, Cheng K, Schultink A, Reca I-B, Pauly M (2011) *O*-Acetylation of *Arabidopsis* hemicellulose xylogucan requires AXY4 or AXY4L proteins with a TBL and DUF231 domain. Plant Cell 23:4041–4053

Gille S, Pauly M (2012) *O*-acetylation of plant cell wall polysaccharides. Front Plant Sci 3:12. doi:10.3389/fpls.2012.00012

Goncalves VMF, Evtuguin DV, Rosario M, Domingues M (2008) Structural characterization of the acetylated heteroxylan from the natural hybrid *Paulownia elongata/Paulownia fortunei*. Carbohydr Res 343:256–266

Goubet F, Jackson P, Deery MJ, Dupree P (2002) Polysaccharide analysis using carbohydrate gel electrophoresis: a method to study plant cell wall polysaccharides and polysaccharide hydrolases. Anal Biochem 300:53–68

Helle S, Cameron D, Lam J, White B, Duff S (2003) Effect of inhibitory compounds found in biomass hydrolysates on growth and xylose fermentation by a genetically engineered strain of *S. cerevisiae*. Enzyme Microb Technol 33:786–792

Higa HH, Butor C, Diaz S, Varki A (1989) *O*-acetylation and de-*O*-acetylation of sialic acids. *O*-acetylation of sialic acids in the rat liver Golgi apparatus involves an acetyl intermediate and essential histidine and lysine residues- a transmembrane reaction? J Biol Chem 264:19427–19434

Hoffmann-Benning S, Kende H (1992) On the role of abscisic acid and gibberellin in the regulation of growth in rice. Plant Physiol 99:1156–1161

Hoije A, Sandstrom C, Roubroeks J, Andersson R, Gohil S, Gatenholm P (2006) Evidence of the presence of 2-O-β-D-xylopyranose-α-L-arabinofuranose side chains in barley husk arabinoxylan. Carbohydr Res 341:2959–2966

Iwai H, Masaoka N, Ishii T, Satoh S (2002) A pectin glucuronyltransferase gene is essential for intracellular attachment in the plant meristem. Proc Natl Acad Sci USA 99:16319–16324

Jensen JK, Kim H, Cocuron J-C, Orler R, Ralph J, Wilkerson CG (2011) The DUF579 domain containing proteins IRX15 and IRX15-L affect xylan synthesis in *Arabidopsis*. Plant J 66:387–400

Johansson MH, Samuelson O (1977) Reducing end groups in birch xylan and their alkaline degradation. Wood Sci Technol 11:261–263

Joseleau J-P, Comtat J, Ruel K (1992) Chemical structure of xylans and their interaction in the plant cell walls. In: Visser J et al (eds) Xylans and xylanases. Elsevier, Amsterdam, pp 1–15

Joshi CP, Mansfield SD (2007) The cellulose paradox-simple molecule, complex biosynthesis. Curr Opin Plant Biol 10:220–226

Jones L, Ennos AR, Turner SR (2001) Cloning and characterization of *irregular xylem4* (*irx4*): a severely lignin-deficient mutant of *Arabidopsis*. Plant J 26:205–216

Kannangara R, Branigan C, Liu Y, Penfield T, Rao V, Mouille G, Hofte H, Pauly M, Riehmann JL, Broun P (2007) The transcription factor WIN1/SHN1 regulates cutin biosynthesis in *Arabidopsis thaliana*. Plant Cell 19:1278–1294

Kiefer L, York W, Darvill A, Albersheim P (1989) Xyloglucan isolated from suspension-cultured sycamore cell walls is *O*-acetylated. Phytochemistry 28:2105–2107

Klein-Marcuschamer D, Oleskowicz-Popiel P, Simmons BA, Blanch HW (2010) Technoeconomic analysis of biofuels: a wiki-based platform for lignocellulosic biorefineries. Biomass Bioenergy 34:1914–1921

Ko J-H, Kim W-C, Han K-H (2009) Ectopic expression of MYB46 identifies transcriptional regulatory genes involved in secondary wall biosynthesis in *Arabidopsis*. Plant J 60:649–665

Kubo M, Udagawa M, Nishikubo N, Horiguchi G, Yamaguchi M, Ito J, Mimura T, Fukuda H, Demura T (2005) Transcription switches for protoxylem and metaxylem vessel formation. Genes Dev 19:1855–1860

Lee C, Zhong R, Richardson EA, Himmelsbach D, Mcphail BT, Ye ZH (2007) The *PARVUS* gene is expressed in cells undergoing secondary wall thickening and is essential for glucuronoxylan biosynthesis. Plant Cell Physiol 48:1659–1672

Lee C, Teng Q, Huang W, Zhong R, Ye ZH (2009) The F8H glycosyltransferase is a functional paralog of FRA8 involved in glucuronoxylan biosynthesis in *Arabidopsis*. Plant Cell Physiol 50:812–827

Lee C, Teng Q, Zhong R, Ye ZH (2011) The four *Arabidopsis* REDUCED WALL ACETYLATION genes are expressed in secondary wall-containing cells and required for acetylation of xylan. Plant Cell Physiol 52:1289–1301

Martinant JP, Cadalen T, Billot A, Chartier S, Leroy P, Bernard M, Saulnier L, Brankard A (1998) Genetic analysis of water-extractable arabinoxylans in bread wheat endosperm. Theor Appl Genet 97:1069–1075

McCarthy RL, Zhong R, Ye Z-H (2009) MYB83 is a direct target of SND1 and acts redundantly with MYB46 in the regulation of secondary cell wall biosynthesis in *Arabidopsis*. Plant Cell Physiol 50:1950–1964

Miyamoto K, Oka M, Ueda J (1997) Update on the possible mode of action of the jasmonates: focus on the metabolism of cell wall polysaccharides in relation to growth and development. Physiol Plant 100:631–638

Mortimer JC, Miles GP, Brown DM, Zhang Z, Segura MP, Weimar T, Yu X, Seffen KA, Stephens E, Turner SR, Dupree P (2010) Absence of branches from xylan in Arabidopsis *gux* mutants reveals potential for simplification of lignocellulosic biomass. Proc Natl Acad Sci USA 40:17409–17414

Nguyen V-L, Huynh B-L, Wallwork H, Stangoulis J (2011) Identification of quantitative trait loci for grain arabinoxylan concentration in bread wheat. Crop Sci 51:1143–1150

Olsen AN, Ernst HA, Leggio LL, Skriver K (2005) NAC transcription factors: structurally distinct, functionally diverse. Trends Plant Sci 10:79–87

Pena MJ, Zhong R, Zhou G-K, Richardson EA, O'Neill MA, Darvill AG, York WS, Ye Z-H (2007) Arabidopsis *irregular xylem*8 and *irregular xylem*9: Implications for the complexity of glucuronoxylan biosynthesis. Plant Cell 19:549–563

Persson S, Caffall KH, Freshour G, Hilley MT, Bauer S, Poindexter P, Hahn MG, Mohnen D, Somerville C (2007) The *Arabidopsis irregular xylem*8 mutant is deficient in glucuronoxylan and homogalacturonan, which are essential for secondary cell wall integrity. Plant Cell 19:237–255

Philippe S, Saulnier L, Guillon F (2006a) Arabinoxylan and (1,3)(1,4)-β-glucan deposition in cell walls during wheat endosperm development. Planta 224:449–461

Philippe S, Tranquet S, Utille J-P, Saulnier L, Guillon F (2006b) Investigation of ferulate deposition in endosperm cell walls of mature and developing wheat grains using a polyclonal antibody. Planta 225:1287–1299

Pjon C-J, Furuya M (1967) Phytochrome function in *Oryza sativa* L. I. Growth responses of etiolated coleoptiles to red, far-red and blue light. Plant Cell Physiol 8:709–718

Riemann M, Muller A, Korte A, Furuya M, Weilwer EW, Nick P (2003) Impaired induction of the jasmonate pathway in the rice mutant *hebiba*. Plant Physiol 133:1820–1830

Selig M, Adney WS, Himmel M, Decker S (2009) The impact of cell wall acetylation on corn stover hydrolysis by cellulolytic and xylanolytic enzymes. Cellulose 16:711–722

Sekimata K, Kimura T, Kaneko I, Nakano T, Yoneyama K, Takeuchi Y, Yoshida S, Asami T (2001) A specific brassinosteroid biosynthesis inhibitor, Brz 2001: evaluation of its effects on *Arabidopsis*, cress, tobacco, and rice. Planta 213:716–721

Scheller HV, Ulvskov P (2010) Hemicelluloses. Annu Rev Plant Biol 61:263–289

Somerville CR (2006) Cellulose synthesis in higher plants. Ann Rev Cell Dev Biol 22:53–78

Sonbol F-M, Fornalé S, Capellades M, Encina A, Touriño S, Torres J-L, Rovira P, Ruel K, Puigdomènech P, Rigau J et al (2009) The maize ZmMYB42 represses the phenylpropanoid pathway and affects the cell wall structure, composition and degradability in *Arabidopsis thaliana*. Plant Mol Bio 70:283–296

Takano M, Kanegae H, Shinomura T, Miyao A, Hirochika H, Furuya M (2001) Isolation and characterization of rice phytochrome A mutants. Plant Cell 13:521–534

Thines B, Katser L, Melotto M, Niu Y, Mandaokar A, Liu G, Nomura K, He S-Y, Howe GA, Browse J (2007) JAZ repressor proteins are targets of the SCF^{COI1} complex during jasmonate signaling. Nature 448:661–666

Toyomasu T, Yamane H, Murofushi N, Nick P (1994) Phytochrome inhibits the effectivness of gibberellins to induce cell elongation in rice. Planta 194:256–263

Turner SR, Somerville CR (1997) Collapsed xylem phenotype of *Arabidopsis* identifies mutants deficient in cellulose deposition in the secondary cell wall. Plant Cell 9:689–701

Ueda J, Miyamoto K, Aoki M (1994) Jasmonic acid inhibits the IAA-induced elongation of oat coleoptile segments: a possible mechanism involving the metabolism of cell wall polysaccharides. Plant Cell Physiol 35:1065–1070

Ueda J, Miyamoto K, Kamisaka S (1995) Inhibition of the synthesis of cell wall polysaccharides in oat coleoptile segments by jasmonic acid: relevance to its growth inhibition. J Plant Growth Reg 14:69–76

Urbanowicz BR, Pena MJ, Ratnaparkhe S, Avci U, Backe J, Steet HF, Foston M, Li H, O'Neill MA, Ragauskas AJ, Darvill AG, Wyman C, Gilbert HJ, York WS (2012) 4-*O*-methylation of glucuronic acid in *Arabidopsis* glucuronoxylan is catalyzed by a domain of unknown function family 579 protein. Proc Natl Acad Sci USA 109:14253–14258

Wilson SM, Burton RA, Doblin MS, Stones BA, Newbigin EJ, Fincher GB, Bacic A (2006) Temporal and spatial appearance of wall polysaccharides during cellularization of barley (*Hordeum vulgare*) endosperm. Planta 224:655–667

Winter D, Vinegar B, Nahal H, Ammar R, Greg V, Wilson GV, Provart NJ (2007) An "electronic fluorescent pictograph" browser for exploring and analyzing large-scale biological data sets. PLoS ONE 2:e718. doi:10.1371/journal.pone.0000718

Wu AM, Rihouey C, Seveno M, Hornblad E, Singh SK, Matsunaga T, Ishii T, Lerouge P, Marchant A (2009) The *Arabidopsis* IRX10 and IRX10-like glycosyltransferases are critical for glucuronoxylan biosynthesis during secondary cell wall formation. Plant J 57:718–731

Wu AM, Hornblad E, Voxeur A, Gerber L, Rihouey C, Lerouge P, Marchant A (2010) Analysis of the *Arabidopsis* IRX9/IRX9-L and IRX14/IRX14-L pairs of glycosyltransferase genes reveals critical contributions to biosynthesis of the hemicellulose glucuronoxylan. Plant Physiol 153:542–554

Yamaguchi M, Ohtani M, Mitsuda N, Kubo M, Ohme-Takagi M, Fukuda H, Demura T (2010) VND-INTERACTING2, a NAC domain transcription factor, negatively regulates xylem vessel formation in *Arabidopsis*. Plant Cell 22:1249–1263

Zablackis E, Huang J, Muller B, Darvill AG, Albersheim P (1995) Structure of plant-cell walls: characterization of the cell-wall polysaccharides of *Arabidopsis thaliana* leaves. Plant Physiol 107:1129–1138

Zeng W, Chatterjee M, Faik A (2008) UDP-xylose-stimulated glucuronyltransferase activity in wheat microsomal membranes: characterization and role in glucurono(arabino)xylan biosynthesis. Plant Physiol 147:78–91

Zeng W, Jiang N, Nadella R, Killen TL, Nadella V, Faik A (2010) A glucurono(arabino)xylan synthase complex from wheat (*Triticum aestivum* L.) contains members of the GT43, 47, and 75 families and functions cooperatively. Plant Physiol 154:78–97

Zhong R, Ye Z-H (2007) Regulation of cell wall biosynthesis. Curr Opin Plant Biol 10:564–572

Zhong R, Lee C, Zhou J, McCarthy RL, Ye Z-H (2008) A battery of transcription factors involved in the regulation of secondary cell wall biosynthesis in *Arabidopsis*. Plant Cell 20:2763–2782

Chapter 11
Towards Redesigning Cellulose Biosynthesis for Improved Bioenergy Feedstocks

Catherine Rayon, Anna T. Olek and Nicholas C. Carpita

Abstract With an estimated 200 billion tons produced annually, cellulose is the most abundant biopolymer on earth. Cellulose is expected to be the principal feedstock for liquid biofuels and bio-based products, but its *para*-crystalline nature results in recalcitrance to deconstruction required for biological and chemical conversion to useful products. Recent work solving the 3D structure of a bacterial cellulose synthase, modeling of plant cellulose synthases, and the 3D contour structure of the catalytic domain of a plant cellulose synthase have contributed new perspectives on the organization of catalytic units in the rosette complex. These discoveries stimulate new approaches to engineer the complex to make altered forms of cellulose for enhancing efficiency of biomass deconstruction for biofuel production or for synthesis of new materials and nanoproducts.

Keywords Lignocellulosic biomass • Cellulose recalcitrance • Cellulose biosynthesis • Cellulose synthase

Lignocellulosic biomass, the thickened secondary cells of vascular plants, is the primary source of reduced carbon for biofuels and bio-based products (Sarkar et al. 2009). Chemically, lignocellulosic cell wall is a complex organization of a *para*-crystalline cellulose microfibrils interlaced with cross-linking glycans, typically xylans in angiosperms and mannans in gymnosperms (Pauly and Keegstra 2008). This polysaccharide framework interacts with a co-extensive matrix of lignin to form architectural structures that provide fiber strength, resistance to extreme tensions in tracheary elements created by transpirational pull of water,

C. Rayon
EA 3900-BIOPI, Université de Picardie Jules Verne, 80039 Amiens, France

A. T. Olek · N. C. Carpita (✉)
Department of Botany and Plant Pathology, Purdue University, 915 West State Street, West Lafayette, IN 47907-2054, USA
e-mail: carpita@purdue.edu

M. C. McCann et al. (eds.), *Plants and BioEnergy*, Advances in Plant Biology 4,
DOI: 10.1007/978-1-4614-9329-7_11, © Springer Science+Business Media New York 2014

and hydrophobic barriers that maintain columns of water required for survival (Boerjan et al. 2003). A wealth of information about how lignin formation can be modified comes from studies vascular and fiber cell identity, where timing and balance of transcription regulation through a cascade of NAC and MYB domain-containing factors and microRNA expression define the lignin and polysaccharide composition of secondary walls during vascularization (Brady et al. 2010; Zhong et al. 2010; Zhao and Dixon 2011). Understanding the fine control of networks offers many opportunities and strategies to engineer biomass composition more precisely.

Considerable research effort is devoted to reducing the recalcitrance of cell walls to conversion to ethanol and advanced biofuels. As described above, much of this recalcitrance is attributed to lignin interactions that prevent hydrolysis by digestive enzymes, but the high crystallinity of cellulose and its interactions with cross-linking glycans and lignin also constitutes additional targets for enhancing efficiency of deconstruction and conversion (Himmel et al. 2007). Our focus here is on progress in understanding the mechanism of synthesis and assembly of cellulose to gain the knowledge needed to develop strategies to redesign this foundational structure of plant cell wall architecture to be an optimal raw material for biofuel production.

11.1 The Construction, Delivery and Turnover of Cellulose Synthase Complexes

Cellulose synthesis begins with the expression of several primary- and secondary wall-specific isoforms of the CesA and their assembly into complexes at the Golgi membrane (Haigler and Brown 1986). We have long known that a continuous supply of these complexes is required to maintain synthesis, indicating that each CSC might synthesize one or a limited number of microfibrils (Herth 1985; Haigler and Brown 1986). Demonstration that cellulose synthases have extremely short half-lives (on the order of 20 min) supports this view (Jacob-Wilk et al. 2006). A major distinction of plant cells is that they move large numbers of vesicle clusters from Golgi to PM during cell wall synthesis (Toyooka et al. 2009). The alternate pathway of the CSC trafficking via cortical microtubules to the PM in contrast to actin-based trafficking of the non-cellulosic polysaccharides offers some opportunities for control of the relative abundance of each in the developing wall. Cytoskeletal elements associated with CesAs including kinesins (Zhong et al. 2010) and dynamin-related proteins (Collings et al. 2008; Hirano et al. 2010; Taylor 2011).

We are just learning how CSC internalization via dynamins is associated with a clathrin-dependent pathway in support of the turnover (Fujimoto et al. 2010).

From in vivo labeling and in vitro biosynthesis dating back to the 1960s, we have known that primary and secondary wall non-cellulosic non-cellulosic cross-linking glycans and pectins are synthesized at the Golgi and transported in large secretory vesicles via actin filaments to the plasma membrane where they are assembled around newly synthesized cellulosic microfibrils (for review, see Carpita 2011). More recently we have learned that cellulose synthase complexes are packaged and exported by a different route involving small vesicles and cortical microtubules (Gutierrez et al. 2009; Crowell et al. 2009). Live-cell imaging shows a cellulose synthase interactive protein (CSI1) co-localizes with CesAs and moves bidirectionally with them (Gu et al. 2010). CSI1 is a microtubule-associated protein that forms a direct bridge to the cellulose synthase complex, and disruption of this connection in the *csi1* mutant results in dissociation of co-alignments of CSCs and the microtubule, suggesting both CSC delivery and guidance functions for microfibril deposition (Li et al. 2012). Both the N-terminal Zn-finger and catalytic domains of CesA have potential phosphorylation sites at Ser residues, and mutation of these sites to mimic and a permanent on or off state results in severe alterations in cell elongation and velocities of tracking of fluorescence-tagged CesAs (Chen et al. 2010). In summary, discovery of the dynamics of cellulose synthases, their special routes of its trafficking to the plasma membrane, and their activation mechanisms through phosphorylation each provide targets for manipulation to enhance rates of synthesis upon delivery to the plasma membrane. However, they do not in themselves provide a means of manipulation of cellulose microfibril structure. For that we need to learn more about the mechanism of synthesis.

11.2 The Cellulose Synthesis at the Plasma Membrane

To redesign the cell wall as an optimal feedstock requires a more complete understanding of the biology of the cellulose synthase complex, from the biochemical mechanism of the polymerization of the (1 → 4)-β-D-glucan chains, to the coordination of the crystallization process, and to the higher order bundling that occurs *in muro*. In plants, individual β-glucan chains of cellulose are synthesized and crystallized at the plasma membrane by large membrane complexes termed 'particle rosettes' (Giddings et al. 1980; Mueller and Brown 1980). Particle rosettes are six-membered hexagonal arrays about 25 nm in diameter and represents the membrane spanning domains and short extracellular loops of roughly three dozen cellulose synthase polypeptides (Fig. 11.1). Cellulose synthases (CesAs) are intrinsic membrane proteins with their catalytic domains extending into the cytoplasm. Membrane foot-printing gives estimated diameters of about 50 nm for the collection of 36 cellulose synthase polypeptides (Bowling and Brown 2008). The 36 β-glucan chains from the rosette complex, i.e. six chains per particle, give an estimated microfibril diameter of around 3.6–3.8 nm for most primary wall cellulose (Kennedy et al. 2007). More recent evidence employed solid-state ^{13}C-nuclear magnetic resonance (NMR) spectroscopy and neutron and x-ray diffraction indicate a smaller microfibril diameters on the order of 2.7–3.0 nm, which indicate that the crystalline portions of the microfibril are only 18–24 glucans (Fernandes et al. 2011; Thomas et al. 2013). As these techniques are relevant for the

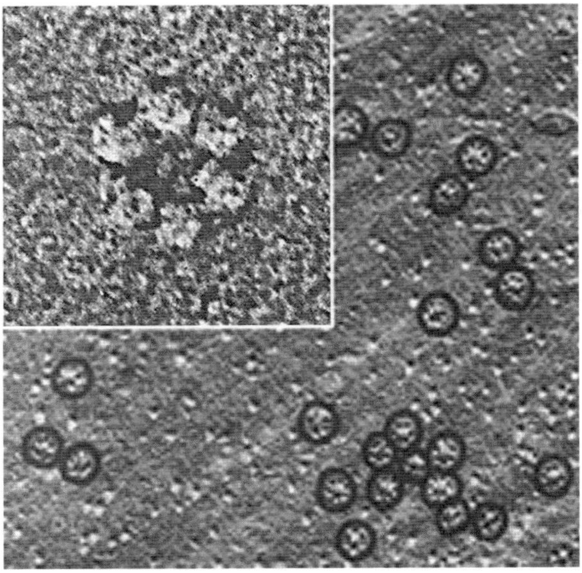

Fig. 11.1 Particle rosette structures of plant cellulose synthase complexes. Freeze-etch images of the P-face of the plasma membrane showing clusters of rosettes associated with the developing of secondary wall spiral thickenings of a Lepidium tracheary element (from Herth 1985). The inset shows the 6-fold symmetry of a single particle rosette from a Zinnia tracheary element developing in vitro (from C. Haigler, unpublished data, as seen in Delmer 1999). A substructure can be observed in each of the particles. In these freeze-etch images, only the membrane-spanning domains and extracellular loops of the CesA proteins can be observed

crystalline domains, the inference is that as many as 18 of the glucan chains surface coat the microfibril in a non-crystalline, flexible way that is capable of interacting tightly with non-cellulosic glycans, such as xyloglucan, glucomannan and xylan, or other microfibrils to form bundles. Bundling of both primary and secondary wall microfibrils can be extensive, creating a higher order of complexity and recalcitrance to deconstruction (Ding and Himmel 2006). Targets of opportunity for altering the microfibril structure include introduction of alternative sugars or linkages into the microfibril during synthesis to disrupt the *para*-crystalline array, altering the number of β-glucan chains per microfibril to minimize diameters, and interference with bundling with altered non-cellulosic glycans that alter nano-scale architecture without radically altering the functional structure of the cell wall.

11.3 The Mechanism of Cellulose Synthesis

Particle rosettes comprise several dozen cellulose synthase polypeptides of about 110 kDa, each with a large, cytoplasmic *N*-terminal region containing a ‘zinc-finger’ (ZnF) domain involved in coupling CesAs (Kurek et al. 2002), and eight

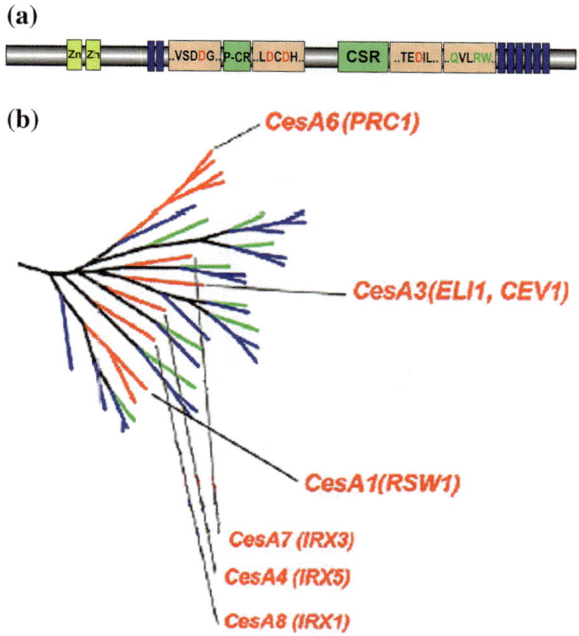

Fig. 11.2 The CesA gene family of Arabidopsis. **a** Domain model for a CesA. Two ZnF domains (*in yellow*) are found in the N terminus before the first membrane-spanning domain (*in blue*). The large central domain contains four highly conserved catalytic motifs (formerly called "U motifs") of D, DxD, (TE) D, and QxxRW, important for substrate binding and catalysis. The class-specific region (CSR) is conserved among orthologs of the same subclade and vary in the number of upstream conserved Cys residues, the number of consecutive basic amino acids, Lys and Arg, and the number of consecutive acidic amino acids, Asp and Glu, downstream from the basic residues (after Carpita and Vergara 1998; Vergara and Carpita 2001). Eight transmembrane domains, two upstream and six downstream of the catalytic domain, are predicted to interact to form a channel through which a single β-glucan chains are secreted to the cell wall. However, the finding that catalytic dimers form the fundamental unit of construction suggests that the membrane-spanning domains might fuse into a larger channel to extrude two chains instead of one. **b** Orthologous relationships of CesAs of Arabidopsis (*red*), rice (*green*), and maize (*blue*). Of the ten Arabidopsis CesA genes, at least three are coexpressed during primary wall formation, and mutations in each of them, AtCesA1 (RSW1; At4g32410), AtCesA6 (PRC1; At5g64740), and AtCesA3 (CEV1 and ELI1; At5g05170), result in cellulose deficiencies and demonstrate non-rdundancy. The *irx* mutants AtCesA8 (IRX1; At4g18780), AtCesA7 (IRX3; At5g17420), and AtCesA4 (IRX5; At5g44030), which are also non-redundant, are deficient in cellulose synthesis specifically in secondary walls (after Penning et al. 2009)

membrane spans, two upstream and six downstream, sandwiching a catalytic domain (CatD) containing four catalytic motifs, formerly called U-motifs, with D, DxD, D, and QxxRW (Fig. 11.2; Delmer 1999). Within the catalytic domain are the Plant-Conserved Sequence (P-CR), located between the first D and DxD motif, and the 'Class-Specific Region' (CSR), for which high similarity across many species was observed among subclasses of orthologous isoforms (Fig. 11.2b; Vergara and Carpita 2001).

Plant CesAs share homology with sequences of bacterial CesA proteins with respect to the four catalytic motifs essential for substrate binding and catalysis (Saxena et al. 1995; Pear et al. 1996). The biochemical mechanism of cellulose synthesis in plants must solve three fundamental physical problems: (a) synthesis of the (1 → 4)-β-D-glucosyl linkage requires that each glucosyl residue is turned 180° with respect to its neighboring sugar, (b) a membrane channel of sufficient size is needed to permit extrusion of the glucan chain, and (c) a mechanism is needed to couple the many synthases into the rosette complex (Carpita 2011). The structure of the *Rhodobacter sphaeroides* CesA (BcsA) solves the first two problems with a single-site mechanism that toggles between two conformations of the non-reducing end acceptor glucosyl residue, spiraling the chain through the 8-membered membrane channel into a second protein, BcsB, that guides the chain through the periplasmic space (Morgan et al. 2013). However, the BcsA synthase functions as a monomer, and plants do not contain the accessory protein but extrude the chains directly to the extracellular surface where they are organized by an unknown mechanism into microfibrils. In contrast, plant CesAs couple into complexes via three additional sequences not present in bacterial CesAs, the Zn-finger domains, the CSR, and/or P-CR. Recently, we showed that recombinant expression of only the large catalytic domains between membrane-spanning domains II and III can be isolated as monomers in the presence of thiol reducing agents but dimerize when the proteins are concentrated or the thiol reducing agent depleted (Olek et al. 2013). Small-angle x-ray scattering (SAXS) of the monomer shows an elongated two-domain structure, with dimers coupled through the smaller domains.

The 3D crystal structure of the BcsA synthase gives a significant conformation of the amino acids that function in UDP-Glc binding, chain termination positioning and catalysis of glycosyl transfer (Morgan et al. 2013). Structure modeling of a cotton secondary wall CesA showed good conservation of the active site defined by the four catalytic motifs and other amino acids with the BcsA (Sethaphong et al. 2013). Olek et al. (2013) used several threading and structure prediction models to show the catalytic motifs give quite similar active site conservation with the BcsA when a rice CesA is truncated to remove P-CR and CSR domains not present in the bacterial protein. However, considerable uncertainty remains concerning the P-CR and CSR domains subtending the active site because of the lack of specific templates. *Ab initio* modeling (Sethaphong et al. 2013) versus composite modeling against chimeric templates (Olek et al. 2013) gave drastically different models for the conformation of the P-CR and CSR.

Sethaphong et al. (2013) extend their model to form a symmetrical hexamer that involves coupling of the CSR and P-CR. In contrast, Olek et al. (2013) predict dimerization through the P-CR and/or CSR domains form the fundamental unit of synthesis, and three dimers couple by complementary Zn-finger domains to form a single particle of the 6-membered particle rosette. Interaction of Zn-fingers is precluded in the symmetrical hexamer. In either model, each monomer synthesizes a single β-glucan chain by a single catalytic mechanism that toggles between the alternating position of *O*-4 of the non-reducing terminal glucose needed to form the (1 → 4)-β-D-linkage. However, based on the contour structure predicted by

SAXS, the fundamental dimer hypothesis predicts the UDP-Glc entry from the outward faces and extrusion of two chains through the cavity formed by the dimer (Olek et al. 2013).

While the 3D crystal structure of a plant CesA is needed to determine the roles of the P-CR and CSR unequivocally, it is reasonable to predict that the mechanism of synthesis and the structure of the active site of the bacterial synthase is conserved in the plant enzyme. Thus, amino acid targets for manipulation of the linkage would be those predicted to be involved with toggling mechanism while maintaining the uridinyl binding domain of UDP-Glc (Morgan et al. 2013). Mapping the differences in the mechanism of (1 → 3)-β-D-glucan vs. (1 → 4)-β-D-glucan synthases could provide some guidance for protein engineering. One could contemplate protein engineering strategies to convert the uridinyl binding domain to a guanidinyl binding domain to accommodate a GDP-Man. Man is a C-2 epimer of Glc and may need some remodeling of the toggling domain to accommodate the axial -OH of Man rather than the equatorial -OH of Glc. Man in place of Glc could potentially introduce slight alterations in the crystal structure of cellulose that would ease deconstruction without modifying the fundamental microfibril function.

The formation of catalytic dimers as the fundamental units of synthesis gives a new perspective on the requirement for multiple isoforms of the CesA associated with primary and secondary wall cellulose formation. As mutants of each impair synthesis, their association with the synthase complex is non-redundant (Taylor 2003), and direct interactions of three distinct CesA polypeptides have been shown in vivo by bimolecular fluorescence complementation (Desprez et al. 2007) and in vitro by affinity pull-down experiments (Taylor et al. 2003; Wang et al. 2008; Atanassov et al. 2009), and yeast 2-hybrid studies (Timmers et al. 2009). The key question becomes whether or not certain isoforms form better homo- or heterodimers, and the role domain swapping or active-site manipulation might play in modulating the number of chains produced.

11.4 Accessory Proteins that Might Function in Cellulose Synthesis

Amor et al. (1995) proposed that a sucrose synthase (SuSy) associated with the plasma membrane constituted a UDP-Glc metabolic channeling mechanism for cellulose synthase. This idea was strengthened by demonstration immunologically of SuSy in the rosette structures (Fujii et al. 2010). However, several lines of conflicting data have cast doubt on a requirement for SuSy in cellulose synthesis. A quadruple mutant that eliminates all detectable SuSy has no effect on rates of cellulose synthesis in Arabidopsis (Barratt et al. 2009). Baroja-Ferandéz et al. (2012) challenged this conclusion by claiming that the reaction conditions for SuSy were not optimized and that even the quadruple mutant had sufficient SuSy activity to support cellulose synthase. Smith et al. (2012) argued that the activity measured was from two additional isoforms that are strictly phloem localized and irrelevant for mesophyll cells of the leaf. Regardless of a facilitative or required role for SuSy as a metabolic channel for UDP-Glc to cellulose synthase, over-expression of SuSy in transgenic poplar results in significant increases in cellulose content (Coleman et al. 2009). Thus, enhancing amounts of SuSy in the absence of over-expression of the components of the cellulose synthase complex is a reasonable strategy to enhance yields of cellulose.

Several interactions of CesAs with other proteins have been inferred from mutants whose phenotypes include disruption of cellulose synthesis. Loss of trichome birefringence (*TBR*), long-recognized as a cellulose deficiency (Potikha and Delmer 1995), was traced to a gene encoding a transmembrane domain-containing protein with a 'domain of unknown function' (DUF231) (Bishoff et al. 2010). TBR is a member of a very large family that also contains genes that encode senescence-related, repression of freezing-tolerance, and resistance to powdery mildew.

KORRIGAN, a transmembrane-containing endo-(1 → 4)-β-D-glucanase required for cell growth and cellulose deposition, co-localizes with CesAs at the plasma membrane but remains without a specific function (Crowell et al. 2010). Members of *COBRA*, a large gene family encoding glycosyl phosphatidylinositol (GPI)-anchored membrane associated protein, are necessary for normal cell development and cell-wall architecture, with different isoforms associated with normal primary- or secondary-wall cellulose biosynthesis in grasses (Schindelman et al. 2001; Roudier et al. 2005; Sato et al. 2010). While the co-localize to the plasma membrane with CesAs, they do not appear to function directly in synthesis of the glucan chains, but in the orientation and patterning of both cellulose and lignin during wall deposition. Their absence gives rise to stem 'brittleness' without change in tensile strength in stress-strain experiments (Sindhu et al. 2007). As some studies indicate that COBRAs might function in crystallization, there is potential to manipulation to alter recalcitrance properties of cellulose. Thus, manipulation of secondary wall COBRAs might have a utility in wall densification or fragmentation during processing.

11.5 Conclusions

Good progress has been made in unraveling the intricacies of the protein structure, synthase assembly, trafficking, and synthase mechanisms of cellulose synthase complexes. At each step of resolution, researchers are given important insights on how the features of microfibril assembly could be altered to improve biomass as a feedstock for biofuels and bio-based products. However, in a broader sense unraveling the complexities of transcriptional networks that are responsible for establishing the diverse lignocellulosic compositions and architectures found in any species holds special promise for redesigning the cell wall for deconstruction to valued end-products.

Acknowledgments This review was completed through support of the Center for Direct Catalytic Conversion of Biomass to Biofuels, an Energy Frontier Research Center funded by the U.S. Department of Energy, Office of Science, Office of Basic Energy Sciences (award no. DE–SC0000997).

References

Amor Y, Haigler CH, Johnson S, Wainscott M, Delmer DP (1995) A membrane-associated form of sucrose synthase and its potential role in synthesis of cellulose and callose in plants. Proc Nat Acad Sci USA 92:9353–9357

Atanassov II, Pittman JK, Turner SR (2009) Elucidating the mechanisms of assembly and subunit interaction of the cellulose synthase complex of Arabidopsis secondary cell walls. J Biol Chem 284:3833–3841

Baroja-Fernández E, Muñoz FJ, Li J, Bahaji A, Almagro G, Montero M, Etxeberria E, Hidalgo M, Sesma MT, Pozueta-Romero J (2012) Sucrose synthase activity in the *sus1/sus2/sus3/sus4 Arabidopsis* mutant is sufficient to support normal cellulose and starch production. Proc Nat Acad Sci USA 109:321–326

Barratt DHP, Derbyshire P, Findlay K, Pike M, Wellner N, Lunn J, Feil R, Simpson C, Maule AJ, Smith AM (2009) Normal growth of Arabidopsis requires cytosolic invertase but not sucrose synthase. Proc Nat Acad Sci USA 106:13124–13129

Bischoff V, Silvia N, Neumetzler L, Schindelasch D, Urbain A, Eshed R, Persson S, Delmer D, Scheible WR (2010) *TRICHOME BIREFRINGENCE* and its homolog At5g01360 encode plant-specific DUF231 proteins required for cellulose biosynthesis in Arabidopsis. Plant Physiol 153:590–602

Boerjan W, Ralph J, Baucher M (2003) Lignin biosynthesis. Annu Rev Plant Biol 54:519–546

Bowling AJ, Brown RM (2008) Cytoplasmic domain of cellulose synthesizing complexes in plants. Protoplasma 233:115–127

Brady SM, Zhang L, Megraw M, Martinez NJ, Jiang E, Yi CS, Liu W, Zeng A, Taylor-Teeples M, Kim D et al (2010) A stele-enriched gene regulatory network in the Arabidopsis root. Mol Syst Biol 7: Art. No. 459

Carpita NC, Vergara CE (1998) A recipe for cellulose. Science 279:672–673

Carpita NC (2011) Update on mechanisms of plant cell wall biosynthesis: how plants make cellulose and other (1 → 4)-β-D-glycans. Plant Physiol 155:171–184

Chen S, Ehrhardt DW, Somerville CR (2010) Mutations of cellulose synthase (CESA1) phosphorylation sites modulate anisotropic cell expansion and bidirectional mobility of cellulose synthase. Proc Nat Acad Sci USA 107:17188–17193

Coleman HD, Yan J, Mansfield SD (2009) Sucrose synthase affects carbon partitioning to increase cellulose production and altered cell wall ultrastructure. Proc Nat Acad Sci USA 106:13118

Collings DA, Gebbie LK, Howles PA, Hurley UA, Birch RJ, Cork AH, Hocart CH, Arioli T, Williamson RE (2008) Arabidopsis dynamin-like protein DRP1A: a null mutant with widespread defects in endocytosis, cellulose synthesis, cytokinesis, and cell expansion. J Exp Bot 59:361–376

Crowell EF, Bischoff V, Desprez T, Rolland A, Stierhof Y-D, Schumacher K, Gonneau M, Höfte H, Vernhettes S (2009) Pausing of Golgi bodies on microtubules regulates secretion of cellulose synthase complexes in Arabidopsis. Plant Cell 21:1141–1154

Crowell EF, Gonneau M, Stierhof Y-D, Höfte H, Vernhettes S (2010) Regulated trafficking of cellulose synthases. Curr Opin Plant Biol 13:700–705

Delmer DP (1999) Cellulose biosynthesis: exciting times for a difficult field of study. Annu Rev Plant Physiol Plant Mol Biol 50:245–276

Desprez T, Juraniec M, Crowell EF, Jouy H, Pochylova Z, Parcy F, Höfte H, Gonneau M, Vernhettes S (2007) Organization of cellulose synthase complexes involved in primary cell wall synthesis in Arabidopsis thaliana. Proc Nat Acad Sci USA 104:15572–15577

Ding S-Y, Himmel ME (2006) The maize primary cell wall microfibril: a new model derived from direct visualization. J Agric Food Chem 54:597–606

Fernandes AN, Thomas LH, Altaner CM, Callow P, Forsyth VT, Apperley DC, Kennedy CJ, Jarvis MC (2011) Nanostructure of cellulose microfibrils in spruce wood. Proc Nat Acad Sci USA 108:E1195–E1203

Fujii S, Hayashi T, Mizuno K (2010) Sucrose synthase is an integral component of the cellulose synthesis machinery. Plant Cell Physiol 51:294–301

Fujimoto M, Arimura S, Ueda T, Takanashi H, Hayashi Y, Nakano A, Tsutsumi N (2010) Arabidopsis dynamin-related proteins DRP2B and DRP1A participate together in clathrin-coated vesicle formation during endocytosis. Proc Nat Acad Sci USA 107:6094–6099

Giddings TH Jr, Brower DL, Staehelin LA (1980) Visualization of particle complexes in the plasma membrane of *Micrasterias denticulata* associated with the formation of cellulose fibrils in primary and secondary cell walls. J Cell Biol 84:327–339

Gu Y, Kaplinsky N, Bringmann M, Cobb A, Carroll A, Sampathkumar A, Baskin TI, Persson S, Somerville C (2010) Identification of a cellulose-synthase-associated protein required for cellulose biosynthesis. Proc Nat Acad Sci USA 107:12866–12871

Gutierrez R, Lindeboom JJ, Paredez AR, Emons AMC, Ehrhardt DW (2009) Arabidopsis cortical microtubules position cellulose synthase delivery to the plasma membrane and interact with cellulose synthase trafficking compartments. Nat Cell Biol 11:797–808

Haigler CH, Brown RM Jr (1986) Transport of rosettes from the Golgi apparatus to the plasma membrane in isolated mesophyll cells of *Zinnia elegans* during differentiation to tracheary elements in suspension culture. Protoplasma 134:111–120

Herth W (1985) Plasma-membrane rosettes involved in localized wall thickening during xylem vessel formation of *Lepidium sativum* L. Planta 164:12–21

Himmel ME, Ding S-Y, Johnson DK, Adney WS, Nimlos MR, Brady JW, Foust TD (2007) Biomass recalcitrance: engineering plants and enzymes for biofuels production. Science 315:804–807

Hirano K, Kotake T, Kamihara K, Tsuna K, Aohara T, Kaneko Y, Takatsuji H, Tsumuraya Y, Kawasaki S (2010) Rice BRITTLE CULM3 (BC3) encodes a classical dynamin OsDRP2B essential for proper secondary cell wall synthesis. Planta 232:95–108

Jacob-Wilk D, Kurek I, Hogan P, Delmer DP (2006) The cotton fiber zinc-binding domain of cellulose synthase A1 from *Gossypium hirsutum* displays rapid turnover in vitro and in vivo. Proc Nat Acad Sci USA 103:12191–12196

Kennedy CJ, Cameron GJ, Sturcová A, Apperley DC, Altaner C, Wess TJ, Jarvis MC (2007) Microfibril diameter in celery collenchyma cellulose: X-ray scattering and NMR evidence. Cellulose 14:235–246

Kurek I, Kawagoe Y, Jacob-Wilk D, Doblin M, Delmer D (2002) Dimerization of cotton fiber cellulose synthase catalytic subunits occurs via oxidation of the zinc-binding domains. Proc Nat Acad Sci USA 99:11109–11114

Li S, Lei L, Somerville CR, Gu Y (2012) CSI1 links microtubules and cellulose synthase complexes. Proc Nat Acad Sci USA 109:185–190

Morgan JLW, Strumillo J, Zimmer J (2013) Crystallographic snapshot of cellulose synthesis and membrane translocation. Nature 493:181–186

Mueller SC, Brown RM Jr (1980) Evidence for an intramembrane component associated with a cellulose microfibril-synthesizing complex in higher plants. J Cell Biol 84:315–326

Olek AT, Rayon C, Makowski L, Kim HR, Ciesielski P, Badger J, Paul LN, Ghosh S, Kihara D, Crowley M, Himmel ME, Bolin JT, Carpita NC (2013) Small-angle x-ray scattering reveals the structure of the catalytic domain of a cellulose synthase and its assembly into dimers. In review

Pauly M, Keegstra K (2008) Cell-wall carbohydrates and their modification as a resource for biofuels. Plant J 54:559–568

Pear JR, Kawagoe Y, Schreckengost WE, Delmer DP, Stalker DM (1996) Higher plants contain homologs of the bacterial celA genes encoding the catalytic subunit of cellulose synthase. Proc Nat Acad Sci USA 93:12637–12642

Penning BW, Hunter CT, Tayengwa R, Eveland AL, Dugard CK, Olek AT, Vermerris W, Koch KE, McCarty DR, Davis MF, Thomas SR, McCann MC, Carpita NC (2009) Genetic resources for maize cell wall biology. Plant Physiol 151:1703–1728

Potikha T, Delmer DP (1995) A mutant of *Arabidopsis thaliana* displaying altered patterns of cellulose deposition. Plant J 7:453–460

Roudier F, Fernandez AG, Fujita M, Himmelspach R, Borner GHH, Schindelman G, Song S, Baskin TI, Dupree P, Wasteneys GO, Benfey PN (2005) COBRA, an Arabidopsis extracellular glycosyl-phosphatidyl inositol-anchored protein, specifically controls highly anisotropic expansion through its involvement in cellulose microfibril orientation. Plant Cell 17:1749–1763

Sarkar P, Bosneaga E, Auer M (2009) Plant cell walls throughout evolution: towards a molecular understanding of their design principles. J Exp Bot 60:3615–3635

Sato K, Suzuki R, Nishikubo N, Takenouchi S, Ito S, Nakano Y, Nakaba S, Sano Y, Funada R, Kajita S, Kitano H, Katayama Y (2010) Isolation of a novel cell wall architecture mutant of rice with defective Arabidopsis COBL4 ortholog BC1 required for regulated deposition of secondary cell wall components. Planta 232:257–270

Saxena IM, Brown RM Jr, Fevre M, Geremia RA, Henrissat B (1995) Multidomain architecture of β-glycosyl transferases: implications for mechanism of action. J Bacteriol 177:1419–1424

Schindelman G, Morikami A, Jung J, Baskin TI, Carpita NC, Derbyshire P, McCann MC, Benfey PN (2001) COBRA encodes a putative GPI-anchored protein, which is polarly localized and necessary for oriented cell expansion in Arabidopsis. Genes Dev 15:1115–1127

Sethaphong L, Haigler CH, Kubicki JD, Zimmer J, Bonetta D, Debolt S, Yingling Y (2013) Tertiary model of a plant cellulose synthase. Proc Nat Acad Sci USA 110:7512–7517

Sindhu A, Langewisch T, Olek A, Multani DS, McCann MC, Vermerris W, Carpita NC, Johal G (2007) Maize Brittle stalk2 encodes a COBRA-like protein expressed in early organ development but required for tissue flexibility at maturity. Plant Physiol 145:1444–1459

Smith AM, Kruger NJ, Lunn JE (2012) Source of sugar nucleotides for starch and cellulose synthesis. Proc Nat Acad Sci USA 109:E776–E776

Taylor NG (2011) A role for Arabidopsis dynamin related proteins DRP2A/B in endocytosis; DRP2 function is essential for plant growth. Plant Mol Biol 76:117–129

Taylor NG, Howells RM, Huttly AK, Vickers K, Turner SR (2003) Interactions among three distinct CesA proteins essential for cellulose synthesis. Proc Nat Acad Sci USA 100:1450–1455

Thomas LH, Forsyth VT, Šturcová A, Kennedy CJ, May RP, Altaner CM, Apperley DC, Wess TJ, Jarvis MC (2013) Structure of cellulose microfibrils in primary cell walls from collenchyma. Plant Physiol 161:465–476

Timmers J, Vernhettes S, Desprez T, Vincken JP, Visser RGF, Trindade LM (2009) Interactions between membrane-bound cellulose synthases involved in the synthesis of the secondary cell wall. FEBS Lett 583:978–982

Toyooka K, Goto Y, Asatsuma S, Koizumi M, Mitsui T, Matsuoka K (2009) A mobile secretory cluster involved in mass transport from the Golgi to the plant cell exterior. Plant Cell 21:1212–1229

Vergara CE, Carpita NC (2001) β-D-Glycan synthases and the CesA gene family: lessons to be learned from the mixed-linkage (1 → 3), (1 → 4)-β-D-glucan synthase. Plant Mol Biol 47:145–160

Wang J, Elliott JE, Williamson RE (2008) Features of the primary wall CESA complex in wild type and cellulose-deficient mutants of *Arabidopsis thaliana*. J Exp Bot 59:2627–2637

Zhao Q, Dixon RA (2011) Transcriptional networks for lignin biosynthesis: more complex than we thought? Trends Plant Sci 16:227–233

Zhong R, Lee C, Ye Z-H (2010) Evolutionary conservation of the transcriptional network regulating secondary cell wall biosynthesis. Trens Plant Sci 15:625–632

Part III
Biomass Processing

Chapter 12
Developing Novel Enzyme Repertoires for the Efficient Deconstruction of Plant Biomass Tailored for the Bioenergy Industry

Harry J. Gilbert

Abstract Plant biomass, commonly referred to as lignocellulose, represents a renewable and thus sustainable substrate for the liquid biofuel and chemical industries, which is carbon dioxide neutral. There is much debate concerning the economic viability of lignocellulose-based liquid biofuels based, primarily, on the cost of the enzymes required to saccharify plant biomass into its component sugars. As a result there has been a substantial investment in enzyme technology targeted towards improving the efficiency of plant cell wall degradation. This Chapter provides an overview of our current knowledge of plant cell wall degrading enzymes at a structural and biochemical level. The article also describes strategies that can be deployed to discover novel, industrially significant, enzyme functions, and how protein engineering can be used to increase the catalytic efficiency of some enzymes, and broaden the substrate specificity of others. Finally the Chapter highlights the emerging importance of polysaccharide oxidases in lignocellulosic deconstruction focussing on the role these enzymes play in opening up the structure of crystalline cellulose, explaining how they are capable of potentiating the activity of glycoside hydrolases (cellulases) against these recalcitrant structures.

Keywords Liquid biofuels • Lignocellulose glycoside hydrolases • Cellulases • Mannanases • Xylanases • Cellulose oxidases • Protein engineering

---

H. J. Gilbert (✉)
The Institute for Cell and Molecular Biosciences, Newcastle University,
Framlington Place, Newcastle upon Tyne NE42 6EB, UK
e-mail: h.j.gilbert@ncl.ac.uk

M. C. McCann et al. (eds.), *Plants and BioEnergy*, Advances in Plant Biology 4,
DOI: 10.1007/978-1-4614-9329-7_12, © Springer Science+Business Media New York 2014

12.1 Introduction

The cell walls of plants are the most abundant source of organic carbon on the planet. This photosynthetically fixed carbon is recycled by microbial enzymes that convert cell wall polysaccharides to mono- and oligosaccharides. Reflecting the growing industrial significance of plant cell wall deconstruction, the enzymes that catalyze this process have been subjected to intense analysis in the last few years. Today, a growing and significant application of these biocatalysts is in the production of second generation lignocellulosic-based biofuels and in the bioprocessing sector, where the synthesis of high-value chemicals from renewable sources, such as plant biomass, is of particular importance (Ragauskas et al. 2006; Bayer et al. 2007; Sticklen 2008; Himmel and Bayer 2009). Plant cell walls are recalcitrant to biological depolymerization as the extensive interactions between polysaccharides, and between polysaccharides and lignin, restrict access to the battery of microbial enzymes (principally glycoside hydrolases) that break down these composite structures (for review, see Mohnen 2008). Since the early 1990s, when the first structures of these enzymes were determined, there has been an explosion of structural information on these proteins. This structural information is starting to inform and direct protein engineering strategies that, potentially, will make a significant contribution to the toolbox of biocatalysts used to deconstruct the plant cell wall. These novel enzymes could therefore increase the economic viability of using lignocellulose as a substrate for the biofuel industry. This review will discuss how such enzymes can be generated either through screening for natural biocatalysts that display novel, and useful, activities, or by engineering established enzymes.

12.2 The Plant Cell Wall

Plant cell walls are divided into two major types; the primary and secondary wall, and are comprised predominantly of polysaccharides (~90 %). Secondary walls, which provide the major source of biomass, contain low amounts of pectin and the major hemicellulose is xylan, although in gymnosperms glucomannans represent the major hemicellulose. In addition to polysaccharides, secondary walls are often rigidified by the impregnation of lignin, a heterogenous aromatic polymer. The structure of the plant cell has been extensively reviewed previously and will be described briefly here (see Harris and Stone 2008; Mohnen 2008; Mohnen et al. 2008, for an overview of plant cell wall structure).

Cellulose is a β-1,4-linked glucose (Glc) molecule that is substantially crystalline. All the hemicellulosic polysaccharides contain a β-linked sugar backbone. In xylans and mannans the backbone sugars are β-1,4-D-Xyl and β-1,4-D-Man residues, while in glucomannan the backbone consists of randomly dispersed β-1,4-Glc and β-1,4-Man sugars. The backbones of hemicellulosic polysaccharides are decorated with a variety of sugars and acetyl groups explaining why these polymers are not crystalline, exemplified by xyloglucan, which comprises a β-1,4-linked glucose backbone decorated with xylose residues that, in turn can also contain additional sugars.

12.3 CAZy

Enzymes that modify complex carbohydrates, together with their accessory non-catalytic carbohydrate binding modules (CBMs), have been grouped into sequence-based families on the continuously updated CAZy database (Cantarel et al. 2009; http://www.cazy.org/). Members of the same enzyme family display a common fold, while the catalytic apparatus and mechanism are similarly conserved. Currently 45 of the 131 glycoside hydrolase families (GHs) contain enzymes that contribute to plant cell wall deconstruction. Of the 65 CBM families, around half of these modules bind to components of the plant cell wall.

Cellulases: Cellulose utilization involves the activities of endo-β-1,4-glucanases, cellobiohydrolases (also called exo-β-1,4-glucanases), and β-glucosidases, that act synergistically to convert crystalline cellulose to glucose (Beguin and Aubert 1994; Tomme et al. 1995). Endo-β-1,4-glucanases attack the cellulose chain at exposed internal glycosidic bonds, while cellobiohydrolases can attack either the non-reducing or reducing end of exposed cellulose chains (Beguin and Aubert 1994). Cellobiose, which can inhibit the cellobiohydrolases, is cleaved by β-glucosidases (Teeri 1997).

The substrate binding site of cellobiohydrolases consists of a tunnel through which single cellulose chains are threaded (Rouvinen et al. 1990; Divne et al. 1994). A subset of GH9 cellulases, those containing a CBM3c, display an endo-processive mode of action (Irwin et al. 1998; Gilad et al. 2003). CBM3cs are atypical family 3 CBMs that bind weakly to cellulose in isolation. The current model for GH9 enzymes that contain a CBM3c is that the cellulase, through its open substrate binding cleft, is able to bind to internal regions of a cellulose chain and thus exhibits an endo-mode of action. However, after bond cleavage the cellulose chain slides along the substrate binding cleft by two residues, assisted by CBM3c, explaining its processive mode of action (Sakon et al. 1997; Irwin et al. 1998; Gilad et al. 2003). Classically cellulose hydrolysis, deploying *Hypocreajecorina* (formerly *Trichoderma reesei*) as the model system, is viewed as a synergistic process between endo-acting cellulases that create new ends from that the exo-acting cellobiohydrolases can release cellobiose from either the reducing (GH7 and GH48) or non-reducing (GH6) end of the cellulose chains (reviewed in Kleywegt et al. 1997; Teeri 1997). This model, however, is inconsistent with several features of cellulose degradative systems. Thus biochemical and structural data indicate that GH6 cellobiohydrolases are not, exclusively, exo acting (Amano et al. 1996; Armand et al. 1997; Varrot et al. 1999). Furthermore, some highly active cellulase systems lack a classic pair of cellobiohydrolases that act from the reducing and non-reducing ends of cellulose chains, respectively (Xie et al. 2007; DeBoy et al. 2008; Weiner et al. 2008). An intriguing report by Tolonen and colleagues (2009) showed that a single endo-processive GH9 cellulase (CfCel9) was essential for cellulose degradation in *Clostridium phytofermentans*. Given the redundancy in cellulase systems, demonstration that a single enzyme is essential for cellulose degradation is rare, and questions the classical synergy model. Indeed, CfCel9 is an extremely exciting target in the development of consolidated bioprocessing

(CBP) systems, in which synthetic biology is deployed to introduce lignocellulosic degradative capacity into a biofuel producing organisms, obviating the need for external enzyme mixtures. Currently, enzyme mixtures used in the saccharification process represents one of the most significant costs in the biofuel industry, and thus the development of CBP organisms offers an attractive approach to generating liquid fuels. However, as detailed above, most cellulase systems consist of a large number of enzymes, and the transfer of this degradative capacity into the biofuel-generating organism represents a significant synthetic biology challenge.

β-**Mannanases**: β-mannanases display a $(\beta/\alpha)_8$ barrel-fold and are located primarily within GH5 and GH26. Substrate recognition by β-mannanases is complicated by the requirement of these enzymes to hydrolyze the heterogenous polymer, glucomannan. Recognition of Man and Glc sugars at subsites distal to the active site (−1 subsite) is highly variable, although some general trends are emerging, which point to a divergence in specificity between GH5 and GH26 mannanases. GH5 mannanases are able to accommodate Glc at the −2 and +1 subsites (Tailford et al. 2009), and are thus able to hydrolyze mannosidic linkages flanked by Man or Glc. A particularly exciting enzyme in the context of glucomannan degradation is Man5A from Caldanaerobius polysaccharolyticus, which is capable of acting as both an endo-glucunase and endo-mannanase (Han et al. 2010). With respect to glucomannan degradation Man5A is capable of fulfilling multiple functions and thus could play an important role in the deconstruction of softwoods.

In contrast, the GH26 mannanases generally display tight specificity for Man at both the −2 subsite and −1 subsites. Indeed, a cohort of GH26 mannanases contain an arginine at the −2 substrate that makes extensive interactions with the substrate, and confers unusually high activity against small mannooligosaccharides (Ducros et al. 2002; Cartmell et al. 2008). Screening genomic databases for other GH26 enzymes that retain this arginine may facilitate the identification of novel mannooligosaccharidases. Currently, the two Cellvibrio enzymes that contain a high-affinity −2 subsite do not possess additional negative binding subsites, which may explain why the high activity displayed against mannotriose and mannotetraose is not translated to the hydrolysis of polysaccharides (Hogg et al. 2001; Cartmell et al. 2008). However, these enzymes provide an excellent structural scaffold for building additional distal subsites that, when coupled with the very high affinity −2 subsite, has the potential to generate mannanases with extremely high catalytic activities.

12.4 Structural Changes that Modulate the Mode of Enzyme Action

The structural basis for the GH6 and GH7 cellobiohydrolases and endoglucanases, reflecting the formation of extended loops that form tunnel-like substrate binding regions, is well established and has been extensively reviewed (Kleywegt et al. 1997; Teeri 1997; Varrot et al. 1999). However, it is also apparent that subtle changes to the distal negative subsites of glycoside hydrolases can be used modulate the mode of enzyme action. Thus, endo-acting GH43 arabinanases contain a substrate binding cleft open at both ends, explaining their endo-activity (Alhassid et al. 2009), The single arabinanase from *Cellvibrio japonicus*, CjArb43A, unusually, displays an endo-processive activity (McKie et al. 1997). Again subtle changes to the distal subsite (compared to endo-acting arabinanases) cause a steric block that prevents extension of the substrate past the −3 subsite (Nurizzo et al. 2002; Proctor et al. 2005). Endo-activity can be introduced into the enzyme through only two amino acid substitutions to the −3 subsite, without compromising catalytic efficiency. The redesigned endo-enzyme remains a more powerful catalyst than unmodified endo-acting arabinanases, demonstrating the feasibility of engineering industrially relevant modes of action into highly active carbohydrate modifying enzymes.

12.5 Xylan Degradation

The xylan backbone is hydrolysed primarily by GH10 and GH11 xylanases, while the Araf side chains are removed by enzymes from various families including GH43, GH51, GH54 and GH62 (for review of xylan degradation, see Gilbert et al. 2008). GH43 arabinofuranosidases may display the highest level of substrate diversity exemplified by the activity of a GH43 enzyme (HiAXHd3) that removes the O_3- Araf side chain from Xyl residues decorated at both O_2 and O_3 with Araf (van den Broek et al. 2005; Sorensen et al. 2006). The crystal structure of this enzyme reveals a cleft that houses the xylan backbone. In the centre of the cleft is the active site pocket. Modification of the rim of the active site pocket introduces endo-xylanase activity, while the resultant enzyme variant, Y165A, retains arabinofuranosidase activity. The crystal structure of Y165A shows that the mutation creates a topology that allows either the xylan backbone or arabinose side chains to enter the active site, explaining the observed multiple catalytic functions of the engineered enzyme, Fig. 12.1 (McKee et al. 2012). These data demonstrate that the active site of HiAXHd3 is tuned to hydrolyse arabinofuranosyl or xylosyl linkages, and it is the topology of the distal regions of the substrate binding surface that confers specificity. The introduction of xylanase activity into an AXHd3 arabinofuranosidase is of considerable biotechnological significance, particularly in industries, such as the biofuel sector, which utilize the plant cell wall as the major substrate. The chemistry of plant cell walls is highly complex, and thus a large number of enzymes, with different substrate specificities, are required to fully saccharify these composite structures. From an industrial perspective, the enzyme component in the bioenergy and bioprocessing sectors is a significant cost, and thus it is important to minimize the number of different glycoside hydrolases deployed. By introducing additional catalytic functions into a biotechnologically significant glycoside hydrolase, the work of McKee et al. demonstrates the feasibility of generating limited enzyme cocktails that display the range of activities

(c)

AXHd3	Xylanase (k_{cat}/K_M)	Arabinofuranosidase (k_{cat}/K_M)
Wild type	No activity	2×10^7 min^{-1} M^{-1}
Y165A	7×10^4 min^{-1} ml mg^{-1}	1×10^5 min^{-1} M^{-1}
Y165A/F493A /E593A	8×10^6 min^{-1} ml mg^{-1}	7×10^6 min^{-1} M^{-1}

(a)

(b)

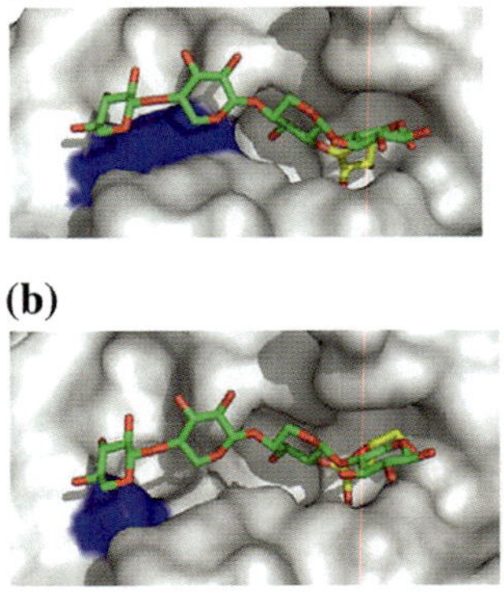

Fig. 12.1 An engineered arabinofuranosidase that is able to degrade xylan. *Panel* **a** Orthogonal views of the active site cleft of the wild type HiAXHd3 arabinofuranosidase shown as a molecular surface with the sugar ligand, coloured *green* (Xyl) or *yellow* (Ara) shown as *sticks*. The position of Y165 is shown in *blue*. *Panel* **b** The active site cleft of The Y166A mutant shown as for *panel* **a**, the removal of the Y166 side-chain increases the accessibility of the active site to xylan. *Panel* **c** shows the catalytic activity of wild type and the Y165A/F493A/E593A mutant of HiAXHd3

required to efficiently degrade plant cell walls. In a more generic sense the engineering of HiAXHd3 demonstrates that GH43 glycoside hydrolases provide a platform for generating bespoke multi-functional enzymes that target industrially significant, chemically complex, substrates, exemplified by the plant cell wall.

12.6 How Do We Look for Novel Plant Cell Wall Degrading Enzymes?

One of the major challenges facing the liquid biofuels industry is the cost of enzymes used in the process. As a result there has been a significant investment in enzyme discovery, in the hope of finding novel and superior (to current enzymes) plant cell wall degrading biocatalysts. Two potential strategies for mining existing genomic data is to identify proteins with a non-catalytic module that indicates a plant cell wall degrading function, but, also contains a domain that is >150 amino acids, which may comprise a novel enzyme. Recently, Bras et al. (2011) characterized a protein that is expressed at highly levels when *Clostridium thermocellum* is cultured on cellulose. The protein contains a ~250 residue sequence with no sequence similarity with any entries in the CAZy database, and a type I dockerin, a domain that integrates proteins into the cellulosome, a highly efficient plant cell wall degrading multienzyme complex. Analysis of the protein showed it to be an endo-β-1,4-glucanase that acts in synergy with the cellulosomal cellobiohydrolase. The crystals structure of the enzyme (Cel124A) in complex with cellotriose showed that the glycanase hydrolyzed glycosidic bonds at the interface between crystalline and amorphous regions of cellulose. Indeed, the specificity of Cel124 for precise structures within cellulose may explain why *C. thermocellulm* expresses such a large number of different endoglucanases. It is possible that the enzymes recognize different substructures of cellulose, which may explain why the cellulosome is one of the most efficient crystalline cellulose degrading systems known.

In addition to searching for non-CAZy sequences, enzymes with novel specificities can also be identified by interrogating the CAZy database for proteins that display features that are atypical of the family in which they are located. This approach has led the Fontes laboratory to identify a GH5 enzyme in which there is departure from the highly conserved amino acids that interact with O_3 of the sugar bound in the active site. The enzyme was characterized and shown to be an arabinoxylan specific xylanase, in which the specificity was conferred by a pocket that was linked β-1,3 to the xylose in the active site (Correia et al. 2011, Montanier et al. 2011). Thus, the productive binding energy to the arabinose, compensated for the loss of interactions with O_3 of the xylose bound in the active site, explaining why the enzyme only hydrolyzes arabinoxylan and not undecorated xylan. Significantly, the distal region of the substrate binding cleft is unusually open suggesting that the enzyme is capable of attacking highly decorated xylans. This has now been confirmed; the enzyme is able to attack corn stem xylan, an extremely complex form of the hemicelluloses, while typical GH10 and GH11 xylanases display no activity against this polysaccharide.

12.7 Plant Cell Wall Degrading Enzymes Display Complex Molecular Architectures

Microbial plant cell wall hydrolases display complex molecular architectures in which the catalytic module is appended, by flexible linker sequences, to one or more CBMs (reviewed in Boraston et al. 2004). CBMs have now been described that bind to the major polysaccharides found in plant cell wall structures (for review see Boraston et al. 2004). Although ubiquitous, the mechanism by which CBMs potentiate catalysis remains unclear, the most likely explanation is that they reduce the “accessibility problem” by simply bringing the appended catalytic modules into intimate association with their target substrate. Proteins that display a CBM-like function, termed Expansins, may offer advantages over classic CBMs. These proteins have been shown to mechanically weaken plant cell walls (McQueen-Mason and Cosgrove 1994), and their use in improving cellulase efficiency has been reported (Han and Chen 2007). Currently, expansins appear to disrupt the cellulose-hemicellulose interface, while the functional importance of Swollenin, another cellulose binding protein, remains opaque.

Type B CBMs generally bind to substrates of the catalytic modules. Exceptions to this rule include the CBM35s appended to three xylan degrading enzymes, which bind to both glucuronic acid (GlcA) and the unsaturated product released

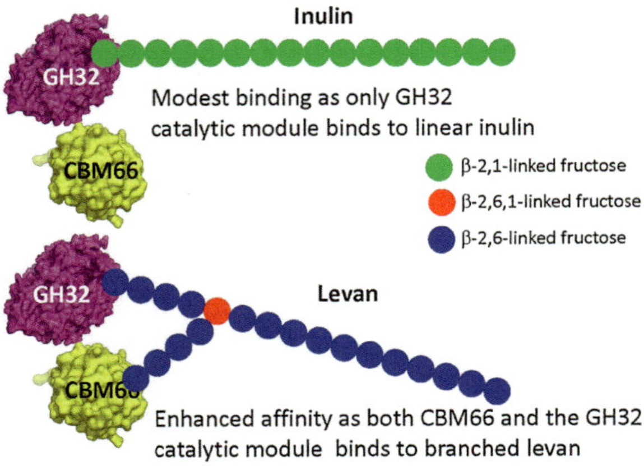

Fig. 12.2 Model for the role of BsCBM66 in exo-acting β-fructosidases. The binding of BsCBM66 to terminal fructofuranose residues enhances the activity of two GH32 broad-acting β-fructosidases against levan, but not against the other major fructan inulin. It is proposed that BsCBM66 facilitates the enzymatic targeting of branched substrates, such as levan, by binding in synergy to the terminal fructose residues of a branch structure leading to increased affinity through avidity effects. This avidity effect does not occur when the enzyme is attacking linear fructans such as inulin

by pectate lyases, but not to 4-O-methyl-D-glucuronic acid (MeGlcA), the more common uronic acid found in xylans. It has been proposed that the rate of GlcA methylation is lower than glucuronoxylan synthesis in rapidly dividing cells (Peña et al. 2007). This has led to the hypothesis that, by targeting unmethylated uronic, the CBM is directing enzymes to more open structures that are particularly susceptible to enzyme degradation. It is possible that this cohort of CBM35s initially direct the xylan degrading apparatus to regions of cell walls that are being actively degraded, for which anhydrogalacturonic is a marker, but, as xylan structures are revealed, the enzyme is shuttled onto the hemicellulosic polysaccharide affording the enzyme access to its target substrate (Montanier et al. 2009), a view consistent with recent data showing that these CBMs enhance degradation of non-methylated glucuronoxylans (Urbanowicz et al. 2012).

Recent studies by Cuskin et al. (2012) have shown that members of a new CBM family, CBM66, are exo binders recognizing the non-reducing end of branched fructan polysaccharides. When an exo-acting glycoside hydrolase that attacks fructans was linked to the CBM66 there was a 100-fold enhancement in activity again levan, a highly branched fructans, but not against inulin, a linear undecorated fructan. These data show that the CBM66 enhances catalytic activity by increasing the affinity of the enzyme for its substrate through an avidity effect (Fig. 12.2). Such a mechanism can only occur if the CBM and the appended catalytic module are binding to the same polysaccharide molecule, explaining why the potentiation only occurs for branched substrates. Within CBM66 are members appended to arabinofuranosidase families (GH43 and GH51), and thus this

CBM-mediated targeting of branched polysaccharides is highly relevant to plant cell wall degradation. Indeed, the technology developed around the CBM66 family provides a platform for engineering enzymes exposed to different evolutionary pressures to CBM-containing glycoside hydrolases.

12.8 Overcoming the Problem of Crystalline Cellulose with Cellulose Oxidases

The major barrier to deconstructing cellulose, the most abundant substrate available to the biofuel industry, is its highly crystalline structure. Glycoside hydrolases distort the sugar in the active site into its transition state conformation, and to achieve this substrate distortion, enzymes need to bind isolated chains of their target polysaccharides. As cellulose chains form highly crystalline structures it is not obvious how cellulases are able to channel isolated cellulose chains into their active sites. It has been shown by Koivula et al. (1998) that a tryptophan, at the entrance to the active site of the pivotal Trichoderma cellobiohydrolase, Cel6A, is essential for activity against crystalline cellulose, but is not required when the enzyme is acting on disordered or soluble substrates. It was proposed that the tryptophan intercalates between a cellulose microfibril and a surface cellulose chain, and the resultant "solubilised" glucan molecule can then be fed into the active site tunnel.

A more generic mechanism for attacking crystalline polysaccharides is the recent discovery of polysaccharide oxidases. These oxidases were first shown to oxidize and thus disrupt the crystalline structure of chitin, making the polysaccharide highly accessible to chitinases (glycoside hydrolases), and thus greatly potentiating the activity of the glycoside hydrolases (Vaaje-Kolstad et al. 2005, 2010). The role of oxidases has now been extended to cellulose where cellulose oxidases mediate a similar disruptive oxidation of the polysaccharide, increasing its access to cellulase action (Forsberg et al. 2011; Quinlan et al. 2011). These enzymes contain a copper binding site, and the redox metal plays a central role in the oxidation reaction (Quinlan et al. 2011; Aachmann et al. 2012; Vaaje-Kolstad et al. 2012). Significantly, the oxidation reaction does not require substrate distortion and thus the enzymes can bind to the planar surface presented by crystalline polysaccharides such as cellulose, explaining their unique activity against high recalcitrant substrates. These oxidases represent a major advance in cellulose deconstruction, and comprises the disruptive C_1 factor proposed by Reese and colleagues (1950). It should be noted that while these oxidases play a critical role in the cellulases systems of both aerobic fungi (Harris et al. 2010; Quinlan et al. 2012) and bacteria (Forsberg et al. 2011), they are absent in the corresponding anaerobic cellulase complexes. Despite the lack of oxidative enzymes anaerobic cellulase consortia, exemplified by the *Clostridium thermocellum* cellulosomes (Fontes and Gilbert 2010), display similar catalytic efficiencies against crystalline substrates to the corresponding aerobic fungal systems (Ding et al. 2012). How the anaerobic cellulase systems overcome the lack of an oxidative enzyme is currently unclear.

12.9 Future Perspectives

It is evident that the explosion of omics technologies, genomic, metagenomic and metatranscriptomic, provides us with an unrivalled opportunity for enzyme discovery programmes. Indeed the power of this information is revealed by the important recent discovery of polysaccharide oxidases. In the next few years the smart analysis of omics data, in conjunction with the further development of oxidases, and the continued use of structure based rational design, is likely to generate highly efficient lignocellulosic degrading enzyme cocktails. This portfolio of biocatalysts is likely to remove the rate limiting step in the use of plant biomass as an economically viable substrate for biofuel production.

References

Aachmann FL, Sorlie M, Skjak-Brak G, Eijsink VGH, Vaaje-Kolstad G (2012) NMR structure of a lytic polysaccharide monooxygenase provides insight into copper binding, protein dynamics, and substrate interactions. Proc Natl Acad Sci USA 109:18779–18784

Alhassid A, Ben-David A, Tabachnikov O, Libster D, Naveht E, Zolotnitsky G, Shoham Y, Shoham G (2009) Crystal structure of an inverting GH 43 1,5-alpha-L-arabinanase from Geobacillus stearothermophilus complexed with its substrate. Biochem J 422:73–82

Amano Y, Shiroishi M, Nisizawa K, Hoshino E, Kanda T (1996) Fine substrate specificities of four exo-type cellulases produced by *Aspergillus niger*, *Trichoderma reesei*, and *Irpex lacteus* on (1 → 3), (1 → 4)-β-D-glucans and xyloglucan. J Biochem 120:1123–1129

Armand S, Drouillard S, Schulein M, Henrissat B, Driguez H (1997) A bifunctionalized fluorogenic tetrasaccharide as a substrate to study cellulases. J Biol Chem 272:2709–2713

Bayer EA, Lamed R, Himmel ME (2007) The potential of cellulases and cellulosomes for cellulosic waste management. Curr Opin Biotechnol 18:237–245

Beguin P, Aubert JP (1994) The biological degradation of cellulose. FEMS Microbiol Rev 13:25–58

Boraston AB, Bolam DN, Gilbert HJ, Davies GJ (2004) Carbohydrate-binding modules: fine-tuning polysaccharide recognition. Biochem J 382:769–781

Bras JL, Cartmell A, Carvalho AL, Verze G, Bayer EA, Vazana Y, Correia MA, Prates JA, Ratnaparkhe S, Boraston AB, Romao MJ, Fontes CM, Gilbert HJ (2011) Structural insights into a unique cellulase fold and mechanism of cellulose hydrolysis. Proc Natl Acad Sci USA 108:5237–5242

Cantarel BL, Coutinho PM, Rancurel C, Bernard T, Lombard V, Henrissat B (2009) The Carbohydrate-Active EnZymes database (CAZy): an expert resource for glycogenomics. Nucl Acids Res 37:233–238

Cartmell A, Topakas E, Ducros VM, Suits MD, Davies GJ, Gilbert HJ (2008) The *Cellvibrio japonicus* mannanase CjMan26C displays a unique exo-mode of action that is conferred by subtle changes to the distal region of the active site. J Biol Chem 283:34403–34413

Correia MA, Mazumder K, Bras JL, Firbank SJ, Zhu Y, Lewis RJ, York WS, Fontes CM, Gilbert HJ (2011) Structure and function of an arabinoxylan-specific xylanase. J Biol Chem 286:22510–22520

Cuskin F, Flint JE, Gloster TM, Morland C, Basle A, Henrissat B, Coutinho PM, Strazzulli A, Solovyova AS, Davies GJ, Gilbert HJ (2012) How nature can exploit nonspecific catalytic and carbohydrate binding modules to create enzymatic specificity. Proc Natl Acad Sci USA 109:20889–20894

DeBoy RT, Mongodin EF, Fouts DE, Tailford LE, Khouri H, Emerson JB, Mohamoud Y, Watkins K, Henrissat B, Gilbert HJ, Nelson KE (2008) Insights into plant cell wall degradation from the genome sequence of the soil bacterium *Cellvibrio japonicus*. J Bacteriol 190:5455–5463

Ding SY, Liu YS, Zeng Y, Himmel ME, Baker JO, Bayer EA (2012) How does plant cell wall nanoscale architecture correlate with enzymatic digestibility? Science 338:1055–1060

Divne C, Stahlberg J, Reinikainen T, Ruohonen L, Pettersson G, Knowles JK, Teeri TT, Jones TA (1994) The three-dimensional crystal structure of the catalytic core of cellobiohydrolase I from *Trichoderma reesei*. Science 265:524–528

Ducros VM, Zechel DL, Murshudov GN, Gilbert HJ, Szabo L, Stoll D, Withers SG, Davies GJ (2002) Substrate distortion by a β-mannanase: snapshots of the Michaelis and covalent-intermediate complexes suggest a B(2,5) conformation for the transition state. Angew Chem Int Ed Engl 41:2824–2827

Fontes CM, Gilbert HJ (2010) Cellulosomes: highly efficient nanomachines designed to deconstruct plant cell wall complex carbohydrates. Annu Rev Biochem 79:655–681

Forsberg Z, Vaaje-Kolstad G, Westereng B, Bunaes AC, Stenstrom Y, MacKenzie A, Sorlie M, Horn SJ, Eijsink VGH (2011) Cleavage of cellulose by a CBM33 protein. Protein Sci 20:1479–1483

Gilad R, Rabinovich L, Yaron S, Bayer EA, Lamed R, Gilbert HJ, Shoham Y (2003) CelI, a noncellulosomal family 9 enzyme from *Clostridium* thermocellum, is a processive endoglucanase that degrades crystalline cellulose. J Bacteriol 185:391–398

Gilbert HJ, Stalbrand H, Brumer H (2008) How the walls come crumbling down: recent structural biochemistry of plant polysaccharide degradation. Curr Opin Plant Biol 11:338–348

Han YJ, Chen HZ (2007) Synergism between corn stover protein and cellulase. Enzyme Microb Technol 41:638–645

Han Y, Dodd D, Hespen CW, Ohene-Adjei S, Schroeder CM, Mackie RI, Cann IK (2010) Comparative analyses of two thermophilic enzymes exhibiting both β-1,4 mannosidic and β-1,4 glucosidic cleavage activities from Caldanaerobius polysaccharolyticus. J Bacteriol 192:4111–4121

Harris PJ, Stone BA (2008) Chemistry and molecular organization of plant cell walls. In: Himmel ME (ed) Biomass recalcitrance: deconstructing the plant cell wall for bioenergy. Blackwell, Oxford, pp 60–93

Harris PV, Welner D, McFarland KC, Re E, Poulsen J-CN, Brown K, Salbo R, Ding H, Vlasenko E, Merino S, Xu F, Cherry J, Larsen S, Lo Leggio L (2010) Stimulation of lignocellulosic biomass hydrolysis by proteins of glycoside hydrolase family 61: structure and function of a large, enigmatic family. Biochemistry 49:3305–3316

Himmel ME, Bayer EA (2009) Lignocellulose conversion to biofuels: current challenges, global perspectives. Curr Opin Biotechnol 20:316–317

Hogg D, Woo EJ, Bolam DN, McKie VA, Gilbert HJ, Pickersgill RW (2001) Crystal structure of mannanase 26A from Pseudomonas cellulosa and analysis of residues involved in substrate binding. J Biol Chem 276:31186–31192

Irwin D, Shin DH, Zhang S, Barr BK, Sakon J, Karplus PA, Wilson DB (1998) Roles of the catalytic domain and two cellulose binding domains of Thermomonospora fusca E4 in cellulose hydrolysis. J Bacteriol 180:1709–1714

Kleywegt GJ, Zou JY, Divne C, Davies GJ, Sinning I, Stahlberg J, Reinikainen T, Srisodsuk M, Teeri TT, Jones TA (1997) The crystal structure of the catalytic core domain of endoglucanase I from *Trichoderma reesei* at 3.6 Å resolution, and a comparison with related enzymes. J Mol Biol 272:383–397

Koivula A, Kinnari T, Harjunpaa V, Ruohonen L, Teleman A, Drakenberg T, Rouvinen J, Jones TA, Teeri TT (1998) Tryptophan 272: an essential determinant of crystalline cellulose degradation by *Trichoderma reesei* cellobiohydrolase Cel6A. FEBS Lett 429:341–346

McKee LS, Pena MJ, Rogowski A, Jackson A, Lewis RJ, York WS, Krogh KB, Vikso-Nielsen A, Skjot M, Gilbert HJ, Marles-Wright J (2012) Introducing endo-xylanase activity into an exo-acting arabinofuranosidase that targets side chains. Proc Natl Acad Sci USA 109:6537–6542

McKie VA, Black GW, Millward-Sadler SJ, Hazlewood GP, Laurie JI, Gilbert HJ (1997) Arabinanase A from Pseudomonas fluorescens subsp. cellulosa exhibits both an endo- and an exo- mode of action. Biochem J 323:547–555

McQueen-Mason S, Cosgrove DJ (1994) Disruption of hydrogen bonding between plant cell wall polymers by proteins that induce wall extension. Proc Natl Acad Sci USA 91:6574–6578

Mohnen D (2008) Pectin structure and biosynthesis. Curr Opin Plant Biol 11:266–277

Mohnen D, Bar-Peled M, Somerville C (2008) Biosynthesis of plant cell walls. In: Himmel ME (ed) Biomass recalcitrance: deconstructing the plant cell wall for bioenergy. Blackwell, Oxford, pp 94–187

Montanier C, van Bueren AL, Dumon C, Flint JE, Correia AM, Prates JA, Firbank SJ, Lewis RJ, Grondin GG, et al (2009) Evidence that family 35 carbohydrate binding modules display conserved specificity but divergent function. Proc Natl Acad Sci USA 106:3065–3070

Montanier CY, Correia MA, Flint JE, Zhu Y, Basle A, McKee LS, Prates JA, Polizzi SJ, Coutinho PM, Lewis RJ, Henrissat B, Fontes CM, Gilbert HJ (2011) A novel, noncatalytic carbohydrate-binding module displays specificity for galactose-containing polysaccharides through calcium-mediated oligomerization. J Biol Chem 286:22499–22509

Nurizzo D, Turkenburg JP, Charnock SJ, Roberts SM, Dodson EJ, McKie VA, Taylor EJ, Gilbert HJ, Davies GJ (2002) *Cellvibrio japonicus* α-L-arabinanase 43A has a novel five-blade β-propeller fold. Nat Struct Biol 9:665–668

Peña MJ, Zhong R, Zhou GK, Richardson EA, O'Neill MA, Darvill AG, York WS, Ye ZH (2007) Arabidopsis irregular xylem8 and irregular xylem9: implications for the complexity of glucuronoxylan biosynthesis. Plant Cell 19:549–563

Proctor MR, Taylor EJ, Nurizzo D, Turkenburg JP, Lloyd RM, Vardakou M, Davies GJ, Gilbert HJ (2005) Tailored catalysts for plant cell-wall degradation: redesigning the exo/endo preference of *Cellvibrio japonicus* arabinanase 43A. Proc Natl Acad Sci USA 102:2697–2702

Quinlan RJ, Sweeney MD, Lo Leggio L, Otten H, Poulsen JC, Johansen KS, Krogh KBRM, Jorgensen CI, Tovborg M, Anthonsen A, Tryfona T, Walter CP, Dupree P, Xu F, Davies GJ, Walton PH (2011) Insights into the oxidative degradation of cellulose by a copper metalloenzyme that exploits biomass components. Proc Natl Acad Sci USA 108:15079–15084

Ragauskas AJ, Williams CK, Davison BH, Britovsek G, Cairney J, Eckert CA, Frederick WJ Jr, Hallett JP, Leak DJ, Liotta CL, Mielenz JR, Murphy R, Templer R, Tschaplinski T (2006) The path forward for biofuels and biomaterials. Science 311:484–489

Reese ET, Siu RG, Levinson HS (1950) The biological degradation of soluble cellulose derivatives and its relationship to the mechanism of cellulose hydrolysis. J Bacteriol 59:485–497

Rouvinen J, Bergfors T, Teeri T, Knowles JK, Jones TA (1990) Three-dimensional structure of cellobiohydrolase II from *Trichoderma reesei*. Science 249:380–386

Sakon J, Irwin D, Wilson DB, Karplus PA (1997) Structure and mechanism of endo/exocellulase E4 from Thermomonospora fusca. Nat Struct Biol 4:810–818

Sorensen HR, Jorgensen CT, Hansen CH, Jorgensen CI, Pedersen S, Meyer AS (2006) A novel GH43 α-L-arabinofuranosidase from *Humicola insolens*: mode of action and synergy with GH51 α-L-arabinofuranosidases on wheat arabinoxylan. Appl Microbiol Biotechnol 73:850–861

Sticklen MB (2008) Plant genetic engineering for biofuel production: towards affordable cellulosic ethanol. Nat Rev Genet 9:433–443

Tailford LE, Ducros VM, Flint JE, Roberts SM, Morland C, Zechel DL, Smith N, Bjornvad ME, Borchert TV, Wilson KS, Davies GJ, Gilbert HJ (2009) Understanding how diverse β-mannanases recognize heterogeneous substrates. Biochemistry 48:7009–7018

Tomme P, Warren RA, Gilkes NR (1995) Cellulose hydrolysis by bacteria and fungi. Adv Microb Physiol 37:1–81

Teeri TT (1997) Crystalline cellulose degradation: New insight into the function of cellobiohydrolases. Trends Biotechnol 15:160–167

Tolonen AC, Chilaka AC, Church GM (2009) Targeted gene inactivation in *Clostridium* phytofermentans shows that cellulose degradation requires the family 9 hydrolase Cphy3367. Mol Microbiol 74:1300–1313

Urbanowicz BR, Pena MJ, Ratnaparkhe S, Avci U, Backe J, Steet HF, Foston M, Li H, O'Neill MA, Ragauskas AJ, Darvill AG, Wyman C, Gilbert HJ, York WS (2012) 4-O-methylation of glucuronic acid in Arabidopsis glucuronoxylan is catalyzed by a domain of unknown function family 579 protein. Proc Natl Acad Sci USA 109:14253–14258

Vaaje-Kolstad G, Horn SJ, van Aalten DMF, Synstad B, Eijsink VGH (2005) The non-catalytic chitin-binding protein CBP21 from Serratia marcescens is essential for chitin degradation. J Biol Chem 280:28492–28497

Vaaje-Kolstad G, Westereng B, Horn SJ, Liu Z, Zhai H, Sorlie M, Eijsink VGH (2010) An oxidative enzyme boosting the enzymatic conversion of recalcitrant polysaccharides. Science 330:219–222

Vaaje-Kolstad G, Bohle LA, Gaseidnes S, Dalhus B, Bjoras M, Mathiesen G, Eijsink VGH (2012) Characterization of the chitinolytic machinery of *Enterococcus faecalis* V583 and high-resolution structure of its oxidative CBM33 enzyme. J Mol Biol 416:239–254

van den Broek LA, Lloyd RM, Beldman G, Verdoes JC, McCleary BV, Voragen AG (2005) Cloning and characterization of arabinoxylan arabinofuranohydrolase-D3 (AXHd3) from *Bifidobacterium adolescentis* DSM20083. Appl Microbiol Biotechnol 67:641–647

Varrot A, Schulein M, Davies GJ (1999) Structural changes of the active site tunnel of *Humicola insolens* cellobiohydrolase, Cel6A, upon oligosaccharide binding. Biochemistry 38:8884–8891

Weiner RM, Taylor LE 2nd, Henrissat B, Hauser L, Land M, Coutinho PM, Rancurel C, Saunders EH, Longmire AG, Zhang H, Bayer EA, Gilbert HJ, Larimer F, Zhulin IB, Ekborg NA, Lamed R, Richardson PM, Borovok I, Hutcheson S (2008) Complete genome sequence of the complex carbohydrate-degrading marine bacterium, *Saccharophagus degradans* strain 2-40 T. PLoS Genet 4:e1000087

Xie G, Bruce DC, Challacombe JF, Chertkov O, Detter JC, Gilna P, Han CS, Lucas S, Misra M, Myers GL, Richardson P, Tapia R, Thayer N, Thompson LS, Brettin TS, Henrissat B, Wilson DB, McBride MJ (2007) Genome sequence of the cellulolytic gliding bacterium Cytophaga hutchinsonii. Appl Environ Microbiol 73:3536–3546

Chapter 13
Using Natural Plant Cell Wall Degradation Mechanisms to Improve Second Generation Bioethanol

Adriana Grandis, Amanda P. de Souza, Eveline Q. P. Tavares and Marcos S. Buckeridge

Abstract Cell wall hydrolysis is one of the key processes needed for development of the technology for second-generation (2G) bioethanol production. Thus, finding and characterizing enzymes that can deal with the complexity of the walls has been the main focus of research. As a result, data on pretreatments of many kinds and performances of enzyme cocktails containing glycosyl hydrolases from microorganisms are becoming quickly available. Here we propose that the efficiency of the 2G process could be increased even further by acquiring control of mechanisms that plants themselves use to degrade their own walls, so that wall loosening provoked by such processes would decrease the energy demand for pretreatments and facilitate hydrolysis. The examined in this chapter are plant-microorganism interaction, cell wall storage mobilization, fruit ripening, abscission, and aerenchyma formation. These systems are seen as having in common the use of *modules* that are coupled sequentially in order to lead to cell wall modification, including hydrolysis, for performance of different biological functions. These modules are (1) target cells perception of a message from the hormonal balance, (2) cell separation, (3) cell expansion, (4) programmed cell death, (5) hemicellulose-cellulose relaxation/hydrolysis and (6) cellulose hydrolysis. We propose that the use of synthetic biology to transform bioenergy feedstocks could be a route to increase the efficiency of 2G processes.

13.1 Introduction

The main bioethanol producers in the world are the US and Brazil, the former producing it from maize starch and the second from sucrose from sugarcane culms. In order to increase bioethanol production without increasing acreage, second

A. Grandis · A. P. de Souza · E. Q. P. Tavares · M. S. Buckeridge (✉)
Laboratory of Plant Physiological Ecology, Department of Botany, Institute of Biosciences, University of São Paulo, São Paulo, Brazil
e-mail: msbuck@usp.br

M. C. McCann et al. (eds.), *Plants and BioEnergy*, Advances in Plant Biology 4, DOI: 10.1007/978-1-4614-9329-7_13,

generation (2G) bioethanol technologies could be used. These technologies consist in the production of free fermentable sugars from cell walls. To access these sugars it is of great importance the production of basic scientific information about what the cell walls are like and how enzymes attack each polymer as well as the entire composite.

The production of bioethanol by means of 2G technologies involves the use of strategies that should modify the organization of cell wall architecture. The so-called 2G route involves biomass pretreatments in which polysaccharides would become available for enzymatic hydrolysis by application of physical and/or chemical treatments (Soccol et al. 2010; Dos Santos et al. 2011).

Moreover, due to the complex nature of the plant cell wall, degradation of lignocellulosic biomass requires the use of different classes of enzymes that have to be used in high proportions in order to produce enough free sugars for efficient fermentation (Verma et al. 2010). The production of 2G bioethanol requires ca. 11 millions of filter paper units (FPU) (i.e. 19 kg) to produce 84 L of bioethanol (Himmel et al. 1997, 1999) or 15–25 kg of cellulase per ton of biomass (Carroll and Somerville 2009; Taylor et al. 2008). Furthermore, due to the different compositions of the cell wall polymers, it would still be necessary to prospect for individually different classes of enzymes for characterization and subsequently combine them into cocktails that should be specific for every type of biomass, including the pretreated ones. Thus, one of the first challenges to be faced in order to turn 2G bioethanol viable commercially is to develop efficient systems of enzymes for biomass degradation (Verma et al. 2010). Due to the high costs and limited capacity for enzyme production, new ways to make biomass available for fermentation could be helpful, since they could hopefully lead to the use of lower proportions of enzymes in the process.

The proposed way to couple pretreatment and hydrolysis has been the use of enzyme cocktails on cell walls of pretreated materials (Fig. 13.1). Such enzyme cocktails are usually artificial assemblies of extracellular proteins produced by microorganisms (Balat 2011). In fact, many initiatives exist to use microorganisms as producers of heterologous glycosyl hydrolases so that enzyme engineering could be used to improve their action.

The sources of enzymes chosen by the scientific community varies, but with the advent of metagenomic techniques, microorganisms and animal genes that encode for glycosyl hydrolases became the main focus for enzyme search in the hope to find new targets that could be somehow more efficient to hydrolyze plant cell walls. This strategy relies mainly on the idea that microorganisms that perform composting of plant residues in the environment and some animal digestive systems are capable to hydrolyze the plant cell walls (Gómez et al. 2008).

Some initiatives exist in which enzymes coding genes from microorganisms have been cloned into plants in order to try to loosen walls and turn them more amenable to hydrolysis (Kaida et al. 2009; Xin et al. 2011). Although expression of the enzymes in vivo have been demonstrated, there is no evidence that these biomasses would be more easily hydrolysed and/or pretreated than biomass derived from untransformed plants.

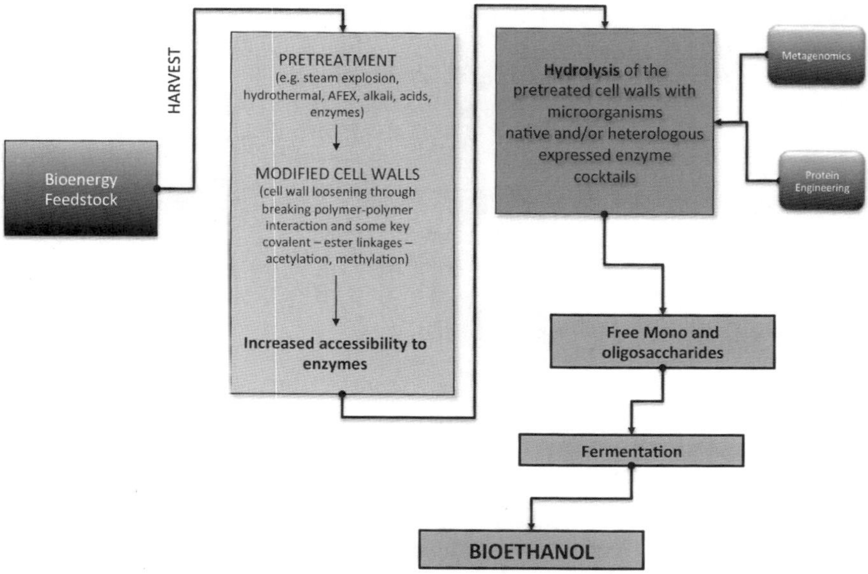

Fig. 13.1 Schematic representation of the second-generation process of bioethanol production, highlighting the different types of pretreatments and hydrolysis

Thus, even with the large effort of the scientific community world wide to improve pretreatment and hydrolysis processes, cell wall recalcitrance to hydrolysis continues to be a barrier to turn the 2G processes economically viable (Himmel et al. 2007). A common mistake has been to attribute recalcitrance solely to the presence of lignin. However, this recalcitrance is also due to the enormous complexity of the architecture of the cell wall with its several polymers integrated in a composite that has been selected in a way to avoid enzyme attack and degradation.

One additional strategy to be used to improve the knowledge about 2G processes that has not been consistently thought of, or experimented with the same intensity, is the use of hydrolytic systems that plants themselves possess. Some of them can be quite efficient to degrade and/or change the architecture of cell walls and might lead to important additions to 2G processes, especially with the advent of the synthetic biology era, in which the capacity of changing biological systems (e.g. turning on/off entire biochemical pathways) is expect to be developed.

Similarly to the strategies currently being designed in which biological engineering is going to be used within microorganisms to produce more efficient enzymes, plant biotechnology could help not only by the introduction of genes of animals and/or microorganisms, but also by finding mechanisms of cell wall hydrolysis that already exist in nature. The main mechanisms of this kind known are: (1) plant-microorganism interaction; (2) seed cell wall storage mobilization; (3) fruit ripening, (4) abscission and (5) aerenchyma formation.

In this chapter, we will provide a review of aspects of some representative mechanisms that could be relevant to improve plants so that they would be better prepared for pretreatment and hydrolysis. Here, this process will be named *biological pretreatment*. The idea is that in the future, plants could have some of their hydrolytic processes redesigned to alter some features of the cell walls in the way of turning them more suitable for use in industry, without however loosing the perspective of the technologies current under development for 2G bioethanol production.

In order to understand how walls can be hydrolysed, some aspects of cell wall composition, structure and architecture will be introduced in the following section. The understanding of the cell wall complexity will allow the reader to appreciate how some natural systems in which walls are modified could be used for biotechnology applications in bioethanol production.

13.2 Composition, Structure and Architecture of the Plant Cell Wall

The cell walls are composed of a mixture of polymers, mainly carbohydrates, proteins and secondary metabolites. These polymers interact through a mixture of covalent and non-covalent linkages to form a "supramolecular complex" that is thought to be a multifunctional structure responsible for controlling the mechanical properties of the cell.

In order to elucidate plant cell wall structure and functions, wall scientists have followed some key steps mainly during the second half of the 20th Century. The first step was to set up methods for extraction, determination of composition and structure of wall components. This step was followed by the development of methods to determine the structure (linkage types) and physico-chemical properties of these polymers. All this work culminated in the proposition of a succession of models that pictured a general assembly for the polymers in the wall (Keegstra et al. 1973; McCann and Roberts 1991; Carpita and Gibeaut 1993). Likewise, many aspects of the biochemistry of the wall were developed, culminating in the characterization of several glycosyl hydrolases (e.g. Fry 2004) that correlate with physiological events in plants, which are controlled by plant hormones and environmental cues. With the advance of cellular and molecular biological techniques, part of the wall research community invested in finding and correlating wall structure with genes associated with its metabolism (e.g. Holland et al. 2000; Lima et al. 2001) and also succeeded to find probes (enzymes and monoclonal antibodies) capable to localise wall components in the cell (e.g. Willats et al. 2001; Pattathil et al. 2010).

McCann and Roberts (1991) coined the term "architecture" of the cell wall in a reference to the existence of three different independent domains in the wall: pectin, cellulose-hemicellulose and proteins. This view implies that the three domains might interact and this sole interaction would have emergent properties. Carpita and Gibeaut (1993) showed that at the level of higher plants, the wall structure has different polysaccharides (e.g. arabinoxylan, β-glucan, xyloglucan and mannan) that may play similar functions. Thus, the "architectural" model of the wall points out that cell wall structure is degenerated and that the domains are changing during evolution, i.e. monocots and ferns changed their main hemicelluloses from xyloglucan to arabinoxylan and mannans respectively, with a concomitant decreased pectin domain (Buckeridge 2006; Sarkar et al. 2009).

In higher plants, cellulose, (1 → 4)-β-D-glucan, is the most abundant biological material on earth and in dry wood correspond around of 40–50 % of dry mass. Hemicellulose is the second polysaccharide representing 25–35 % of dry mass and their composition is a mixture of various polymerised monosaccharides, such as xylose and arabinose or glucose, mannose and galactose whose proportions change according to the plant tissue and the species.

The principal hemicelluloses found in nature are xyloglucan, galactomannan or galactoglucomannan, β-glucan and glucuronoarabinoxylan (GAX). Xyloglucan is a major hemicellulose in the primary walls of flowering plants, all non-monocots and about one-half of monocot species. Some seeds store this polysaccharide in large amounts as a resource for the embryo after germination (Buckeridge et al. 2000a). Xyloglucans typically comprise (1 → 4)-β-D-glucan backbones, with three of every four glucose residues substituted with (1 → 6)-α-D-xylopyranosides, and some of the xylose residues are further substituted with (1 → 2)-β-D-galactopyranosides. The most widespread technique used for studies on xyloglucan has been the digestion with a *Trichoderma* endo-β-glucanase (Anderson and Stone 1975; Tiné et al. 2006). This enzyme hydrolyses xyloglucan polymers only on unbranched glucosyl residues.

The cell walls from grasses (Poales), which are the main plants used in bioenergy, have a distinct architecture when compared with non-monocots and non-Commelinid monocots (Carpita and Gibeaut 1993). Besides the fact that low levels of pectin are present in their walls, grasses contain as their main hemicelluloses the mixed-linked (1 → 3), (1 → 4)-β-D-glucan (β-glucan) and glucurono-arabinoxylan (GAX).

β-Glucan interlaced by GAX of low degree of substitution and glucomannans, is thought to be tightly associated with the cellulose microfibrils. GAX of higher degree of arabinosyl substitution and some glucomannan constitutes the major pore-determining interstitial material between the microfibrils (see below).

One of the principal enzymes involved in the depolymerization of the endosperm β-glucan is the (1 → 3), (1 → 4)-β-D-glucan endohydrolase, whose activity parallels that of the *B. subtilis* enzyme in that hydrolysis is restricted to (1 → 4)-β-linked glucosyl residues adjacent to (1 → 3)-β-linked residues on the non-reducing side (Parrish et al. 1960). Like the *B. subtilis* endohydrolase, hydrolysis of β-glucan with these enzymes yields mostly cellobiosyl- and cellotriosyl-(1 → 3)-glucosyl oligosaccharides and smaller amounts of larger cellodextrin-(1 → 3)-glucose oligomers (Anderson and Stone 1975).

In GAX, the arabinose residues are attached at the *O*-3 positions along the (1 → 4)-β-linked xylan backbone, and the glucuronic acids are attached to the *O*-2 positions (Carpita 1996). Arabinoxylans are widespread in the walls of all flowering plants, but in nongramineous species the polymer is of much lower abundance, and the arabinose residues are attached mostly at the *O*-2 rather than the *O*-3 of the xylosyl units.

A highly substituted GAX (HS-GAX), with six of seven xylosyl branched units is associated with the maximum growth rate of coleoptiles of the grass species *Lolium multiflorum* and a sequence-dependent xylanase was found that requires branching with GlcA to cleave the neighboring (1 → 4)-α-D-xylosyl linkage (Nishitani and Nevins 1991). When maize GAXs are depleted of arabinosyl units by mild-acid hydrolysis, this endo-β-D-xylanase releases a homogeneous group of deca- or undecamers of glucuronoxylan. The xylosyl units are also substituted with acetyl groups at the positions *O*-2 and *O*-3 and some methyl groups have been detected (Carpita 1996).

Porosity of the GAX domain could be determined by the extent of removal of the appendant units. Some highly substituted GAX remain intercalated in the small amount of pectins that are also found in the primary wall. Because β-glucans are thought to be tightly associated with individual cellulose microfibrils (Carpita et al. 2001), GAX might be the main molecule in the interstitial space. However, Buckeridge and Crivellari (unpublished) have recently performed experiments with sugarcane walls that did not support this hypothesis. When isolated walls (alcohol insoluble residues) of leaves and culms of sugarcane were treated with lichenase (a glycosyl hydrolase that attacks specifically β-glucan), most of the polymer present in the walls was hydrolyzed and, at the same time, its retrieval did not interfere in the action of endo β-xylanase on the wall. Also, De Souza et al. (2013) observed that β-glucans in sugarcane walls are lightly associated with cellulose and/or the cellulose-hemicellulose matrix in the wall. These data suggests that in sugarcane, although β-glucan is present in some association with other polymers in the wall, they seem to be free for enzyme hydrolysis whereas GAXs are likely to be strongly bound to other polymers, probably cellulose.

Another distinctive feature of Poales and their relatives is the enrichment of aromatic substances in nonlignified walls (Carpita 1996). A large portion of the aromatic substances is esters of hydroxycinnamates (Harris and Hartley 1980). The GAXs are cross-linked in walls by both esterified and etherified hydroxycinnamates and by other phenolic substances, as ferulic acid (Iiyama et al. 1994).

On the basis of the knowledge of the structure of the cell wall polysaccharides, the mode of action of the glycosyl hydrolases and also the genes that encode for these enzymes, it is now possible to design routes that can deal with the complex hydrolysis mechanisms of either the less complex seed polysaccharide degrading systems or even to access more complex wall systems such as the ones from grasses. The latter is not an easy task as relatively little knowledge exists about the composition, structure and architecture of the wall of most bioenergy feedstocks.

The cell walls of one of the most important bioethanol feedstocks, sugarcane, has been recently analysed by De Souza et al. (2013), who demonstrated that hemicelluloses are quantitatively the main components of the cell walls. Regarding carbohydrates alone, whereas hemicelluloses account for ca. 60 % of the walls, cellulose makes around 30 %, and pectins about 10 %. Hemicelluloses are of three types: arabinoxylan, β-glucan and xyloglucan, the former being about

50 % of the hemicelluloses with the other two types sharing the rest in equal proportion. These authors highlighted the elevated complexity of cell walls and proposed a model for enzyme hydrolysis in which phenylpropanoids in the wall would have to be retrieved first in order to open the way for carbohydrate specific enzymes, pectins being the first to be attacked, followed by the members of the cellulose-hemicellulose matrix.

In summary, cell wall hydrolysis for bioenergy production purposes can be thought of as a procedure capable to disassemble cell wall architecture so that individual polysaccharides would be available to hydrolases. The goal of the process is the production of fermentable monosaccharides. On the other hand, even if the cell wall architecture can be dismantled, the polysaccharides released would still offer the challenge related to their branching patterns, which are different for every polysaccharide, resulting in a rather large diversity of combinations that are named fine structures.

Although the structure of cell wall polymers have been studied for several species, less attention is given to fine structure studies. Since the fine structure of polysaccharides could involve a formation of a glycomic code and interfere in hydrolysis efficiency (De Souza and Buckeridge, unpublished), the understanding of branching patterns and its combinations could be an important issue for 2nd generation processes (see topic 13.4).

In order to illustrate how natural systems could be helpful in bioenergy sector, in the following sections we will discuss five major systems existent in nature, which include cell walls modifications by enzymes produced for specific situations. It must be highlighted here that these items do not intend to review every biological process exhaustively, but only give key information that are relevant for the mechanisms of cell wall modifications (including hydrolysis) that could be used to transform plants and make them more suitable as bioenergy crops.

13.3 Cell Wall Degradation During Plant-Microorganism Interaction

When a pathogen invades a plant tissue, a biochemical conflict takes place, in which host and invader will produce a series of substances, the former to defend itself and the latter to penetrate in the cells (Garcia-Brugger et al. 2006). Among the strategies used by both organisms there are cell wall hydrolases (Walton 1994). In this way, the process of plant-microorganism interaction offers an opportunity to find ways to efficiently disassemble the plant cell wall. Both, saprophytic and pathogenic microorganisms produce extracellular enzymes that can degrade plant polysaccharides. These microorganisms, including bacteria and fungi, and also invertebrates as nematodes can digest cell wall polymers in order to obtain sugars and energy for their growth as well as to penetrate and colonize the cells in a plant tissue (Walton 1994; Annis and Goodwin 1997). In most cases, plants will respond with a reaction named hypersensitivity, which includes programmed cell death and with that can restrict the infectious region to a small area surrounding the infection. However, in these cases, pathogens can use this opportunity to feed on dead cells and gain nutrients and with that they may become even more invasive (Greenberg and Yao 2004).

The first class of cell wall hydrolases produced by the pathogen is the pectinases (Tomassini et al. 2009). They are thought to loosen the tissue through digestion of the middle lamella and consequently kill the plant cells that they attack (Cooper 1983). The pectinases known to be involved in fungal and bacterial invasion are endo- and exo-polygalacturonases, pectate-lyases, and pectin methylesterases (Walton 1994; Lagaert et al. 2009). It is also believed that even saprophytes need to use cell wall degradation strategies in order to establish interaction with the plant and for that they also produce pectinases (e.g. Marques et al. 2006).

Pectins are polysaccharides that control cell wall porosity. Some studies demonstrated that the porosity of the substrate walls is the main limiting factor in the enzymatic hydrolysis of lignocellulosic biomass (Chandra et al. 2007). Thus, the knowledge about how microorganisms use the pectinases to invade the plant could be a way to improve the biomass hydrolysis to 2G bioethanol.

Futhermore, several other enzymes that are able to hydrolyse cell walls can be produced by microorganisms. Recently, the need for enzymes to be used in the production of bioethanol from biomass led many groups to produce a large amount of scientific information about glycosyl hydrolases from microorganisms. The isolation and characterization of enzymes as well as cloning their genes, open a way to engineering those derived from microorganisms, i.e. modify them to perform hydrolysis more efficiently (Serpa and Polikarpov 2011; Ward 2011). Then, new cocktails of enzymes could be designed that would efficiently produce free fermentable sugars for bioethanol production. A review of the enzymes from microorganisms used for bioenergy has been recently published (Polizeli et al. 2011).

13.4 Seed Cell Wall Storage Mobilization

Cell Wall Storage Polysaccharides (CWSP) are found as the principal storage compound in seeds of many taxonomically important groups of plants. These groups developed extremely efficient biochemical mechanisms to disassemble cell walls and use the products of hydrolysis for growth. During evolution, several different groups of polymers have been selected as reserves (e.g. mannans, galactomannans, galactoglucomannans, arabinogalactans and xyloglucans) (Buckeridge et al. 2000a, 2000b; Buckeridge 2010). The selective pressures that ended up producing CWSP were directed towards an increase their proportion by introducing a storage cell wall between already existent primary walls (Tiné, Braga, Pattathil, Hahn and Buckeridge, unpublished).

The CWSP have different chemical structures that are related with different biological functions. Mannans and galacto- and glucomannans that are present in palm and coffee seeds and lettuce and tomato respectively are insoluble in water and display strong intermolecular interaction. Because of these characteristics, their biological function is usually associated with conferring hardness to plant tissues. Endo-β-mannanase is the principal enzyme involved in mannan hydrolysis in all species studied and several genes have been cloned whose expression is specific to the seed (Bewley et al. 1997; Lisboa et al. 2006; Gong and Bewley 2007).

In legumes, the main function of the endospermic cell walls appear to be storage, with the yield of galactomannan reaching more than 30 % of the seed dry weight in many species (Buckeridge et al. 2000a). These walls are thickened with galactomannan and in certain cases (e.g. *Trigonella foenum-graecum* and *Schyzolobium parayba*) the protoplasm disappears, giving place to the storage wall. In these cases, the endosperm is non-living and degradation is performed by three enzymes (α-galactosidase, endo-β-mannanase and exo-β-mannosidase) made in the aleurone layer (Reid 1971; Buckeridge and Dietrich 1996).

Differently from being stored in the endosperm, arabinogalactan and xyloglucans are stored in cotyledons. This is extremely relevant because the cotyledon is, evolutionarily speaking, an adapted leaf and the integration of the metabolism of these polysaccharides with the entire plant is likely to preserve several features of the leaf cell walls.

Pectin polymers are also found as cell wall storage polysaccharides, notably arabinogalactan (Buckeridge et al. 2000b). The cotyledons of lupins (especially *Lupinus angustifolius*) have been used as a model to study mobilization of this polymer.

Lupinus angustifoluis seeds accumulate high proportions of galactan (Buckeridge and Reid 1994). These authors showed that an exo-β-(1,4)-galactanase purified from *L. angustifolius* seed acts specifically on this type of galactan. This enzyme is analogous to galactanases found in fruits (TBG4 in tomato for example—see below) where it is thought to play a major function in the changes of porosity in cell walls (Brummell 2006).

Of the CWSP known, xyloglucan seems to be the more complex in most senses. Its basic structure is similar to the primary wall xyloglucans. They have a backbone composed of (1 → 4)-β-linked glucan (like cellulose) with regular branching with (1 → 6)-α-linked xylosyl residues that can be branched further with (1 → 2)-β-linked galactosyl residues. There is no case reported where fucose was detected in storage xyloglucan. Four enzymes responsible for storage xyloglucan degradation have been detected, purified and characterized (α-xylosidase, β-galactosidase, β-glucosidase and xyloglucan-endo-β-glucanase, latter on renamed xyloglucan transglucosylase/hydrolase (XTH) (Buckeridge et al. 2000a; Buckeridge 2010).

In *Hymenaea*, Buckeridge et al. (1997) discovered that storage xyloglucan contains two types of constitutive blocks in the main chain, i.e., XXXG and XXXXG. Later on, Tiné et al. (2006) showed that these blocks are not combined randomly in the polymer molecules, which led to the proposition that storage xyloglucans might have a combination of branching patterns that forms a code that have to be decrypted by the enzymatic system in order to be hydrolysed (Buckeridge 2010). Indeed, the fact that branching of cell wall polymers might not be random has been also observed early on for galactomannans (Reid and Edwards 1995).

As this kind of information is not available in most cases, research should be urgently directed to that in order to complement the efforts in the search for new enzymes as well as enzyme engineering.

13.5 Fruit Ripening

During ripening, the cells of fruits display changes in cell walls so that the final result will afford the release of seeds for dispersion. Several species have been adapted to human consumption and became sources of food, being nowadays extremely important in agriculture. That is the case of tomato, apple, pear, melon, papaya, grape to name but a few. Due to the economic importance of fruits, development and ripening have been intensively studied, with tomato being one of the most studied fruit regarding studies of cell wall changes related to texture (Brummell and Harpster 2001; Brummell 2006). Changes in fruit texture that occur during ripening in many cases lead to softening, this process being slightly different for each species (Harker et al. 1997; Brummell 2006). Nonetheless, in all cases the texture is related to changes in cell walls that are associated with the production of several hydrolases. Due to the fact that enzymes can act to hydrolyse fruit cell wall polysaccharides and/or to allow structural changes that modify the mechanical properties of plant tissues, fruit systems can be considered as an opportunity to find mechanisms of cell wall degradation in plants that could help developing biotechnology for bioenergy.

In fruit systems, the most studied degradation process is the activity of endopolygalacturonases, pectin methylesterases and β-galactosidases on pectic matrix of the walls. Together with the detection of enzyme activities, changes in cell walls of tomato are consistent with a large decrease in uronic acid as well as even larger decreases in galactose in wall fractions (Ahmed and Labavitch 1980; Rose et al. 1998; Brummell and Harpster 2001; Brummell 2006).

It has been demonstrated that pectinases are associated with ripening and softening through degradation of endopolygalacturonases present in the middle lamella. The action of this enzyme is thought to lead to cell separation, a phenomenon that is linked to the subsequent softening of the fruits and change in their texture and taste (Fray and Grierson 1993; Barsan et al. 2010).

The xyloglucan-cellulose network has also been pointed out as an important target in ripening (Rose and Bennett 1999; Brummell 2006). These authors highlight the role of expansions and XTHs in the process and the former authors even compare the process of fruit ripening with the cell expansion system in hypocotyls. Brummell (2006) reviewed fruit ripening from the viewpoint of the wall changes, putting forward the idea that whereas pectin degradation varies among different fruit species (Brummell 2006), the xyloglucan-cellulose network seems to undertake changes that are much more consistent among species. However, the author also highlights the fact that, contrarily to what has been observed for pectins, xyloglucan small fragments have not been detected. Thus, the effects of expansin and enzymes on the xyloglucan-cellulose matrix seem to be more towards a relaxation of this matrix rather than hydrolysis. This seems to lead to cell wall swelling that happens concomitantly with pectin solubilization.

The increase in pore sizes, thought to be related to the attack of exo-β-galactanase in pectin matrix (Brummell 2006), probably opens the way for expansions, XTHs, cellulases, as well as other debranching enzymes, that will attack the cellulose-hemicellulose matrix of the wall, finally changing wall texture as a whole. Indeed, according to Carpita et al. (1979), pore sizes of the walls in plant tissues would be in the range of 35–40 Å, whereas the glycosyl hydrolases stoke radius [e.g. a 20 kDa protein would be approximately 2.5 nm according to Carroll and Somerville (2009)] is in general higher than the pore sizes, so that this is a clear limitation for penetration of enzymes in the cell wall matrix.

Although endo-β-glucanases have been found to increase during fruit ripening, very little evidence has been produced that cellulose would be hydrolysed. However, it has been speculated that cellulases would be capable to act on the surface of microfibrils (Brummell 2006) without, however, any assignment of a function for this event.

In spite of the fact of hydrolysis, in a broad sense, being the main process in course during fruit ripening, other steps such as cell separation and expansion, have been observed (Brummell 2006). On the other hand, differently from some other cell wall degradation processes such as aerenchyma formation and abscission (see below), fruit ripening seems do not include programmed cell death.

From a general point of view, fruit ripening is a process in which walls are firstly attacked by pectinases that increases wall porosity, probably allowing the action of other enzymes like XTHs and endo-β-glucanases that will transform the cellulose-hemicellulose matrix, promoting swelling and increasing even more wall porosity. The extensins that bind to the cellulose-hemicellulose matrix of the wall also could participate of these processes, being probably the first proteins to interact with the wall matrix. Thus, the cell wall alteration plus a higher hydration capacity given by the presence of pectins with lower molecular weight modifies fruits softening.

All this knowledge is extremely useful for bioenergy purposes, as it makes possible to understand details of how plants can manipulate their own walls. These events could be compared with the ones that occur during biomass pretreatment process used to achieve the 2G bioethanol. The possibility of applying such knowledge in bioenergy crops to produce similar effects in grass stems cell walls (e.g.) could lead to the establishment of a biological pretreatment that could be implemented using synthetic biology tools in the future.

13.6 Abscission

Abscission is a general term applied to the processes that takes place during the detachment of leaves, fruits, seeds and also cotyledons, lateral branches and many reproductive structures. In all cases studied, extremely precise biochemical pathways lead to cell wall modification and breakdown in order to form a fracture line known as the abscission zone. The events that occur during abscission and dehiscence processes are cleary related to cell wall modification, in a very similar way to the other ones reported in this chapter.

Roberts et al. (2002), highlight the facts that endopolygalacturonases and endo-β-glucanases are closely related to the processes of cell separation. In fruits like tomato, silencing of abscission-related polygalacturonases (e.g. TAPG1) led to a delay in abscission and increased the force needed to break the abscission zone in explants treated with ethylene (Jiang et al. 2008).

In the abscission zone, cortex cells of different plant tissues are thought to be positionally differentiated target cells that will receive the signal from hormones (ethylene, auxin, abscisic acid, gibberellin) in such a balance that the cross talk among these will lead to transdifferentiation and finally to abscission. McManus et al. (1998) demonstrated the existence of transdifferentiation in mature cortical cells of bean abscission tissues. According these authors, this means that cells from the cortex of some plant tissues have the flexibility to differentiate into functionally competent ethylene-responsive cells that exhibit a gene expression compatible with an abscission cell.

Although not reported in most cases, cell wall degradation is thought to occur during abscission, since detection of activity of endo-β-glucanase has been reported for many abscission systems (Roberts et al. 2002). Also, accumulation of mRNA related with genes encoding Cel1 and Cel4 in tomato fruit and flower abscission zones have been reported (Gonzáles-Carranza et al. 1998).

The similarity of this process with the ones described above is remarkable. Abscission is widespread in plants, occurring in many different tissues of different organs in plants. It includes wall separation and expansion, which are associated with pectin degradation as well as modifications in the cellulose hemicellulose matrix. However, programmed cell death has not been reported for abscission processes, according to the literature revised for this chapter. Thus, there seems to be a great variety of combinations of action of hydrolases to perform cell separation and hydrolysis so that it could be used to design technologies associated to 2G bioethanol production.

13.7 Aerenchyma Formation

Aerenchyma comprises a series of interconnected gas chambers developing on parenchymatic tissue (cortex) of shoots and roots as a result of sequencial events, showing typical features of a programmed cell death process. Its formation is regarded as a response to oxygen shortage (Gunawardena et al. 2001a), exogenous ethylene (Gunawardena et al. 2001b), nitrogen, phosphorus and/or sulphur starvation (He et al. 1994; Siyiannis et al. 2011), mechanical impedance (He et al. 1996b), oxidative stress (Steffens et al. 2011) and osmotic stress (Karahara et al. 2012).

In opposition to the inducible aerenchyma, the constitutive formation is commonly observed in aquatic species as *Juncus effusus* (Visser and Bögemann 2006) and *Sagittaria lancifolia* (Schussler and Longstreth 2000), and also in maize relatives (Mano et al. 2007) and in rice wetland species (Justin and Armstrong 1991), independently of environmental stimuli. In both cases, its development lies upon a cell separation process termed schizogeny (thought of as a cell separation process that does not include cell wall hydrolysis). Alternatively, for the lysogenic process (i.e. cell separation followed by cell wall hydrolysis) to take place, the cells targeted to form aerenchyma clearly undergo a programmed cell death. It is essential to mention that both processes rely on polysaccharide hydrolysis, whereas the cell separation in schizogenous aerenchyma formation depends on middle lamella degradation. As for the lysogenous process, cell wall modifications occur and for that a wide array of enzymes would be required.

Concerning cell wall modifications, most of the aerenchyma-related data is restricted to the lysigenic and inducible formation in maize roots. As reported for other cell degradation events, in this model the inductive role is played by ethylene (He et al. 1992, 1996a; Gunawardena et al. 2001a). One of the earliest signs of aerenchyma formation is the ethylene accumulation, followed by cell death and endo-β-glucanases activity (Kawase 1979; He et al. 1996a). Indeed, inhibitors of cell death or ethylene biosynthesis block the enhancement of endo-β-glucanases activity (He et al. 1994, 1996a), suggesting the coupling between cell wall degradation, ethylene signalling and programmed cell death.

Another potentially degradative enzyme associated with aerenchyma formation is endo-β-xylanase. The rising activity levels of this enzyme were obtained in waterlogged maize roots, coinciding with the rising of ethylene levels (Bragina et al. 2001). Also concerning ethylene induction, the role of XTH, a putative cell wall loosening enzyme, was evaluated during aerenchyma formation. Transcripts of certain *XTH* accumulated in flooded maize roots forming aerenchyma, reaching its maximum within 12 h and remaining high for 144 h during flooding. On the other hand, the inhibition of ethylene synthesis blocked any rise of this transcript accumulation, while exogenous ethylene led to *XTH* transcript accumulation even under aerobic conditions (Saab and Sachs 1996). A transcriptional analysis corroborated the induction of XTH expression during the aerenchyma formation in maize roots, as well as for endo-β-glucanase (Rahji et al. 2011).

Even though the enzymatic or transcriptional approaches have been used to suggest the link between a set of glycosil hydrolases and aerenchyma formation, to our knowledge, data regarding cell wall modifications in situ have been reported only by Gunawardena et al. (2001b), who documented the distinctive levels of methyl esterification changing in the middle lamella in cross sections of the cortex of the maize root along aerenchyma formation. These authors observed that those changes initiated before ultrastructural modifications typical of cells undergoing programmed cell death that have been reported in a previous work (Gunawardena et al. 2001a). Indeed, rise in the expression of pectin methyl esterases, pectate lyase and an endopolygalacturonase have been observed during aerenchyma formation (Rahji et al. 2011). These data not only corroborate the existence of homogalacturonan modifications but also suggest that more severe changes on the cell wall structure and not only on the middle lamella can occur. We have observed a similar pattern of events in sugarcane roots, not only with the pectinases, but also with the observation of the rise in gene expression of several genes that encode hemicellulases (e.g. lichenase, endo-β-xylanase, XTH, β-glucosidase—unpublished results).

Gunawardena et al. (2001a) reported the formation of vesicles and vacuole rupture while the cell wall appeared intact. It is reasonable to hypothesize that the release of hydrolytic enzymes in this step could be responsible for the cell wall degradation (Bouranis et al. 2007).

The data produced that are related with aerenchyma formation to date indicates that this phenomenon is similar to the other processes discussed above in the sense that cell walls are modified, including cell separation, hemicellulose-cellulose matrix relaxation and perhaps some hydrolysis of cellulose. Thus, the knowlegde about aerenchyma formation can also be considered as a possible target for use in biotechnology that includes cell wall modifications for bioenergy purposes. Recently we have produced a consistent set of results describing events during the aerenchyma formation in sugarcane roots. We found that the events are essencially similar to maize (e.g.), with the exception that in sugarcane it is constitutively activated. Our results (manuscripts in preparation) show very clearly that cell separation, programmed cell death, hemicellulose-cellulose changes and some attack to cellulose occurs in sugarcane. We are now searching for the transcription factors responsible for regulating this processes aiming at possibly inducing them in culms so that this process might be used as a biological pretreatment in the future.

13.8 Cell Wall Modifications in Natural Systems in the Context of Bioenergy

As mentioned above, in order to establish 2G bioethanol production, the retrieval of energy from wall polymers would have to include cell wall modification. From the different systems discussed along this chapter, it is possible to think of combining some of them in order to help increasing the production of bioethanol from bioenergy feedstocks that are being studied for use in 2G processes.

In Table 13.1 the different modules that plants apparently use to transform their cell walls in order to perform different biological functions are shown in perspective. Six events seem to be common to most of the natural processes that clearly involve modifications in cell walls. Some of these events have been investigated more deeply and others are implied, but not proven, as no direct experimental evidences have been produced for them yet. In Table 13.1, the events are placed in an order (from left to right) that they seem to occur in all systems. However, some steps are missing in some of the processes (e.g. cell separation in storage cell wall mobilization and programmed cell death in fruit ripening).

Table 13.1 Correlation among situations that include cell wall modifications and natural events in plant biology where these situations are key to biological function

Processes	Target cells	Cell separation	Cell expansion	Programmed cell death	Hemicellulose hydrolysis	Cellulose hydrolysis
Plant microorganism interaction	?	YES	?	YES	YES	YES
Storage cell wall mobilization	?	NO	YES	?	YES	?
Fruit ripening	YES	YES	YES	NO	YES	YES/NO
Abscission	YES	YES	YES	?	YES	YES
Aerenchyma formation	YES	YES	YES	YES	YES	?

YES = there is literature showing that it occurs in most (or all) systems studied; NO = not observed in the systems studied to date; YES/NO = observed only in some of the systems studied to date; ? = not reported to occur to date

The combination of events is consistent with a chain of processing steps that would follow a sequence of modifications in the walls that are apparently similar in the five processes reviewed in this chapter. The general idea is that in a parenchymatic tissue, a given group of cells would be targeted to start the process (as seen in abscission according to Roberts et al. (2002) and also in aerenchyma formation). The following step would be cell separation that probably involves the action of endo-polygalacturonase(s) on the pectins of the middle lamella. This might be involved with the release of calcium that could be responsible for the signaling related to programmed cell death. The latter seems to occur in parallel with cell expansion and subsequent production and release glycosyl hydrolases that would attack hemicelluloses and cellulose. Every one of these events can be regarded as a *module* that can or not be used in the complete process that leads to partial or complete cell wall modification.

There are two possible ways to use this information for biotechnology purposes: (1) using them for production of transformed plants; or (2) express some of the hydrolase genes heterologously in order to use them as additions to enzyme cocktails already in use in industry.

For the first option, it should be possible to use the modern tools of synthetic biology in order to induce cell separation and expansion along with some modification of the hemicellulose-cellulose matrix. In this case, plants would produce biomass with the characteristics expected to be analogous to a pretreated biomass material, i.e. materials that would be easily hydrolyzed when being processed in industry, especially during the hydrolysis process. For advancing the second option, some of the enzymes known to be present during hydrolysis processes in bioenergy feedstocks, such as sugarcane, miscanthus, maize, poplar, willow and others should be studied regarding their specificity. Some of these enzymes may be heterologously expressed for production and added to existent cocktails.

In both cases, there is a relatively long way to go in terms of production of the proof of concepts that will be necessary in order to develop the necessary technologies. However, in some cases, such processes are already present in plants and could be improved using transformation, possibly of transcription factors that trigger the whole process.

It is important to note here that wood feedstocks would probably have to be treated differently regarding the strategies discussed in this chapter. Wood feedstocks have large proportions of dead cells composed of high proportions of cellulose and also with a much higher degree of lignification. For this type of bioenergy feedstock, an addition to the strategy would be the modification of lignin composition to facilitate access to the cell wall polysaccharides, which in fact is currently under investigation (Vanholme et al. 2013). As grasses can have some lignin and phenypropanoids in their cell walls, as well as the cells of the vascular system, whose walls bear many features in common with the wood tissues, the use of the strategies proposed in this chapter for tissues containing large proportion or primary cell walls, along with the ones being developed for decreasing lignin interference, would probably have to be coupled in the future in order to improve bioenergy production even more.

Based on the examples discussed in this chapter, it is very likely that natural processes that involve cell wall modifications in plants will be useful as additions to 2G bioethanol technologies, via the use of modern molecular techniques associated to systems and synthetic biology.

Acknowledgments This work is part of the production of the Instituto Nacional de Ciência e Tecnologia do Bioetanol-INCT do Bioetanol (FAPESP 2008/57908-6 and CNPq 574002/2008-1) and of the Centro de Processos Biológicos e Industriais para Biocombustíveis-CeProBIO (FAPESP 2009/52840-7 and CNPq 490022/2009-0). Financial support by FAPESP Projects 2010/17104-5, 2010/17070-3, 2011/07586-5 and 2011/02344-3.

References

Ahmed AE, Labavitch JM (1980) Cell wall metabolism in ripening fruit 1. Cell wall changes in ripening "Bartlett" pears. Plant Physiol 65:1009–1013

Anderson MA, Stone BA (1975) A new substrate for investigating the specificity of glucan hydrolases. FEBS Lett 52:202–207

Annis SL, Goodwin PH (1997) Recent advances in the molecular genetics of plant cell wall-degrading enzymes produced by plant pathogenic fungi. Eur J Plant Pathol 103:1–14

Balat M (2011) Production of bioethanol from lignocellulosic materials via the biochemical pathway: a review. Energy Converse Manag 52:858–875

Barsan C, Sanchez-Bel P, Rombaldi C, Egea I, Rossignol M, Kuntz M et al (2010) Characteristics of the tomato chromoplast revealed by proteomic analysis. J Exp Bot 61:2413–2431

Bewley JD, Burton RA, Morohashi U, Fincher GB (1997) Molecular cloning of a cDNA encoding a (1 → 4)-β-mannan endohydrolase from the seeds of germinated tomato (*Lycopersicon esculentum*). Planta 203:454–459

Bouranis DL, Chorianopoulou SN, Siyiannis VF, Protonotarios VE, Hawkesford MJ (2007) Lysigenous aerenchyma development in roots: triggers and cross-talks for a cell elimination program. Int J Plant Dev Biol 1:127–140

Bragina TV, Martinovich LI, Rodionova NA, Bezborodov AM, Grineva GM (2001) Ethylene-induced activation of xylanase in adventitious roots of maize as a response to the stress effect of root submersion. Appl Biochem Microbiol 37:722–725

Brummell DA (2006) Cell wall disassembly in ripening fruit. Funct Plant Biol 33:103–119

Brummell DA, Harpster MH (2001) Cell wall metabolism in fruit softening and quality and its manipulation in transgenic plants. Plant Mol Biol 47:311–340

Buckeridge MS (2006) Implications of emergence, degeneracy and redundancy for the modeling of the plant cell wall. In: Hayashi T (ed) The science and the lore of the plant cell wall: biosynthesis, structure and function. BrownWalker Press, Boca Raton, pp 41–47

Buckeridge MS (2010) Seed cell wall storage polysaccharides: models to understand cell wall biosynthesis and degradation. Plant Physiol 154:1–7

Buckeridge MS, De Souza AP (2013) The fine structure of cell wall polysaccharides might be related to the existence of a glycomic code in plant cell walls. In: Lam E (ed) Compendium of bioenergy plants: sugarcane (submitted)

Buckeridge MS, Dietrich SMC (1996) Mobilisation of the raffinose family oligosaccharides and galactomannan in germinating seeds of *Sesbania marginata* Benth (Leguminosae-Faboideae). Plant Sci 117:33–43

Buckeridge MS, Reid JSG (1994) Purification and properties of a novel β-galactosidase or exo-(1 → 4)-β-D-galactanase from the cotyledons of germinated *Lupinus angustifolius* L. seeds. Planta 192:502–511

Buckeridge MS, Crombie HJ, Mendes CJM, Reid JSG, Gidley MJ, Vieira CJ (1997) A new family of xyloglucan oligosaccharides from cotyledons of *Hymenaea courbaril*: structure determination of the oligosaccharide XXXXG by enzymatic sequencing and NMR. Carbohydr Res 303:233–237

Buckeridge MS, Dietrich SMC, Lima DU (2000a) Galactomannans as the reserve carbohydrate of legume seeds. In: Gupta AK, Kaur N (eds) Developments in crop science. Elsevier Science BV, Amsterdam, pp 283–316

Buckeridge MS, Santos HP, Tiné MAS (2000b) Mobilisation of storage cell wall polysaccharides in seeds. Plant Physiol Biochem 38:141–156

Buckeridge MS, De Souza AP, Arundale RA, Anderson-Teixeira KJ, DeLucia E (2012) Ethanol from sugarcane in Brazil: a midway strategy for increasing ethanol production while maximizing environmental benefits. GCB Bioenergy 4:119–126

Carpita NC (1996) Structure and biogenesis of the cell walls of grasses. Ann Rev Plant Physiol Plant Mol Biol 47:445–476

Carpita NC, Gibeaut DM (1993) Structural models of primary cell walls in flowering plants: consistency of molecular structure with the physical properties of the walls during growth. Plant J 3:1–30

Carpita NC, Sabularse D, Montezinos D, Delmer DP (1979) Determination of the pore size of cell walls of living plant cells. Science 205:1144–1147

Carpita N, Tierney M, Campbell M (2001) Molecular biology of the plant cell wall: searching for the genes that define structure, architecture and dynamics. Plant Mol Biol 47:1–5

Carroll A, Somerville C (2009) Cellulosic biofuels. Ann Rev Plant Biol 60:165–182

Chandra RP, Bura R, Mabee WE, Berlin A, Pan X, Saddler JN (2007) Substrate pretreatment: the key to effective enzymatic hydrolysis of lignocellulosics? Biofuels Adv Biochem Eng Biotechnol 108:67–93

Cooper RM (1983) The mechanisms and significance of enzymatic degradation of host cell walls by parasites. In: Callow JA (ed) Biochemical plant pathology. Wiley, Chichester, pp 101–137

De Souza AP, Leite DCC, Pattathil S, Hahn MG, Buckeridge MS (2013) Composition and structure of sugarcane cell wall polysaccharides: implications for second-generation bioethanol production. Bioenerg Res 6:564–579

Dos Santos WD, Gómez EO, Buceridge MS (2011) Bioenergy and the sustainable revolution. In: Buckeridge MS, Goldman GH (eds) Routes to cellulosic ethanol. Springer, New York, pp 15–26

Fray RG, Grierson D (1993) Molecular genetics of tomato fruit ripening. Trends Genet 9:438–443

Fry S (2004) Primary cell wall metabolism: tracking the careers of wall polymers in living plant cells. New Phytol 161:641–675

Garcia-Brugger A, Lamotte O, Vandelle E, Bourque S, Lecourieux D et al (2006) Early signaling events induced by elicitors of plant defenses. Mol Plant Microbe Interact 19:711–724

Gómez LD, Steele-King CG, McQueen-Mason SJ (2008) Sustainable liquid biofuels from biomass: the writing's on the walls. New Phytol 178:473–485

Gong X, Bewley JD (2007) Shorting out the LeMANs: endo-β-mannanase genes and their encoded proteins in tomato. Seed Sci Tech 17:143–154

Gonzáles-Carranza ZH, Lozoya-Gloria E, Roberts JA (1998) Recent developments in abscission: shedding light on the shedding process. Trends Plant Sci 3:10–14

Greenberg JT, Yao N (2004) The role and regulation of programmed cell death in plant-pathogen interactions. Cell Microbiol 6:201–211

Gunawardena A, Lan H, Pearce DM, Jackson MB, Hawes CR, Evans DE (2001a) Characterization of programmed cell death during aerenchyma formation induced by ethylene or hypoxia in roots of maize (*Zea mays* L.). Planta 212:205–214

Gunawardena A, Pearce DM, Jackson MB, Hawes CR, Evans DE (2001b) Rapid changes in cell wall pectic polysaccharides are closely associated with early stages of aerenchyma formation, a spatially localized form of programmed cell death in roots of maize (*Zea mays* L.) promoted by ethylene. Plant Cell Environ 24:1369–1375

Harker FR, Redgwell RJ, Hallet IC, Murray SH, Carter G (1997) Texture of fresh fruit. Hortic Rev 20:121–124

Harris PJ, Hartley RD (1980) Phenolic constituents of the cell walls of monocotyledons. Biochem Syst Ecol 8:153–160

He CJ, Morgan P, Drew MC (1992) Enhanced sensitivity to ethylene in nitrogen of phosphate-starved roots of *Zea mays* L. during aerenchyma formation. J Plant Physiol 98:137–142

He CJ, Drew M, Morgan PW (1994) Induction of enzymes associated with lysigenous aerenchyma formation in roots of *Zea mays* during hypoxia or nitrogen starvation. J Plant Physiol 105:861–865

He CJ, Pinlayson SA, Drew MC, Jordan WR, Morgan PW (1996a) Ethylene biosynthesis during aerenchyma formation in roots of *Zea mays* subjected to mechanical impedance and hypoxia. J Plant Physiol 112:1679–1685

He CJ, Morgan PW, Drew MC (1996b) Transduction of an ethylene signal is required for cell death and lysis in the root cortex of maize during aerenchyma formation induced by hypoxia. J Plant Physiol 112:463–472

Himmel ME, Adney WS, Baker JO, Elander R, McMillan JD, Nieves RA, Sheehan JJ, Thomas SR, Vinzant TB, Zhang M (1997) Advanced bioethanol production technologies: a prospective. In: Woodward J, Saha B (eds) ACS symposium series, vol 666. American Chemical Society, Washington, DC, pp 2–45

Himmel ME, Ruth MF, Wyman CE (1999) Cellulase for commodity products from cellulosic biomass. Curr Opin Biotechnol 10:358–364

Himmel ME, Ding SY, Johnson DK, Adney WS, Nimlos MR, Brady JW, Foust TD (2007) Biomass recalcitrance: engineering plants and enzymes for biofuels production. Science 315:804–807

Holland N, Holland D, Helentjaris T, Dhugga KS, Xoconostle-Cazares B, Delmer DP (2000) A comparative analysis of the plant cellulose synthase (CesA) gene family. Plant Physiol 123:1313–1323

Iiyama K, Lam TBT, Stone BA (1994) Covalent crosslinks in the cell wall. Plant Physiol 104:315–320

Jiang CZ, Lu F, Imsabai W, Meir S, Reid MS (2008) Silencing polygalacturonase expression inhibits tomato petiole abscission. J Exp Bot 59:973–979

Justin SHFW, Armstrong W (1991) Evidence for the involvement of ethene in aerenchyma formation in adventitious roots of rice (*Oryza sativa* L.). New Phytol 118:49–62

Kaida R, Kaku T, Baba K, Oyadomari M, Watanabe T, Nishida K, Kanaya T, Shani Z, Shoseyov O, Hayashi T (2009) Loosening xyloglucan accelerates the enzymatic degradation of cellulose in wood. Mol Plant 2:904–909

Karahara I, Umemura K, Soga Y, Akai Y, Bando T, Ito Y, Tamaoki D, Uesugi K, Abe J, Yamauchi D, Mineyuki Y (2012) Demonstration of osmotically dependent promotion of aerenchyma formation at different levels in the primary roots of rice using a 'sandwich' method and X-ray computed tomography. Ann Bot 110:503–509

Kawase M (1979) Role of cellulase in aerenchyma development in sunflower. Amer J Bot 66:183–190

Keegstra K, Talmadge K, Baurer WD, Albersheim O (1973) Structure of the plant cell walls. Plant Physiol 51:188–196

Lagaert S, Beliën T, Volckaert G (2009) Plant cell walls: protecting the barrier from degradation by microbial enzymes. Stem Cell Dev Biol 20:1064–1073

Lima DU, Santos HP, Tiné MAS, Molle FD, Buckeridge MS (2001) Patterns of expression of cell wall related genes in sugarcane. Gen Mol Biol 24:191–198

Lisboa CGS, Tiné MAS, Buckeridge MS (2006) Endo-β-mannanase from the endosperm of seeds of *Sesbania virgata* (Cav.) Pers. (Leguminosae): purification, characterisation and importance for germination and early seedling growth. Braz J Plant Physiol 18:269–280

Mano Y, Omori F, Takamizo T, Kindiger B, Bird R, Loaisiga C, Takahashi H (2007) QTL mapping of root aerenchyma formation in seedlings of a maize e rare teosinte "*Zea nicaraguensis*" cross. Plant Soil 295:103–113

Marques MR, Buckeridge MS, Braga MR, Dietrich SMC (2006) Characterization of an extracellular endopolygalacturonase from saprobe *Mucor ramosissimus* Samutsevitsch and its action as trigger of defensive response in tropical plants. Mycopathologia 162:337–346

McCann MC, Roberts K (1991) Architecture of the primary cell wall. In: Lloyd CW (ed) The cytoskeletal basis of plant growth and form. Academic Press, London, pp 109–129

McManus MT, Thompson DS, Merriman C, Lyne L, Osborne DI (1998) Transdifferentiation of mature cortical cells to functional abscission cells in bean. Plant Physiol 116:891–899

Nishitani K, Nevins DJ (1991) Glucuronoxylan xylanohydrolase: a unique xylanase with the requirement for appendant glucuronosyl units. J Biol Chem 266:6539–6543

Ong RG, Chundawat SPS, Hodge DB, Keskar S, Dale B (2013) Linking plant biology and pretreatment: understanding the structure and organization of the plant cell wall and interactions with cellulosic biofuel production. In: Carpita NC, Buckeridge MS, McCann MC (eds) Plants and bioenergy. Springer, New York

Parrish FW, Perlin AS, Reese ET (1960) Selective enzymolysis of poly-β-D-glucans, and the structures of the polymers. Can J Chem 38:2094–2104

Pattathil S, Avci U, Baldwin D, Swennes AG, McGill JA, Popper Z et al (2010) A comprehensive toolkit of plant cell wall glycan-directed monoclonal antibodies. Plant Physiol 153:514–525

Polizeli MLTM, Corrêa ECP, Polizeli AM, Jore JA (2011) Hydrolases from microorganisms used for degradation of plant cell wall and bioenergy. In: Buckeridge MS, Goldman GH (eds) Routes to cellulosic ethanol. Springer, New York, pp 117–136

Rahji I, Yamauchi T, Takahashi H, Nishiuchi S, Shiono K, Watanabe R, Mliki A, Nagamura Y, Tsutsumi N, Nishizawa NK, Nakazono M (2011) Identification of genes expressed in maize root cortical cells during lysigenous aerenchyma formation using laser microdissection and microarray analyses. New Phytol 190:351–368

Reid JSG (1971) Reserve carbohydrate metabolism in germinating seeds of *Trigonella foenungraecum* L. (Legum). Planta 100:131–142

Reid JSG, Edwards M (1995) Galactomannans and other cell wall storage polysaccharides in seeds. In: Stephen AM (ed) Food polysaccharides and their applications. Marcel Dekker, New York, pp 155–186

Roberts JA, Elliott KA, Gonzalez-Carranza Z (2002) Abscission, dehiscence, and other cell separation processes. Ann Rev Plant Biol 53:131–158

Rose JKC, Bennett AB (1999) Cooperative disassembly of the cellulose-xyloglucan network of plant cell walls: parallels between cell expansion and fruit ripening. Trends Plant Sci 4:176–183

Rose JKC, Hadfield KA, Labavitch JM, Bennett AB (1998) Temporal sequence of cell wall disassembly in rapidly ripening melon fruit. Plant Physiol 117:345–361

Saab IN, Sachs MM (1996) A folding-induced xyloglucan endo-transglycosylase homolog in maize is responsive to ethylene and associated with aerenchyma. J Plant Physiol 112:385–391

Sarkar P, Bosneaga E, Auer M (2009) Plant cell walls throughout evolution: towards a molecular understanding of their design principles. J Exp Bot 60:3615–3635

Schussler EE, Longstreth DJ (2000) Changes in cell structure during the formation of root aerenchyma in *Sagittaria lancifolia* (Alismataceae). Amer J Bot 87:12–19

Serpa VI, Polikarpov I (2011) Enzymes in bioenergy. In: Buckeridge MS, Goldman GH (eds) Routes to cellulosic ethanol. Springer, New York, pp 99–116

Siyiannis VF, Protonotarios VE, Zechmann B, Chorianopoulou SN, Müller M, Hawkesford MJ, Bouranis DL (2011) Comparative spatiotemporal analysis of root aerenchyma formation processes in maize due to sulphate, nitrate or phosphate deprivation. Protoplasma 249:671–686

Soccol CR, Vandenberghe LPS, Medeiros ABP, Karp SG, Buckeridge MS, Ramos LP, Pitarelo AP, Ferreira-Leitão V, Gottschalk LMF, Ferrara MA (2010) Bioethanol from lignocelluloses: status and perspectives in Brazil. Biores Technol 101:4820–4825

Steffens B, Geske T, Sauter M (2011) Aerenchyma formation in the rice stem and its promotion by H_2O_2. New Phytol 190:369–378

Taylor LE II, Dai Z, Decker SR, Brunecky R, Adney WS, Ding SY, Himmel ME (2008) Heterologous expression of glycosyl hydrolases in planta: a new departure for biofuels. Trends Biotechnol 26:413–424

Tiné MAS, Silva CO, Lima DU, Carpita NC, Buckeridge MS (2006) Fine structure of a mixed-oligomer storage xyloglucan from seeds of *Hymenaea courbaril*. Carbohydr Polym 66:444–454

Tomassini A, Sella L, Raiola A, D'Ovidio R, Favaronl F (2009) Characterization and expression of *Fusarium graminearum* endo-polygalacturonases in vitro and during wheat infection. Plant Pathol 58:556–564

Vanholme B, Cesarino I, Goeminne G, Kim H, Marroni F, Van Acker R, Vanholme R, Morreel K, Ivens B, Pinosio S, Morgante M, Ralph J, Bastien C, Boerjan W (2013) Breeding with rare defective alleles (BRDA): a natural *Populus nigra* HCT mutant with modified lignin as a case study. New Phytol 198:765–776

Verma D, Kanagaraj A, Jin S, Singh ND, Kolattukudy PE, Daniell H (2010) Chloroplast-derived enzyme cocktails hydrolyse lignocellulosic biomass and release fermentable sugars. Plant Biotechnol J 8:332–350

Visser EJ, Bögemann GM (2006) Aerenchyma formation in the wetland plant *Juncus effusus* is independent of ethylene. New Phytol 171:305–314

Walton JD (1994) Deconstructing the cell wall. Plant Physiol 104:1113–1118

Ward R (2011) Cellulase engineering for biomass saccharification. In: Buckeridge MS, Goldman GH (eds) Routes to cellulosic ethanol. Springer, New York, pp 137–153

Willats WGT, McCartney L, Mackie W, Knox P (2001) Pectin: cell biology and prospects for functional analysis. Plant Mol Biol 47:9–27

Xin Z, Watnabe N, Lam E (2011) Improving efficiency of cellulosic fermentation via genetic engineering to create "smart plants" for biofuel production. Enzymes in bioenergy. In: Buckeridge MS, Goldman GH (eds) Routes to cellulosic ethanol. Springer, New York, pp 183–199

Chapter 14
Linking Plant Biology and Pretreatment: Understanding the Structure and Organization of the Plant Cell Wall and Interactions with Cellulosic Biofuel Production

Rebecca Garlock Ong, Shishir P. S. Chundawat, David B. Hodge, Sai Keskar and Bruce E. Dale

Abstract In order to more economically process cellulosic feedstocks using a biochemical pathway for fuel production, it is necessary to develop a detailed understanding of plant cell wall characteristics, pretreatment reaction chemistry, and their complex interactions. However given the large number of thermochemical pretreatment methods that are currently being researched and the extreme diversity of plant cell wall structure and composition, this prospect is extremely challenging. Here we present the current state of research at the interface between plant biology and pretreatment chemistry. The first two sections discuss the chemistry of the secondary plant cell wall and how different pretreatment methods alter the overall cell wall structure. The third section addresses how the characteristics of the cell wall and pretreatment efficacy are impacted by different factors such as

R. G. Ong (✉) · S. P. Chundawat · D. B. Hodge · S. Keskar · B. E. Dale
Department of Chemical Engineering and Materials Science, Michigan State University, 3815 Technology Blvd, Lansing, Lansing, MI 48910, USA
e-mail: garlock1@msu.edu

S. P. Chundawat
e-mail: chundawa@egr.msu.edu

R. G. Ong · S. P. Chundawat · D. B. Hodge · S. Keskar · B. E. Dale
DOE Great Lakes Bioenergy Research Center, Madison, WI, USA

S. P. Chundawat
Department of Biochemistry, University of Wisconsin–Madison, Madison, WI, USA

D. B. Hodge
Department of Biosystems and Agricultural Engineering, Michigan State University, XXX, XXX

D. B. Hodge
Department of Civil, Environmental and Natural Resources Engineering, Luleå University of Technology, XXX, XXX

M. C. McCann et al. (eds.), *Plants and BioEnergy*, Advances in Plant Biology 4,
DOI: 10.1007/978-1-4614-9329-7_14, © Springer Science+Business Media New York 2014

plant maturity, classification, and plant fraction. The fourth section summarizes current directions in the development of novel plant materials for improved biochemical conversion. And the final section discusses the use of chemical pretreatments as a screening and analysis tool for rapid identification of amenable plant materials, and for expansion of the fundamental understanding of plant cell walls.

Keywords Enzymatic digestibility • Lignocellulose • Plant breeding and transgenesis • Plant cell wall • Pretreatment chemistry • Screening tools

Abbreviations

AFEX™	Ammonia fiber expansion
BMIMCl	1-butyl-3-methylimidazolium chloride
CBM	Carbohydrate binding module
EMIMAc	1-ethyl-3-methylimidazolium acetate
EMIMCl	1-ethyl-3-methylimidazolium chloride
G	Guaiacyl
GAX	Glucuronoarabinoxylan
H	*p*-hydroxyphenyl
IL	Ionic liquid
S	Syringyl
TAGs	Triacylglycerols

14.1 Introduction

Lignocellulosic materials are a promising source of biofuels because of their abundance and availability. One potential conversion pathway is the biochemical route, through enzymatic hydrolysis and fermentation of cell wall carbohydrates. The difficulty is that although plant cell walls are permeable to small molecules, such as water, carbon dioxide, sugars, and amino acids (Ivakov and Persson 2012), while enzymes, with a diameter of around 51Å (Ishizawa et al. 2007), are too large to penetrate. Therefore to obtain access to polysaccharides embedded within the cell wall in an industrially relevant time scale, some form of chemical or physical pretreatment is needed to disrupt the cell wall structure. A large number of pretreatments are currently being researched (da Costa Sousa et al. 2009; Zhao et al. 2012), corresponding to a wide range of chemistries and modes of action. In addition there is enormous diversity of plant cell walls in terms of their structure and organization (Cosgrove 2005). The chemical and physical interactions between variables related to the feedstock (Fig. 14.1a), and pretreatment (Fig. 14.1b), determines the specific types and magnitudes of effects on cell wall structure (Fig. 14.1c), and ultimately the extent of enzymatic deconstruction.

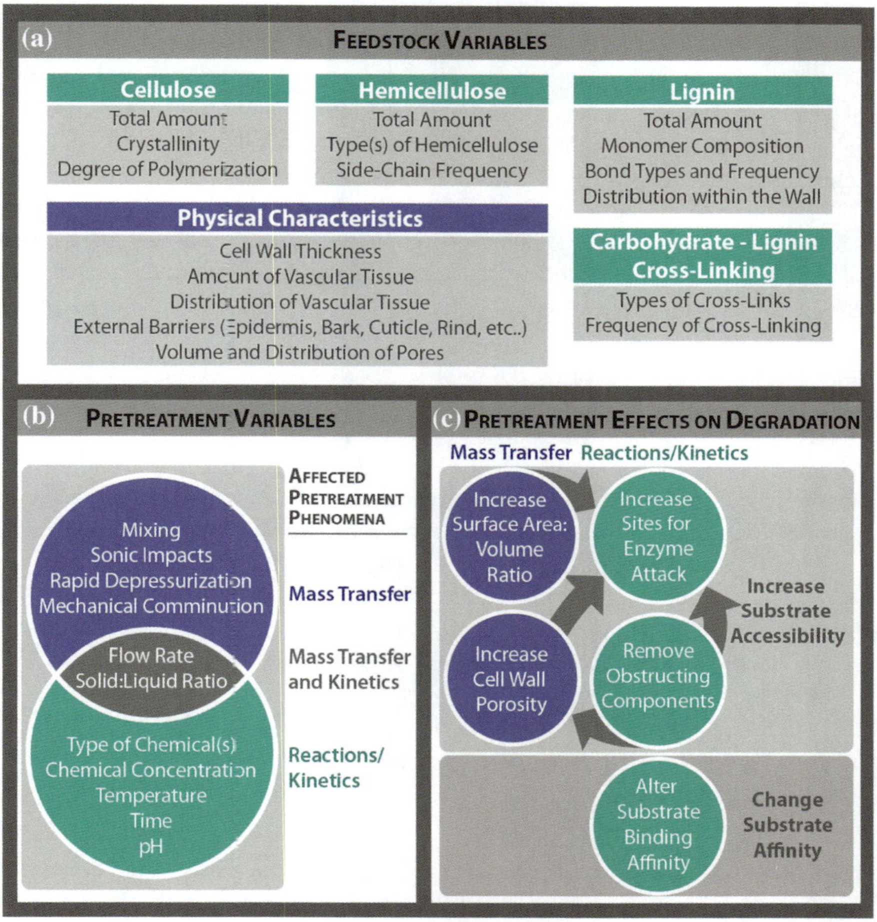

Fig. 14.1 Feedstock variables (**a**), pretreatment variables (**b**), and resulting modes of action (**c**) for improved enzymatic degradation of plant cell wall carbohydrates. Gray arrows in part C represent the potential for a direct impact of one mode of action on another

14.2 Secondary Cell Wall Chemistry

Higher plants have two main cell wall types, with different functions and compositions. Primary walls are laid down during cell growth and elongation, and secondary walls are laid down after cessation of cell growth (Cosgrove 2005; Ivakov and Persson 2012). The middle lamella is located between adjacent cells and binds them together (Cosgrove 2005). After growth stops, lignin deposition begins in the middle lamella and cell corners and progresses to the primary and secondary walls (Ralph et al. 2007; Ivakov and Persson 2012). In woody plants, the primary wall is degraded before secondary wall deposition (Jarvis 2012), however for herbaceous plants the secondary wall is deposited directly inside the primary wall (Wilson and Hatfield

1997; Engels and Jung 1998). Not all types of cells have secondary cell walls, mainly those requiring greater strength or rigidity (Cosgrove 2005), and some secondary walls never lignify (Engels and Jung 1998). But because of their greater thickness, secondary walls make up the bulk of lignocellulosic biomass and cell volume, especially in woody materials (Wilson and Hatfield 1997; Ivakov and Persson 2012). Cellulose, hemicelluloses, and lignin are the major components of the secondary cell wall. Cellulose forms the scaffolding of the cell wall and comprises β-(1 → 4)-linked glucan chains arranged in crystalline microfibrils. The hemicelluloses are a diverse class of amorphous carbohydrates that cross-link cellulose microfibrils and lignin within the cell wall. All hemicelluloses have β-(1 → 4)-linked backbones of glucose (glucans), mannose (mannans), glucose and mannose (glucomannans), or xylose (xylans), and may be substituted with sugars, uronic acids, and acetyl groups. Lignin is an amorphous phenylpropanoid polymer that fills in most of the remaining space and is comprised of three different subunits that are differentiated by the number of methoxyl groups on the phenyl ring: syringyls (S) have 2; guaiacyls (G) have 1; and *p*-hydroxyphenyls (H) have 0. Pectins are another class of carbohydrate and represent a major portion of the dicot and gymnosperm primary wall, however they are comparatively easy to extract from the cell wall or degrade (Willför et al. 2005a, b; DeMartini et al. 2011a). For more detailed explanations on cell wall composition and structure of the polymers please refer to a number of reviews (Carpita and Gibeaut 1993; Ralph et al. 2007; Scheller and Ulvskov 2010; Ivakov and Persson 2012).

14.2.1 Variation in Chemistry Due to Classification, Cell Type, and Location

Bioenergy plants are grouped in three classes based on their cell wall composition: grass-like (commelinid monocots), dicot-like (non-commelinid monocots, herbaceous dicots, and hardwoods), and gymnosperm (softwoods). Grass-like secondary cell walls contain glucuronoarabinoxylan (GAX) as the main hemicellulose substituted with arabinose and some glucuronic acid, and lignin comprised of S and G subunits with low levels of H subunits; dicot-like secondary cell walls predominantly contain glucuronoxylan substituted with 4-*O*-methyl-glucuronic acid and infrequently with arabinose, and lignin comprised of similar levels of S and G subunits and trace H subunits; and gymnosperm secondary walls contain slightly more galactoglucomannan than glucuronoarabinoxylan, and lignin comprised mostly of G subunits and low levels of H subunits (Ralph et al. 2007; Scheller and Ulvskov 2010). The type and distribution of the polymers varies within the cell and between cell types. For all plant classifications, the cell corners and middle lamella generally have the highest lignin content compared to the primary and secondary wall (Singh et al. 2009; Siqueira et al. 2011; Sun et al. 2011). For corn stover, cell types can be arranged in order of decreasing lignin and cellulose content: sclerenchyma and tracheids > epidermis > bundle sheath > parenchyma (Sun et al. 2011). Sugarcane follows a similar trend with lignin concentrated in the vessels followed by fiber and parenchyma cells (Siqueira et al. 2011). Lignin in herbaceous dicots is concentrated in the vascular ring (Wilson and Hatfield 1997; Engels and Jung 1998), but pith parenchyma cells, though thin, are also lignified (Engels and Jung 1998).

14.2.2 Covalent Linkages

The S/G ratio determines the types of inter-unit cross-linking that occur within the lignin matrix. The β-O-4 (β-aryl ether) linkage is the most frequent linkage and one of the most easily cleaved chemically (Ralph et al. 2007) and is more common in lignin containing more S subunits (Kishimoto et al. 2009). 4-*O*-5 linkages are more common with a 1:1 S/G ratio, and the β-β linkage is more common with a greater proportion of S subunits (Kishimoto et al. 2009; Rencoret et al. 2011). Lignins with a greater proportion of G subunits tend to be more branched, and also contain more chemically and thermally resistant structures (β-5 and 5-5) (Ralph et al. 2007; Kishimoto et al. 2009; Rencoret et al. 2011). As a result, hardwood lignin is easier to degrade and has a lower glass transition temperature compared to softwood lignin, which contains no syringyl subunits (Lundquist and Lundgren 1972; Lundquist 1973; Awal and Sain 2011).

Lignin is also covalently linked to hydroxycinnamic acids, with *p*-coumaric acids forming ester-linked terminal residues. Ferulic acids, which are also able to form oligomers, are ether-linked to lignin and ester-linked to carbohydrates, either pectins in certain dicots, or GAX arabinose side-chains in grass and dicot secondary walls (Iiyama et al. 1990; Harris and Trethewey 2010), though the frequency is lower for dicots due to significantly lower arabinose substitution (Scheller and Ulvskov 2010; Chiniquy et al. 2012). Ferulate cross-links limit enzymatic degradation (Grabber et al. 1998), but the ester link with hemicellulose is easily cleaved by most pretreatments. In addition to covalent cross-linking through hydroxycinnamic bridges, a variety of direct cross-links have also been proposed between lignin subunits and cell wall carbohydrates (Imamura et al. 1994; Karlsson et al. 2004; Lawoko et al. 2006).

14.2.3 Non-Covalent Interactions

A great deal of interaction between cell wall polymers is in the form of hydrogen bonding and van der Waals forces. In higher plants the glucan chains in the cellulose microfibril are present predominantly in the I_β crystal conformation (Atalla and Vanderhart 1984; Stone 2005). The microfibrils may interact with each other and other cell wall polysaccharides through non-covalent interactions (Altaner and Jarvis 2008; Ivakov and Persson 2012) and through these form aggregate- or bundle-like structures (Donaldson 2007; Abe and Yano 2009). Glucomannans bind more strongly to cellulose and are more resistant to extraction compared to glucuronoxylans (Clayton and Phelps 1965; Åkerholm and Salmén 2001, 2004; Zhang et al. 2011a). Strength of hemicellulose binding is likely related to interactions between the specific sugars in the hemicellulose backbone and cellulose, and a recent modeling study showed fewer hydrogen bonds but greater bond strength between cellulose and glucomannan compared to between cellulose and xylan (Zhang et al. 2011a). However stronger binding of glucomannan may also be related to lower side-chain substitution compared to xylan (Clayton and Phelps 1965). For the same class of hemicellulose, those with lower substitution bind more strongly to cellulose (Whitney et al. 1998; Kabel et al. 2007; Dammström et al. 2009), and the pattern of substitution also appears to have an impact (de Lima and Buckeridge 2001). In addition to the sugar side-chains, most mannans and xylans are acetylated (Scheller and Ulvskov 2010), which likely reduces binding affinity towards cellulose (Altaner and Jarvis 2008). It has also been hypothesized that hemicelluloses may be covalently linked to or embedded within cellulose microfibrils (Cosgrove 2005).

14.3 Pretreatment Chemistry

Thermochemical pretreatments alter the cell wall through chemical reactions that cleave covalent bonds and/or disrupt non-covalent interactions between cell wall polymers (Fig. 14.2) as well as through thermal softening and solubilization of biomass components. These chemical changes in combination with the physical removal and/or relocalization of cell wall components cause structural changes that improve enzymatic digestibility. Most pretreatments can be grouped based on their general effect on cell wall structure: those that remove lignin (alkaline/oxidative), those that remove hemicellulose and relocalize lignin (acidic), and those that fractionate cell wall components (ionic liquid, organosolv, and phosphoric acid) (Fig. 14.2). Most pretreatments, except for biological pretreatments and ionic liquids (ILs), can also be arranged in a continuum based on the nucleophilicity/electrophilicity of their main reactants (Fig. 14.3). Though less precise, the continuum can also be thought of in terms of pH (Pedersen and Meyer 2010; Garlock et al. 2011). Almost all of these pretreatments cleave some fraction of acetyl groups from the hemicellulose backbone (Maloney et al. 1985; Kumar et al. 2009; Shi et al. 2011) and use conditions that break α-ether linkages in lignin (Saake and Lehnen 2007).

The main mode of action for alkaline and oxidative pretreatments is through nucleophilic substitution and/or oxidation of esters and β-ethers within lignin and between cell wall polymers (Tarkow and Feist 1969; Iiyama et al. 1990; Sewalt et al. 1996). At very high alkali concentrations, carbohydrate monomers can be removed via peeling reactions and converted to acids (e.g. lactic acid) (Knill and Kennedy 2003). As reactant concentration and temperature decrease, peeling reactions become less likely to occur and fewer β-aryl-ether bonds are broken. Ammonia, a weaker nucleophile, does not cleave β-ethers but is known to cleave ester-linkages between hemicellulose and hydroxycinnamic acids (Wang et al. 1964; Azarpira et al. 2011). In contrast, acidic pretreatments mainly act through electrophilic hydrolysis of ester cross-links, β-ether bonds, and glycosidic

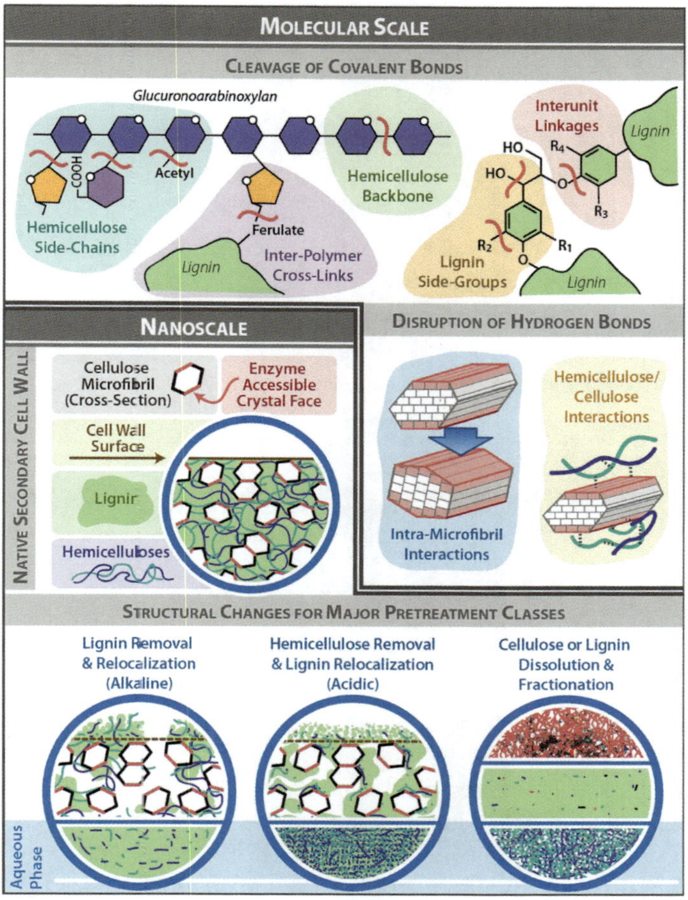

Fig. 14.2 Main molecular scale (chemical) impacts to plant cell wall components by thermochemical pretreatments and, in conjunction with mass transfer of biomass components, the resulting nanoscale (structural) changes for the three main classes of pretreatment

linkages, and they can also catalyze the dehydration of monomeric sugars. Room temperature acid treatment is able to break ether linkages between hydroxycinnamic acids and lignin/hemicelluloses (Wallace et al. 1995); while higher temperatures are needed to hydrolyze esters (Sannigrahi et al. 2009). Though β-ether bonds can be broken by strong acidic pretreatments, they are more readily hydrolyzed by alkali (Saake and Lehnen 2007). The key feature of acidic pretreatments is the hydrolysis of glycosyl linkages that allows for extraction of hemicellulose-derived oligomers and monomers. Xylans are more easily hydrolyzed than mannans (McGee and April 1982; Tunc and van Heiningen 2008; Várnai et al. 2010), and for side-chains, arabinose is more easily removed than galactose, and galactose than 4-*O*-methyl-glucuronic acid (McGee and April 1982; Sun and Cheng 2005). During hydrothermal pretreatments, hydronium ion concentration is

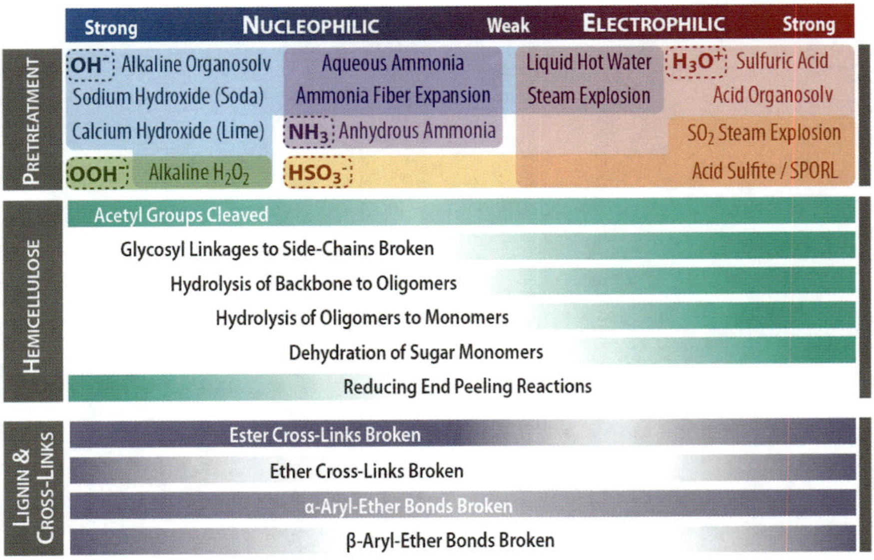

Fig. 14.3 Thermochemical pretreatments arranged in order of reactant nucleophilicity and effect on cell wall covalent linkages

initially governed by water autoionization and later by release of weak, biomass-derived acids (Garrote et al. 1999), and for these weakly acidic pretreatments, most hemicellulose is released as oligomers, only the most amenable side-chains are cleaved, and most lignin inter-unit linkages remain intact (Garlock et al. 2011).

A number of pretreatments (AFEX™[1], liquid hot water, dilute acid, and acid-catalyzed organosolv) have also been shown to deposit lignin-rich globules on the surface of the cell wall (Donohoe et al. 2008; Chundawat et al. 2011; Donohoe et al. 2011; Koo et al. 2012). For acidic pretreatments, particularly those catalyzed by sulfuric acid, lignin can condense and form new bonds (Xiao et al. 2013; Lundquist 1973; Karlsson et al. 1988). Most pretreatments also generate degradation compounds that influence downstream processes, and the specific compounds that are formed are determined by the interaction of plant cell wall chemistry (grass, hardwood, or softwood) with pretreatment chemistry (Chundawat et al. 2010; Du et al. 2010).

In addition to cleavage of covalent bonds, some pretreatments (sodium hydroxide, liquid ammonia, phosphoric acid, and ILs) disrupt hydrogen bonding within cellulose microfibrils and generate more digestible forms of cellulose (amorphous > II, III > I). The main mode of action for ionic liquids is the disruption of hydrogen bonding and decrystallization of cellulose to the extent that fractionation and lignin removal may not be necessary for high enzymatic conversions (Wu et al. 2011). There are indications that IL reactivity is related to both the ability of the

[1] AFEX™ is a registered trademark of MBI International, Lansing, MI.

anion to accept hydrogen bonds (Tadesse and Luque 2011; Gericke et al. 2012; King et al. 2012) and the length of the alkyl substituent chain on the cation, with shorter chain lengths leading to more effective cellulose dissolution (Zhang et al. 2005). Combined, the anion and cation form an electron donor–acceptor matrix with the cellulose hydroxyl groups that facilitates dissolution (Tadesse and Luque 2011; Xu et al. 2012). In ILs, pH is a measure of dissociation between the anion and cation (MacFarlane et al. 2006) and anions and cations can be classified as acidic, basic, or neutral. For example, the imidazolium ring has acidic properties that are believed to result in acid catalytic effects (MacFarlane et al. 2006). An IL with an acidic cation and basic anion like 1-ethyl-3-methylimidazolium acetate (EMIMAc) has a larger degree of dissociation (~pH 11) (Singh et al. 2009; Muhammad et al. 2012), which is likely related to its ability to both decrystallize cellulose and dissolve lignin. Imidiazolium ILs with a weakly basic anion like 1-butyl-3-methylimidizolium chloride (BMIMCl) and 1-ethyl-3-methylimidizolium chloride (EMIMCl) (~pH 6) are more selective for dissolving cellulose (Zhang et al. 2013b). The IL anion may also enhance catalytic reactions, and ILs with anions that are less basic than water (e.g. Cl^{-}) turn strong acids into weaker acids, however ILs with anions that are more basic than water (e.g. acetate) turn weak acids (like water and acetic acid) into stronger acids (MacFarlane et al. 2006). This may be one reason for the beneficial effect of water observed in EMIMAc, though this may also be related to reductions in viscosity (Fu and Mazza 2011). IL viscosity, which is much higher than conventional solvents, impacts cellulose dissolution through mass transfer and this can be difficult to separate from kinetic impacts (Gericke et al. 2012). The effectiveness of an IL is also dependent on temperature. Pure cellulose dissolves in imidazolium ILs between 80 and 100 °C (Zhang et al. 2005; Vitz et al. 2009), but pretreatment of whole biomass requires higher temperatures (~130 °C) for significant decrystallization of undissolved fractions (Kimon et al. 2011), which might be related to the glass transition temperature of lignin (Keskar et al. 2011; Li et al. 2011).

14.4 Impacts of Plant Characteristics on Cell Wall Degradation

14.4.1 Plant Classification

Different pretreatments process certain classifications of plants more effectively than others (Wyman et al. 2013), however, for the same pretreatment method, plant materials can almost always be arranged in the following order, either with regard to digestibility for the same conditions, or severity of conditions required for equivalent digestibility: grasses > herbaceous dicots > hardwoods > softwoods (Arantes and Saddler 2011; DeMartini and Wyman 2011a; Garlock et al. 2012b). This order is largely due to four factors that increasingly hinder pretreatment reaction kinetics and mass transfer: (1) increase in proportion of recalcitrant covalent linkages within the cell wall (esters - > ethers - > carbon–carbon bonds); (2) increase in strength of hydrogen-bonding of major hemicellulose sugars to cellulose; (3) increase in average cell wall thickness and proportion of the cell volume occupied by cell wall; and (4) increase in the proportion of lignin versus cellulose, although if cellulose accessibility is sufficiently increased, the actual presence of lignin during hydrolysis is not a major issue (Jeoh et al. 2007; Chundawat et al. 2011; Rollin et al. 2011; Wiman et al. 2012).

14.4.2 Plant Varieties

A handful of studies have looked at differences in digestibility and yields for cultivars within the same species. Upland and lowland switchgrass when harvested around the same time in the same location had similar sugar yields for most pretreatment methods (Kim et al. 2011) and similar optimal pretreatment conditions and enzyme loading (Garlock et al. 2012a). Results for wheat straw were varied, with one study that indicated sugar yields (g/g dry biomass) from hydrothermally pretreated wheat straw were not influenced by cultivar (Larsen et al. 2012), while two other studies found a significant variation in sugar yields across all cultivars (Lindedam et al. 2010; Lindedam et al. 2012).

14.4.3 Plant Cell Types and Tissues

Herbaceous feedstocks can show major differences in conversion between different portions of the plant or different cell types that may influence practical considerations such as harvesting methods and fractionation prior to pretreatment. In general, pith tends to be more digestible than the vascular bundles and the rind/epidermis. One study found that sugar yields follow the same pattern of digestibility for both hydrothermally pretreated and untreated materials (pith > leaves > rind) (Zeng et al. 2012). Pith cells of sugar cane were highly digestible by enzymes even without pretreatment and following chlorite treatment the rind cells became significantly more digestible (Siqueira et al. 2011).

For herbaceous botanical fractions, the general trend is that stems are easier to digest than leaves. For AFEX™-pretreatment, corn fractions were more digestible in order of decreasing lignin content (husk > leaf > stem > cob) (Garlock et al. 2009). For sodium hydroxide pretreatment, corn stover fractions released the most glucan in order of husks, cob, and leaves > upper stem > lower stem (Duguid et al. 2009) and corn stover and wheat straw fractions that contained more lignin showed a greater improvement with a higher catalyst loading (Duguid et al. 2007, 2009). Hydrothermally pretreated grasses and legume stems had lower percent sugar conversions than leaves, but higher total sugars released (DeMartini and Wyman 2011a). Miscanthus fractions showed decreasing cellulose conversion with: leaves > sheath > stem (Le Ngoc Huyen et al. 2010).

14.4.4 Harvest Date and Maturity

For herbaceous crops that have annual growth cycles, harvest date significantly impacts composition and biomass yields. As the plant approaches full maturity and senescence, the relative proportion of lignin and structural carbohydrates increase with a simultaneous decrease in soluble sugars, protein, and minerals (Dien et al. 2006). Harvest during the growing season can result in a highly digestible material, but one that also has significant nitrogen and ash content (Bals et al. 2010), which can impact farm economics, sustainability, and conversions. Some studies have shown little impact on total sugars released due to maturity (Dien et al. 2006; Garlock et al. 2009). However, there is a consistent decrease in digestibility and biomass yields when harvest is delayed from fall to winter or spring, largely due to loss of leaves and other fragile, digestible portions of the plant (Pordesimo et al. 2005; Adler et al. 2006; Le Ngoc Huyen et al. 2010; Kim et al. 2011). With regard to woody materials, one paper examined sugar yields from different growth rings and found no significant variation between mature wood and juvenile wood, despite an increase in lignin content with age of the ring (DeMartini and Wyman 2011b).

14.4.5 Composition

The most common trend reported for the effect of biomass composition on hydrolysis yields, is that glucan digestibility is negatively correlated to total lignin content (Davison et al. 2006; Dien et al. 2006; Rock et al. 2009; Garlock et al. 2012b). Lignin monomer composition may also be important, as a decrease in the S/G ratio leads to more recalcitrant linkages, and pretreatments that can break them would be expected to show a higher digestibility compared to those that do not. However, based on a number of studies S/G ratio may or may not be correlated to improved digestibility, depending on other plant cell wall properties and whether and how the plant was pretreated (Chen et al. 2002; Mechin et al. 2005; Davison et al. 2006; Li et al. 2010; Studer et al. 2011b; Zhang et al. 2011b).

14.5 Designing Improved Feedstocks

A number of strategies for developing "plants designed for deconstruction" have been reviewed in recent years (Carpita 2012; Jung et al. 2012; Abramson et al. 2013). These strategies can be grouped broadly as altering lignin (content, monolignol composition, and degree of polymerization), increasing and/or altering polysaccharides (content, composition, or crystallinity), expressing cell wall-degrading or modifying enzymes *in planta*, or producing oils in vegetative tissues.

14.5.1 Alterations to Lignin

Initial studies on plants with altered lignin contents began with the "brown midrib" mutations (Barrière et al. 2004) for improved ruminant digestibility. Plant lines have subsequently been engineered with decreased and altered lignin content by changing the expression of monolignol biosynthetic enzymes. Decreasing expression of one or more of the monolignol synthesis enzymes has been shown to decrease total lignin content and improve the enzymatic digestibility of alfalfa following hot water pretreatment (Chen and Dixon 2007). However, decreasing the total lignin content of the cell wall also impairs the overall fitness of the plant and can lead to dwarfed plants and failure to accumulate biomass (Casler et al. 2002; Voelker et al. 2011). As a consequence of this, more recent strategies have been focused on altering the ratio of monolignols, and increasing the S/G ratio in hybrid poplar has been shown to improve alkaline delignification (Stewart et al. 2009) and digestibility following alkaline and dilute acid pretreatment, though there was no significant difference following AFEX™ treatment (Ong 2011). Increasing S/G in *Arabidopsis* was shown to improve the enzymatic release of glucose following hot water pretreatment (Li et al. 2010). A recent study found that decreasing total lignin content concurrently with decreasing S/G in switchgrass improved the enzymatic glucose yield following dilute acid pretreatment, as well as decreasing pretreatment severity and cellulase loadings, and increasing ethanol yield (Fu et al. 2011).

Another strategy has been to introduce novel monolignols or proteins that make the cell wall more amenable to chemical deconstruction without impacting total lignin content or plant fitness. These approaches, all of which have been shown to increase digestibility and/or lignin removal to some extent include adding monolignols that shorten the degree of polymerization (*p*- hydroxybenzyaldehydes) (Eudes et al. 2012), monolignols that incorporate alkali-labile ester linkages within the lignin matrix, e.g. novel ester-based di-lignols as lignin precursors (Grabber et al. 2008; Simmons et al. 2010), and glycoproteins that participate in cross-couplings with lignin, such as tyrosine-rich hydroxyproline-rich glycoprotein (Liang et al. 2008).

14.5.2 Alterations to Polysaccharides

Altering cell wall polysaccharides is another method to reduce cell wall recalcitrance or increase the amount of substrate. One strategy is to decrease cellulose crystallinity by overexpressing cellulose synthases with impaired functionality (Harris et al. 2009) or by overexpressing a membrane-bound endoglucanase, KORRIGAN (Maloney and Mansfield 2010). Another strategy is to increase the carbohydrate content of the plant cell wall. Cellulose content and crystallinity increased in poplar by over-expressing a sucrose synthase gene (Coleman et al. 2006) and various amorphous polysaccharides have been targeted for accumulation, including starch (Chuck et al. 2011 and mixed-linkage β-glucans (Pauly et al. 2011). In contrast, reductions in glucuronoxylan content in poplar showed an increase in digestibility by enzymes alone (Lee et al. 2009). In rice, loss of activity for a xylosyltransferase thought be responsible for arabinosyl substitution of the xylan backbone resulted in a slight increase in arabinose substitution and decrease in hydroxycinnamic acid content, resulting in increased extractability of xylan and enzymatic digestibility (Chiniquy et al. 2012). Alteration of *O*-acetylation of hemicelluloses may also lead to a decrease in acetate content for reduced inhibition of fermentation or altered capacity of hemicelluloses to hydrogen bond with other cell wall polymers (Gille and Pauly 2012). Other work has demonstrated improved enzymatic digestibility of plant cell walls by preventing de-methyl esterification in the pectin homogalacturonan (Lionetti et al. 2010), which limits the ability to form Ca^{2+}-mediated cross-links, increasing primary cell wall porosity and decreasing rigidity and cell-to-cell adhesion in primary cell walls.

14.5.3 Transcription Factors for Secondary Cell Wall Formation

Regulatory networks have recently been identified comprising several transcription factors that act as "master switches" responsible for controlling the temporal and spatial regulation of collections of genes involved in the secondary cell wall synthesis, assembly, and thickening (Shen et al. 2012). One study ectopically overexpressed a MYB transcription factor in switchgrass to down-regulate the genes associated with monolignol biosynthetic pathways and identified phenotypic outcomes of reduced lignin and reduced *p*-coumarate to ferulate ratios that resulted in a tripling of enzymatic sugar release (Shen et al. 2012). Other work identified a mutation in WRKY transcription factors to be responsible for secondary cell wall thickening and significantly increased cellulose, hemicellulose, and lignin deposition in the pith cells of model dicots, increasing the overall plant density, and potentially providing a route for increasing accumulation of fermentable sugars in plant cell walls (Verma et al. 2010).

14.5.4 Expression of Cell Wall Degrading Enzymes in Planta

The high cost and doses of enzymes required for cellulosic biofuels are critical economic barriers for commercialization. Expression of thermophilic cellulases in the apoplast (Sticklen 2006) or mesophilic cellulases in chloroplasts (Verma et al. 2010) are one possible route for generating some of the cellulolytic enzymes in situ. Cellulolytic enzymes can be generated *in planta* to supplement other enzymes during hydrolysis, however even mild pretreatment of the biomass can significantly lower their activity (Teymouri et al. 2004). Expression of feruloyl esterases in grasses which cleave ferulate ester cross-links has been found to improve both enzymatic and in vitro ruminant digestibilities (Buanafina et al. 2008). Expression of plant cellulolytic enzymes that are active under plant physiological conditions (Hartati et al. 2008) or cellulose binding modules (CBMs) (Shoseyov et al. 2006) in the apoplast have been found to increase growth and biomass accumulation, presumably by increased cell wall loosening, but with the potential disadvantage of impaired plant fitness.

14.5.5 Production of Oils in Vegetative Tissues

One way to increase the energy content of lignocellulosic biomass is to modify plants to produce oils, fatty acids, or triacylglycerols (TAGs) in vegetative tissues (Durrett et al. 2008). In one study triacylglycerols were accumulated in senescing *Arabidopsis* leaves by either blocking fatty acid breakdown, or by ectopically expressing the LEC2 seed development transcription factor in leaves (Slocombe et al. 2009). Another study successfully shifted the carbon flux in *Arabidopsis* leaves from starch biosynthesis to the production and accumulation of triacylglycerols by simultaneously reducing the expression of a catalytic subunit of ADP-glucose pyrophosphorylase and ectopically expressing the WRINKLED1 transcription factor that is involved in seed oil biosynthesis (Sanjaya et al. 2011).

14.6 Pretreatment as a Screening and Analysis Tool: Expanding Our Understanding of the Plant Cell Wall

Re-engineering plants to provide phenotypic traits desirable of an ideal biofuel energy crop is an area of intense research, as highlighted previously. However, it is vitally important to evaluate processing capabilities of new materials as they are being generated, as biomass recalcitrance may not favorably correlate with the traits selected for during transgenesis or breeding. With recent advances in high-throughput analytical techniques, it is now feasible to quickly screen for desirable traits from very large libraries of biomass phenotypes, while requiring only small sample quantities for detailed analyses. In addition to screening, high-throughput techniques are also helping to further understanding of the relationship between biomass conversion and plant cell wall characteristics. For example, high throughput composition analysis techniques allowed for screening of thousands of poplar samples for lignin content and S/G ratios, and from this a fairly large subset was further tested using a high-throughput pretreatment and enzymatic hydrolysis method in order to determine the relative impacts of lignin and S/G ratio on sugar yields (Studer et al. 2011b).

As in the example above, high-throughput pretreatments can now be carried out in custom-designed microplate-based reactors that have been developed for both acidic and alkaline pretreatments (Santoro et al. 2010; Selig et al. 2010; Studer et al. 2011a). Rapid, small-scale compositional analysis methods are able to determine cell wall composition, both before and after pretreatment (DeMartini et al. 2011b; Selig et al. 2011). These techniques can be coupled to medium/high-throughput analyses using LC–MS/MS and 2D-NMR for more detailed elucidation of changes in cell wall structure, composition, and degradation (Chundawat et al. 2008; Kim and Ralph 2010;

Morreel et al. 2010). Semi-automated (medium/low throughput) electron micrography and immunolabeling based techniques have also been used in recent years to characterize the complex interplay of pretreatment severity and cell wall ultra-structural modifications (Donohoe et al. 2009; Pattathil et al. 2010; Chundawat et al. 2011; Zhang et al. 2013a). To this end a bio-analytic toolkit was developed, comprising more than 200 glycan-directed monoclonal antibodies that recognize distinct epitopes present on various categories of plant cell wall polysaccharides (Pattathil et al. 2010). This microplate-based, quantitative assay has provided insights into the relationship between pretreatment severity and cell wall polysaccharide accessibility and extraction, and the molecular architecture of the plant cell wall (Alonso-Simón et al. 2010; DeMartini et al. 2011a). As indicated by Moller et al. (2007), monoclonal antibodies directed against cell wall glycans provides complementary compositional data that could be used to optimize pretreatment conditions and enzyme cocktails necessary for more efficient degradation of lignocellulose.

The effectiveness of pretreatments on bioconversion has been evaluated using micro-scale based rapid enzymatic hydrolysis (Chundawat et al. 2008; Banerjee et al. 2010; Gomez et al. 2010; Jäger et al. 2011; Riedlberger and Weuster-Botz 2012) and microbial fermentation based assays (Funke et al. 2010; Riedlberger and Weuster-Botz 2012). These assays can be coupled with microplate-based pretreatments to facilitate rapid screening of several hundred biomass specimens (Studer et al. 2010). Additionally, with developments in micro-scale cell-free protein expression systems it is possible to selectively optimize enzyme combinations necessary for different pretreatments and biomass types (Chandrasekaran et al. 2010).

14.7 Conclusions

In recent years understanding of the chemistry and structure of the plant cell wall has progressed rapidly. Pretreatment research has contributed to understanding of the distribution and composition of various cell wall polysaccharides within the many different classes of cell walls. Future work will continue to delve more deeply into the complex relationships between cell wall and pretreatment chemistry to improve and develop novel conversion methods for release of cell wall sugars and to improve biomass characteristics for conversion to biofuels. High-throughput analytical techniques and tools that allow for rapid analysis of small quantities of samples will allow for more efficient comparisons in the development of new feedstocks and processing methods, and improved understanding of the fundamental relationships between cell wall chemistry and structure and pretreatment chemistry.

References

Abe K, Yano H (2009) Comparison of the characteristics of cellulose microfibril aggregates of wood, rice straw and potato tuber. Cellulose 16:1017–1023

Abramson M, Shoseyov O, Hirsch S, Shani Z (2013) Genetic modifications of plant cell walls to increase biomass and bioethanol production. In: Lee JW (ed) Advanced biofuels and bio-products. Springer, New York, pp 315–338

Adler PR, Sanderson MA, Boateng AA, Weimer PJ, Jung H-JG (2006) Biomass yield and biofuel quality of switchgrass harvested in fall or spring. Agron J 98:1518–1525

Åkerholm M, Salmén L (2001) Interactions between wood polymers studied by dynamic ft-ir spectroscopy. Polym 42:963–969

Åkerholm M, Salmén L (2004) Softening of wood polymers induced by moisture studied by dynamic FTIR spectroscopy. J Appl Polym Sci 94:2032–2040

Alonso-Simón A, Kristensen JB, Øbro J, Felby C, Willats WGT, Jørgensen H (2010) High-throughput microarray profiling of cell wall polymers during hydrothermal pre-treatment of wheat straw. Biotechnol Bioeng 105:509–514

Altaner CM, Jarvis MC (2008) Modelling polymer interactions of the `molecular velcro' type in wood under mechanical stress. J Theor Biol 253:434–445

Arantes V, Saddler J (2011) Cellulose accessibility limits the effectiveness of minimum cellulase loading on the efficient hydrolysis of pretreated lignocellulosic substrates. Biotechnol Biofuel 4(3):1–16

Atalla RH, Vanderhart DL (1984) Native cellulose: a composite of two distinct crystalline forms. Science 223:283–285

Awal A, Sain M (2011) Spectroscopic studies and evaluation of thermorheological properties of softwood and hardwood lignin. J Appl Polym Sci 122:956–963

Azarpira A, Lu F, Ralph J (2011) Reactions of dehydrodiferulates with ammonia. Org Biomol Chem 9:6779–6787

Bals B, Rogers C, Jin MJ, Balan V, Dale B (2010) Evaluation of ammonia fibre expansion (AFEX) pretreatment for enzymatic hydrolysis of switchgrass harvested in different seasons and locations. Biotechnol Biofuel 3(1):1–11

Banerjee G, Car S, Scott-Craig J, Borrusch M, Walton J (2010) Rapid optimization of enzyme mixtures for deconstruction of diverse pretreatment/biomass feedstock combinations. Biotechnol Biofuel 3:22

Barrière Y, Ralph J, Méchin V, Guillaumie S, Grabber JH, Argillier O et al (2004) Genetic and molecular basis of grass cell wall biosynthesis and degradability. II. Lessons from brown-midrib mutants. C R Biol 327:847–860

Buanafina MMdO, Langdon T, Hauck B, Dalton S, Morris P (2008) Expression of a fungal ferulic acid esterase increases cell wall digestibility of tall fescue (*Festuca arundinacea*). Plant Biotechnol J 6:264–280

Carpita NC (2012) Progress in the biological synthesis of the plant cell wall: New ideas for improving biomass for bioenergy. Curr Opin Biotechnol 23:330–337

Carpita NC, Gibeaut DM (1993) Structural models of primary cell walls in flowering plants: Consistency of molecular structure with the physical properties of the walls during growth. Plant J 3:1–30

Casler MD, Buxton DR, Vogel KP (2002) Genetic modification of lignin concentration affects fitness of perennial herbaceous plants. Theor Appl Genet 104:127–131

Chandrasekaran A, Bharadwaj R, Park JI, Sapra R, Adams PD, Singh AK (2010) A microscale platform for integrated cell-free expression and activity screening of cellulases. J Proteome Res 9:5677–5683

Chen F, Dixon RA (2007) Lignin modification improves fermentable sugar yields for biofuel production. Nat Biotechnol 25:759–761

Chen L, Auh C, Chen F, Cheng XF, Aljoe H, Dixon RA et al (2002) Lignin deposition and associated changes in anatomy, enzyme activity, gene expression, and ruminal degradability in stems of tall fescue at different developmental stages. J Agric Food Chem 50:5558–5565

Chiniquy D, Sharma V, Schultink A, Baidoo EE, Rautengarten C, Cheng K et al (2012) XAX1 from glycosyltransferase family 61 mediates xylosyltransfer to rice xylan. PNAS 109:17117–17122

Chuck GS, Tobias C, Sun L, Kraemer F, Li C, Dibble D, et al (2011) Overexpression of the maize Corngrass1 microRNA prevents flowering, improves digestibility, and increases starch content of switchgrass. PNAS, http://www.pnas.org/content/early/2011/10/04/1113971108.abstract. (Ahead of Print)

Chundawat SPS, Balan V, Dale BE (2008) High-throughput microplate technique for enzymatic hydrolysis of lignocellulosic biomass. Biotechnol Bioeng 99:1281–1294

Chundawat SPS, Donohoe BS, Sousa L, Elder T, Agarwal UP, Lu F et al (2011) Multi-scale visualization and characterization of plant cell wall deconstruction during thermochemical pretreatment. Energ Environ Sci 4:973–984

Chundawat SPS, Vismeh R, Sharma LN, Humpula JF, da Costa Sousa L, Chambliss CK et al (2010) Multifaceted characterization of cell wall decomposition products formed during ammonia fiber expansion (AFEX) and dilute acid based pretreatments. Bioresour Technol 101:8429–8438

Clayton DW, Phelps GR (1965) The sorption of glucomannan and xylan on α-cellulose wood fibers. J Polym Sci: Part C 11:197–220

Coleman HD, Ellis DD, Gilbert M, Mansfield SD (2006) Up-regulation of sucrose synthase and udp-glucose pyrophosphorylase impacts plant growth and metabolism. Plant Biotechnol J 4:87–101

Cosgrove DJ (2005) Growth of the plant cell wall. Nat Rev Mol Cell Biol 6:850–861

da Costa Sousa L, Chundawat SPS, Balan V, Dale BE (2009) `Cradle-to-grave' assessment of existing lignocellulose pretreatment technologies. Curr Opin Biotechnol 20:339–347

Dammström S, Salmén L, Gatenholm P (2009) On the interactions between cellulose and xylan, a biomimetic simulation of the hardwood cell wall. BioResources 4:3–14

Davison BH, Drescher SR, Tuskan GA, Davis MF, Nghiem NP (2006) Variation of S/G ratio and lignin content in a *Populus* family influences the release of xylose by dilute acid hydrolysis. Appl Biochem Biotechnol 129–132:427–435

de Lima DU, Buckeridge MS (2001) Interaction between cellulose and storage xyloglucans: the influence of the degree of galactosylation. Carbohydr Polym 46:157–163

DeMartini J, Wyman C (2011a) Composition and hydrothermal pretreatment and enzymatic saccharification performance of grasses and legumes from a mixed-species prairie. Biotechnol Biofuel 4(52):1–10

DeMartini JD, Pattathil S, Avci U, Szekalski K, Mazumder K, Hahn MG et al (2011a) Application of monoclonal antibodies to investigate plant cell wall deconstruction for biofuels production. Energ Environ Sci 4:4332–4339

DeMartini JD, Studer MH, Wyman CE (2011b) Small-scale and automatable high-throughput compositional analysis of biomass. Biotechnol Bioeng 108:306–312

DeMartini JD, Wyman CE (2011b) Changes in composition and sugar release across the annual rings of *Populus* wood and implications on recalcitrance. Bioresour Technol 102:1352–1358

Dien BS, Jung H-JG, Vogel KP, Casler MD, Lamb JFS, Iten L et al (2006) Chemical composition and response to dilute-acid pretreatment and enzymatic saccharification of alfalfa, reed canarygrass, and switchgrass. Biomass Bioenerg 30:880–891

Donaldson L (2007) Cellulose microfibril aggregates and their size variation with cell wall type. Wood Sci Technol 41:443–460

Donohoe B, Decker S, Tucker M, Himmel M, Vinzant T (2008) Visualizing lignin coalescence and migration through maize cell walls following thermochemical pretreatment. Biotechnol Bioeng 101:913–925

Donohoe BS, Selig MJ, Viamajala S, Vinzant TB, Adney WS, Himmel ME (2009) Detecting cellulase penetration into corn stover cell walls by immuno-electron microscopy. Biotechnol Bioeng 103:480–489

Donohoe BS, Vinzant TB, Elander RT, Pallapolu VR, Lee YY, Garlock RJ et al (2011) Surface and ultrastructural characterization of raw and pretreated switchgrass. Bioresour Technol 102:11097–11104

Du BW, Sharma LN, Becker C, Chen S-F, Mowery RA, van Walsum GP et al (2010) Effect of varying feedstock–pretreatment chemistry combinations on the formation and accumulation of potentially inhibitory degradation products in biomass hydrolysates. Biotechnol Bioeng 107:430–440

Duguid KB, Montross MD, Radtke CW, Crofcheck CL, Shearer SA, Hoskinson RL (2007) Screening for sugar and ethanol processing characteristics from anatomical fractions of wheat stover. Biomass Bioenerg 31:585–592

Duguid KB, Montross MD, Radtke CW, Crofcheck CL, Wendt LM, Shearer SA (2009) Effect of anatomical fractionation on the enzymatic hydrolysis of acid and alkaline pretreated corn stover. Bioresour Technol 100:5189–5195

Durrett TP, Benning C, Ohlrogge J (2008) Plant triacylglycerols as feedstocks for the production of biofuels. Plant J 54:593–607

Engels FM, Jung HG (1998) Alfalfa stem tissues: cell-wall development and lignification. Ann Bot 82:561–568

Eudes A, George A, Mukerjee P, Kim JS, Pollet B, Benke PI et al (2012) Biosynthesis and incorporation of side-chain-truncated lignin monomers to reduce lignin polymerization and enhance saccharification. Plant Biotechnol J 10:609–620

Fu C, Mielenz JR, Xiao X, Ge Y, Hamilton CY, Rodriguez M, et al (2011) Genetic manipulation of lignin reduces recalcitrance and improves ethanol production from switchgrass. PNAS, http://www.pnas.org/content/early/2011/02/04/1100310108.abstract. (Ahead of Print)

Fu D, Mazza G (2011) Aqueous ionic liquid pretreatment of straw. Bioresour Technol 102:7008–7011

Funke M, Buchenauer A, Mokwa W, Kluge S, Hein L, Muller C et al (2010) Bioprocess control in microscale: scalable fermentations in disposable and user-friendly microfluidic systems. Microb Cell Fact 9(86):1–13

Garlock R, Chundawat S, Balan V, Dale B (2009) Optimizing harvest of corn stover fractions based on overall sugar yields following ammonia fiber expansion pretreatment and enzymatic hydrolysis. Biotechnol Biofuel 2(29):1–14

Garlock RJ, Balan V, Dale BE (2012a) Optimization of AFEX™ pretreatment conditions and enzyme mixtures to maximize sugar release from upland and lowland switchgrass. Bioresour Technol 104:757–768

Garlock RJ, Balan V, Dale BE, Ramesh Pallapolu V, Lee YY, Kim Y et al (2011) Comparative material balances around pretreatment technologies for the conversion of switchgrass to soluble sugars. Bioresour Technol 102:11063–11071

Garlock RJ, Bals B, Jasrotia P, Balan V, Dale BE (2012b) Influence of variable species composition on the saccharification of AFEX™ pretreated biomass from unmanaged fields in comparison to corn stover. Biomass Bioenerg 37:49–59

Garrote G, Domínguez H, Parajó JC (1999) Hydrothermal processing of lignocellulosic materials. Eur J Wood Wood Prod 57:191–202

Gericke M, Fardim P, Heinze T (2012) Ionic liquids—promising but challenging solvents for homogeneous derivatization of cellulose. Molecules 17:7458–7502

Gille S, Pauly M (2012) *O*-acetylation of plant cell wall polysaccharides. Front Plant Sci 3(12):1–7

Gomez L, Whitehead C, Barakate A, Halpin C, McQueen-Mason S (2010) Automated saccharification assay for determination of digestibility in plant materials. Biotechnol Biofuel 3(23):1–12

Grabber JH, Hatfield RD, Lu FC, Ralph J (2008) Coniferyl ferulate incorporation into lignin enhances the alkaline delignification and enzymatic degradation of cell walls. Biomacromolecules 9:2510–2516

Grabber JH, Ralph J, Hatfield RD (1998) Ferulate cross-links limit the enzymatic degradation of synthetically lignified primary walls of maize. J Agric Food Chem 46:2609–2614

Harris D, Stork J, Debolt S (2009) Genetic modification in cellulose-synthase reduces crystallinity and improves biochemical conversion to fermentable sugar. GCB Bioenergy 1:51–61

Harris P, Trethewey J (2010) The distribution of ester-linked ferulic acid in the cell walls of angiosperms. Phytochem Rev 9:19–33

Hartati S, Sudarmonowati E, Park YW, Kaku T, Kaida R, Baba Ki et al (2008) Overexpression of poplar cellulase accelerates growth and disturbs the closing movements of leaves in sengon. Plant Physiol 147:552–561

Iiyama K, Lam TBT, Stone BA (1990) Phenolic acid bridges between polysaccharides and lignin in wheat internodes. Phytochem 29:733–737

Imamura T, Watanabe T, Kuwahara M, Koshijima T (1994) Ester linkages between lignin and glucuronic acid in lignin-carbohydrate complexes from *Fagus crenata*. Phytochem 37:1165–1173

Ishizawa CI, Davis MF, Schell DF, Johnson DK (2007) Porosity and its effect on the digestibility of dilute sulfuric acid pretreated corn stover. J Agric Food Chem 55:2575–2581

Ivakov A, Persson S (2012) Plant cell walls. eLS. Wiley, pp 1–17 doi: 10.1002/9780470015902.a0001682.pub2

Jäger G, Wulfhorst H, Zeithammel EU, Elinidou E, Spiess AC, Büchs J (2011) Screening of cellulases for biofuel production: online monitoring of the enzymatic hydrolysis of insoluble cellulose using high-throughput scattered light detection. Biotechnol J 6:74–85

Jarvis MC (2012) Sclerenchyma. eLS. Wiley, pp 1–3 doi: 10.1002/9780470015902.a0002082.pub2

Jeoh T, Ishizawa CI, Davis MF, Himmel ME, Adney WS, Johnson DK (2007) Cellulase digestibility of pretreated biomass is limited by cellulose accessibility. Biotechnol Bioeng 98:112–122

Jung H-JG, Samac DA, Sarath G (2012) Modifying crops to increase cell wall digestibility. Plant Sci 185–186:65–77

Kabel MA, van den Borne H, Vincken J-P, Voragen AGJ, Schols HA (2007) Structural differences of xylans affect their interaction with cellulose. Carbohydr Polym 69:94–105

Karlsson O, Ikeda T, Kishimoto T, Magara K, Matsumoto Y, Hosoya S (2004) Isolation of lignin–carbohydrate bonds in wood. Model experiments and preliminary application to pine wood. J Wood Sci 50:141–150

Karlsson O, Lundquist K, Meuller S, Westlid K (1988) On the acidolytic cleavage of arylglycerol β-aryl ethers. Acta Chem Scand B 42:48–51

Keskar SS, Edye LA, Doherty WOS, Bartley JP (2011) The chemistry of acid catalyzed delignification of sugarcane bagasse in the ionic liquid trihexyl tetradecyl phosphonium chloride. J Wood Chem Technol 32:71–81

Kim H, Ralph J (2010) Solution-state 2D NMR of ball-milled plant cell wall gels in DMSO-d6/pyridine-d5. Org Biomol Chem 8:576–591

Kim Y, Mosier NS, Ladisch MR, Ramesh Pallapolu V, Lee YY, Garlock R et al (2011) Comparative study on enzymatic digestibility of switchgrass varieties and harvests processed by leading pretreatment technologies. Bioresour Technol 102:11089–11096

Kimon KS, Leslie Alan E, William Orlando Sinclair D (2011) Enhanced saccharification kinetics of sugarcane bagasse pretreated in 1-butyl-3-methylimidazolium chloride at high temperature and without complete dissolution. Bioresour Technol 102:9325–9329

King AWT, Parviainen A, Karhunen P, Matikainen J, Hauru LKJ, Sixta H et al (2012) Relative and inherent reactivities of imidazolium-based ionic liquids: the implications for lignocellulose processing applications. RSC Adv 2:8020–8026

Kishimoto T, Chiba W, Saito K, Fukushima K, Uraki Y, Ubukata M (2009) Influence of syringyl to guaiacyl ratio on the structure of natural and synthetic lignins. J Agric Food Chem 58:895–901

Knill CJ, Kennedy JF (2003) Degradation of cellulose under alkaline conditions. Carbohydr Polym 51:281–300

Koo B-W, Min B-C, Gwak K-S, Lee S-M, Choi J-W, Yeo H et al (2012) Structural changes in lignin during organosolv pretreatment of *Liriodendron tulipifera* and the effect on enzymatic hydrolysis. Biomass Bioenerg 42:24–32

Kumar R, Mago G, Balan V, Wyman CE (2009) Physical and chemical characterizations of corn stover and poplar solids resulting from leading pretreatment technologies. Bioresour Technol 100:3948–3962

Larsen SU, Bruun S, Lindedam J (2012) Straw yield and saccharification potential for ethanol in cereal species and wheat cultivars. Biomass Bioenerg 45:239–250

Lawoko M, Henriksson G, Gellerstedt G (2006) Characterisation of lignin-carbohydrate complexes (LCCs) of spruce wood (*Picea abies* L.) isolated with two methods. Holzforschung 60:156–161

Le Ngoc Huyen T, Rémond C, Dheilly RM, Chabbert B (2010) Effect of harvesting date on the composition and saccharification of *Miscanthus x giganteus*. Bioresour Technol 101:8224–8231

Lee C, Teng Q, Huang W, Zhong R, Ye Z-H (2009) Down-regulation of PoGT47C expression in poplar results in a reduced glucuronoxylan content and an increased wood digestibility by cellulase. Plant Cell Physiol 50:1075–1089

Li W, Sun N, Stoner B, Jiang X, Lu X, Rogers RD (2011) Rapid dissolution of lignocellulosic biomass in ionic liquids using temperatures above the glass transition of lignin. Green Chem 13:2038–2047

Li X, Ximenes E, Kim Y, Slininger M, Meilan R, Ladisch M et al (2010) Lignin monomer composition affects *Arabidopsis* cell-wall degradability after liquid hot water pretreatment. Biotechnol Biofuel 3(27):1–7

Liang H, Frost CJ, Wei X, Brown NR, Carlson JE, Tien M (2008) Improved sugar release from lignocellulosic material by introducing a tyrosine-rich cell wall peptide gene in poplar. CLEAN–Soil, Air, Water 36:662–668

Lindedam J, Andersen SB, DeMartini J, Bruun S, Jørgensen H, Felby C et al (2012) Cultivar variation and selection potential relevant to the production of cellulosic ethanol from wheat straw. Biomass Bioenerg 37:221–228

Lindedam J, Bruun S, Jorgensen H, Felby C, Magid J (2010) Cellulosic ethanol: Interactions between cultivar and enzyme loading in wheat straw processing. Biotechnol Biofuel 3(25):1–10

Lionetti V, Francocci F, Ferrari S, Volpi C, Bellincampi D, Galletti R et al (2010) Engineering the cell wall by reducing de-methyl-esterified homogalacturonan improves saccharification of plant tissues for bioconversion. PNAS 107:616–621

Lundquist K (1973) Acid degradation of lignin. Part VIII. Low molecular weight phenols from acidolysis of birch lignin. Acta Chem Scand 27:2597–2606

Lundquist K, Lundgren R (1972) Acid degradation of lignin. Part VII. The cleavage of ether bonds. Acta Chem Scand 26:2005–2023

MacFarlane DR, Pringle JM, Johansson KM, Forsyth SA, Forsyth M (2006) Lewis base ionic liquids. Chem Commun, 1905–1917, doi: 10.1039/B516961P

Maloney MT, Chapman TW, Baker AJ (1985) Dilute acid hydrolysis of paper birch: kinetics studies of xylan and acetyl-group hydrolysis. Biotechnol Bioeng 27:355–361

Maloney VJ, Mansfield SD (2010) Characterization and varied expression of a membrane-bound endo-β-1,4-glucanase in hybrid poplar. Plant Biotechnol J 8:294–307

McGee JK, April GC (1982) Chemicals from renewable resources: hemicellulose behavior during organosolv delignification of southern yellow pine. Chem Eng Commun 19:49–56

Mechin V, Argillier O, Rocher F, Hebert Y, Mila I, Pollet B et al (2005) In search of a maize ideotype for cell wall enzymatic degradability using histological and biochemical lignin characterization. J Agric Food Chem 53:5872–5881

Moller I, Sørensen I, Bernal AJ, Blaukopf C, Lee K, Øbro J et al (2007) High-throughput mapping of cell-wall polymers within and between plants using novel microarrays. Plant J 50:1118–1128

Morreel K, Dima O, Kim H, Lu F, Niculaes C, Vanholme R et al (2010) Mass spectrometry-based sequencing of lignin oligomers. Plant Physiol 153:1464–1478

Muhammad N, Omar WN, Man Z, Bustam MA, Rafiq S, Uemura Y (2012) Effect of ionic liquid treatment on pyrolysis products from bamboo. Ind Eng Chem Res 51:2280–2289

Ong RG (2011) Interactions between biomass feedstock characteristics and bioenergy production: from the landscape to the molecular scale, PhD., Michigan State University, USA

Pattathil S, Avci U, Baldwin D, Swennes AG, McGill JA, Popper Z et al (2010) A comprehensive toolkit of plant cell wall glycan-directed monoclonal antibodies. Plant Physiol 153:514–525

Pauly M, Hake S, Kraemer FJ (2011) Maize variety and method of production. US Patent 13/152,219, Filed 2 Jun 2011

Pedersen M, Meyer AS (2010) Lignocellulose pretreatment severity—relating pH to biomatrix opening. New Biotechnol 27:739–750

Pordesimo LO, Hames BR, Sokhansanj S, Edens WC (2005) Variation in corn stover composition and energy content with crop maturity. Biomass Bioenerg 28:366–374

Ralph J, Brunow G, Boerjan W (2007) Lignins. eLS. Wiley, pp 1–10, doi: 10.1002/9780470015902.a0020104

Rencoret J, Gutiérrez A, Nieto L, Jiménez-Barbero J, Faulds CB, Kim H et al (2011) Lignin composition and structure in young versus adult *Eucalyptus globulus* plants. Plant Physiol 155:667–682

Riedlberger P, Weuster-Botz D (2012) New miniature stirred-tank bioreactors for parallel study of enzymatic biomass hydrolysis. Bioresour Technol 106:138–146

Rock K, Thelemann R, Jung H-J, Tschirner U, Sheaffer C, Johnson G (2009) Variation due to growth environment in alfalfa yield, cellulosic ethanol traits, and paper pulp characteristics. Bioenerg Res 2:79–89

Rollin JA, Zhu Z, Sathitsuksanoh N, Zhang YHP (2011) Increasing cellulose accessibility is more important than removing lignin: a comparison of cellulose solvent-based lignocellulose fractionation and soaking in aqueous ammonia. Biotechnol Bioeng 108:22–30

Saake B, Lehnen R (2007) Lignin. In: Ullmann's encyclopedia of industrial chemistry. Wiley-VCH, doi: 10.1002/14356007.a15_305.pub3

Sanjaya Durrett TP, Weise SE, Benning C (2011) Increasing the energy density of vegetative tissues by diverting carbon from starch to oil biosynthesis in transgenic *Arabidopsis*. Plant Biotechnol J 9:874–883

Sannigrahi P, Ragauskas AJ, Miller SJ (2009) Lignin structural modifications resulting from ethanol organosolv treatment of loblolly pine. Energ Fuel 24:683–689

Santoro N, Cantu S, Tornqvist C-E, Falbel T, Bolivar J, Patterson S et al (2010) A high-throughput platform for screening milligram quantities of plant biomass for lignocellulose digestibility. Bioenerg Res 3:93–102

Scheller HV, Ulvskov P (2010) Hemicelluloses. Annu Rev Plant Biol 61:263–289

Selig M, Tucker M, Law C, Doeppke C, Himmel M, Decker S (2011) High throughput determination of glucan and xylan fractions in lignocelluloses. Biotechnol Lett 33:961–967

Selig MJ, Tucker MP, Sykes RW, Reichel KL, Brunecky R, Himmel ME et al (2010) Lignocellulose recalcitrance screening by integrated high-throughput hydrothermal pretreatment and enzymatic saccharification. Ind Biotechnol 6:104–111

Sewalt VJH, Fontenot JP, Allen VG, Glasser WG (1996) Fiber composition and in vitro digestibility of corn stover fractions in response to ammonia treatment. J Agric Food Chem 44:3136–3142

Shen H, He X, Poovaiah CR, Wuddineh WA, Ma J, Mann DGJ et al (2012) Functional characterization of the switchgrass (*Panicum virgatum*) R2R3-MYB transcription factor PvMYB4 for improvement of lignocellulosic feedstocks. New Phytol 193:121–136

Shi J, Ebrik MA, Yang B, Garlock RJ, Balan V, Dale BE et al (2011) Application of cellulase and hemicellulase to pure xylan, pure cellulose, and switchgrass solids from leading pretreatments. Bioresour Technol 102:11080–11088

Shoseyov O, Shani Z, Levy I (2006) Carbohydrate binding modules: biochemical properties and novel applications. Microbiol Mol Biol R 70:283–295

Simmons BA, Loqué D, Ralph J (2010) Advances in modifying lignin for enhanced biofuel production. Curr Opin Plant Biol 13:312–319

Singh S, Simmons BA, Vogel KP (2009) Visualization of biomass solubilization and cellulose regeneration during ionic liquid pretreatment of switchgrass. Biotechnol Bioeng 104:68–75

Siqueira G, Milagres A, Carvalho W, Koch G, Ferraz A (2011) Topochemical distribution of lignin and hydroxycinnamic acids in sugar-cane cell walls and its correlation with the enzymatic hydrolysis of polysaccharides. Biotechnol Biofuel 4:7

Slocombe SP, Cornah J, Pinfield-Wells H, Soady K, Zhang Q, Gilday A et al (2009) Oil accumulation in leaves directed by modification of fatty acid breakdown and lipid synthesis pathways. Plant Biotechnol J 7:694–703

Stewart JJ, Akiyama T, Chapple C, Ralph J, Mansfield SD (2009) The effects on lignin structure of overexpression of ferulate 5-hydroxylase in hybrid poplar. Plant Physiol 150:621–635

Sticklen M (2006) Plant genetic engineering to improve biomass characteristics for biofuels. Curr Opin Biotechnol 17:315–319

Stone B (2005) Cellulose: Structure and distribution. eLS. Wiley, pp 1–8, doi: 10.1038/npg.els.0003892

Studer M, Brethauer S, DeMartini J, McKenzie H, Wyman C (2011a) Co-hydrolysis of hydrothermal and dilute acid pretreated populus slurries to support development of a high-throughput pretreatment system. Biotechnol Biofuel 4(19):1–10

Studer MH, DeMartini JD, Brethauer S, McKenzie HL, Wyman CE (2010) Engineering of a high-throughput screening system to identify cellulosic biomass, pretreatments, and enzyme formulations that enhance sugar release. Biotechnol Bioeng 105:231–238

Studer MH, DeMartini JD, Davis MF, Sykes RW, Davison B, Keller M et al (2011b) Lignin content in natural *Populus* variants affects sugar release. PNAS 108:6300–6305

Sun L, Simmons BA, Singh S (2011) Understanding tissue specific compositions of bioenergy feedstocks through hyperspectral raman imaging. Biotechnol Bioeng 108:286–295

Sun Y, Cheng JJ (2005) Dilute acid pretreatment of rye straw and bermudagrass for ethanol production. Bioresour Technol 96:1599–1606

Tadesse H, Luque R (2011) Advances on biomass pretreatment using ionic liquids: an overview. Energ Environ Sci 4:3913–3929

Tarkow H, Feist WC (1969) A mechanism for improving the digestibility of lignocellulosic materials with dilute alkali and liquid ammonia. In: Hajny GJ, Reese ET (eds) Cellulases and their applications. American Chemical Society, Washington, D. C., pp 197–218

Teymouri F, Alizadeh H, Laureano-Pérez L, Dale B, Sticklen M (2004) Effects of ammonia fiber explosion treatment on activity of endoglucanase from acidothermus cellulolyticus in transgenic plant. Appl Biochem Biotechnol 116:1183–1191

Tunc MS, van Heiningen ARP (2008) Hemicellulose extraction of mixed southern hardwood with water at 150°C: effect of time. Ind Eng Chem Res 47:7031–7037

Várnai A, Siika-aho M, Viikari L (2010) Restriction of the enzymatic hydrolysis of steam-pretreated spruce by lignin and hemicellulose. Enzym Microb Technol 46:185–193

Verma D, Kanagaraj A, Jin S, Singh ND, Kolattukudy PE, Daniell H (2010) Chloroplast-derived enzyme cocktails hydrolyse lignocellulosic biomass and release fermentable sugars. Plant Biotechnol J 8:332–350

Vitz J, Erdmenger T, Haensch C, Schubert US (2009) Extended dissolution studies of cellulose in imidazolium based ionic liquids. Green Chem 11:417–424

Voelker SL, Lachenbruch B, Meinzer FC, Kitin P, Strauss SH (2011) Transgenic poplars with reduced lignin show impaired xylem conductivity, growth efficiency and survival. Plant, Cell Environ 34:655–668

Wallace G, Russell WR, Lomax JA, Jarvis MC, Lapierre C, Chesson A (1995) Extraction of phenolic-carbohydrate complexes from graminaceous cell walls. Carbohydr Res 272:41–53

Wang PY, Bolker HI, Purves CB (1964) Ammonolysis of uronic ester groups in birch xylan. Can J Chem 42:2434–2439

Whitney SEC, Brigham JE, Darke AH, Reid JSG, Gidley MJ (1998) Structural aspects of the interaction of mannan-based polysaccharides with bacterial cellulose. Carbohydr Res 307:299–309

Willför S, Sundberg A, Hemming J, Holborn B (2005a) Polysaccharides in selected industrially important softwood species. In: 59th appita annual conference and exhibition. Auckland, New Zealand, pp 415–422

Willför S, Sundberg A, Pranovich A, Holborn B (2005b) Polysaccharides in some industrially important hardwood species. Wood Sci Technol 39:601–617

Wilson JR, Hatfield RD (1997) Structural and chemical changes of cell wall types during stem development: consequences for fibre degradation by rumen microflora. Aust J Agric Res 48:165–180

Wiman M, Dienes D, Hansen MAT, van der Meulen T, Zacchi G, Lidén G (2012) Cellulose accessibility determines the rate of enzymatic hydrolysis of steam-pretreated spruce. Bioresour Technol 126:208–215

Wu H, Mora-Pale M, Miao J, Doherty TV, Linhardt RJ, Dordick JS (2011) Facile pretreatment of lignocellulosic biomass at high loadings in room temperature ionic liquids. Biotechnol Bioeng 108:2865–2875

Wyman CE, Dale BE, Balan V, Elander RT, Holtzapple MT, Ramirez RS, et al (2013) Comparative performance of leading pretreatment technologies for biological conversion of corn stover, poplar wood, and switchgrass to sugars, 1st edn. In: Wyman CE (ed) Aqueous Pretreatment of Plant Biomass for Biological and Chemical Conversion to Fuels and Chemicals. Wiley, London, pp 245–265

Xiao L-P, Shi Z-J, Xu F, Sun R-C (2013) Characterization of lignins isolated with alkaline ethanol from the hydrothermal pretreated *Tamarix ramosissima*. Bioenerg Res 6:1–14

Xu H, Pan W, Wang R, Zhang D, Liu C (2012) Understanding the mechanism of cellulose dissolution in 1-butyl-3-methylimidazolium chloride ionic liquid via quantum chemistry calculations and molecular dynamics simulations. J Comput Aided Mol Des 26:329–337

Zeng M, Ximenes E, Ladisch MR, Mosier NS, Vermerris W, Huang C-P et al (2012) Tissue-specific biomass recalcitrance in corn stover pretreated with liquid hot-water: enzymatic hydrolysis (Part 1). Biotechnol Bioeng 109:390–397

Zhang H, Fangel JU, Willats WGT, Selig MJ, Lindedam J, Jørgensen H, et al (2013a) Assessment of leaf/stem ratio in wheat straw feedstock and impact on enzymatic conversion. GCB Bioenergy, doi: 10.1111/gcbb.12060, (Ahead of Print)

Zhang H, Wu J, Zhang J, He J (2005) 1-allyl-3-methylimidazolium chloride room temperature ionic liquid: a new and powerful nonderivatizing solvent for cellulose. Macromolecules 38:8272–8277

Zhang X, Yang W, Blasiak W (2011a) Modeling study of woody biomass: interactions of cellulose, hemicellulose, and lignin. Energ Fuel 25:4786–4795

Zhang Y, Culhaoglu T, Pollet B, Melin C, Denoue D, Barrière Y et al (2011b) Impact of lignin structure and cell wall reticulation on maize cell wall degradability. J Agric Food Chem 59:10129–10135

Zhang Z, O'Hara IM, Doherty WOS (2013b) Effects of pH on pretreatment of sugarcane bagasse using aqueous imidazolium ionic liquids. Green Chem 15:431–438

Zhao X, Zhang L, Liu D (2012) Biomass recalcitrance. Part II: Fundamentals of different pretreatments to increase the enzymatic digestibility of lignocellulose. Biofuel Bioprod Biorefin 6:561–579

Chapter 15
Lignocellulosic Biorefineries: Concepts and Possibilities

Kenneth F. Reardon

Abstract To date, research, development, and commercialization within the bioenergy industry has focused on the production of biofuels, with any unconverted biomass used for production of electricity, biogas, animal feed, or fertilizer. However, both the economics and the environmental impacts of biofuel production could be improved by developing processes to obtain a wider range of chemicals (with higher value) from biomass. Example products range from commodity chemicals such as dicarboxylic acids to nutraceuticals. In this article, the concept of a biorefinery will be explored, especially in comparison to a petroleum refinery. Various products and options to produce non-fuel chemicals from plants biomass are outlined. Such processes would lead to a more diverse and sustainable biorefinery.

Keywords Biorefinery • Biorenewables • Biomass • Lignocellulosic • Cellulosic • Biofuels

15.1 Introduction

In the past decade, the term "biorefinery" has increased dramatically in usage (Fig. 15.1), and a Google search on this term now returns more than 655,000 results. Biorefineries have been the subject of a report from the US National Research Council (2000) and numerous conferences, books, and review articles.

The general concept of a biorefinery—a facility that produces not only a biofuel but also other chemical products and power from biomass—is intended to draw an analogy to a petroleum refinery (here, "petrorefinery" will be used to as a matched shorthand term). Both bio- and petrorefineries have chemically complex

K. F. Reardon (✉)
Department of Chemical and Biological Engineering,
Colorado State University, Fort Collins, CO 80523-1370, USA
e-mail: Kenneth.Reardon@colostate.edu

M. C. McCann et al. (eds.), *Flants and BioEnergy*, Advances in Plant Biology 4,
DOI: 10.1007/978-1-4614-9329-7_15,

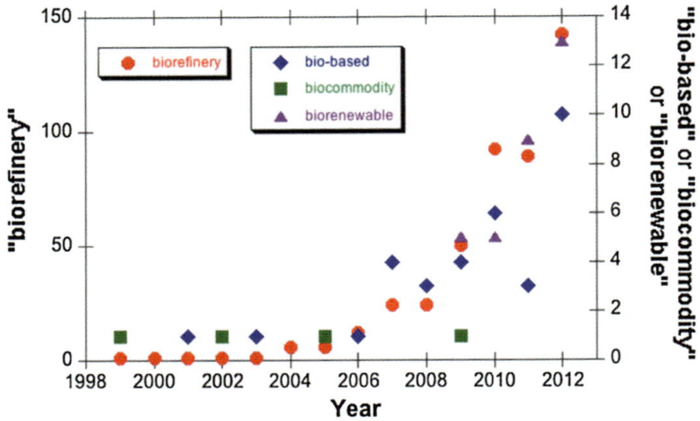

Fig. 15.1 PubMed search results for the number of publications with terms related to biorefineries in the article title or abstract. Publications with "biorefinery" or "biorefineries" are the most numerous (*left axis*) but the use of "bio-based" and "biorenewable" is also growing rapidly

feedstocks and the capability of producing numerous products, which provide important benefits to the overall refinery economics.

But to what extent is this analogy valid? The goal of this article is to consider that question and to focus in particular on the options available for co-products (non-fuel products) that could be produced in a biorefinery. Two general types of biorefineries will be considered: those based on thermochemical conversion (gasification or pyrolysis) and those based biochemical conversion (pretreatment-deconstruction-fermentation). Other platforms, especially those that combine thermochemical and biological conversion steps, are under development and some are moving rapidly toward commercialization (e.g., Zeachem's biological–chemical ethanol process). However, many of the same concepts will apply to those processes. The feedstock to be considered for the biorefineries is lignocellulosic biomass (e.g., grass, wood, corn stover).

15.2 Petroleum Refineries

Petrorefineries began operation in the mid-1800s, primarily to process crude petroleum into kerosene for use in lamps and heaters. With the expansion of the automobile industry, production shifted to emphasize gasoline and diesel as products. Today, many different products are obtained from crude petroleum (Fig. 15.2). Modern refineries may be extremely large (>90,000 m^3/day crude oil processed), are highly integrated, and rely extensively on thermal and chemical processes (Fig. 15.3, Table 15.1). Major unit operations in petrorefineries include separation processes (e.g., desalting, distillation, evaporation) and reaction processes (e.g., thermal, catalytic, and steam cracking; catalytic reforming; isomerization; alkylation). Petrorefineries shift their product output over the course of a year, primarily to respond to increased gasoline demands in the summer and increased heating oil needs in the winter (Suenaga and Smith 2011).

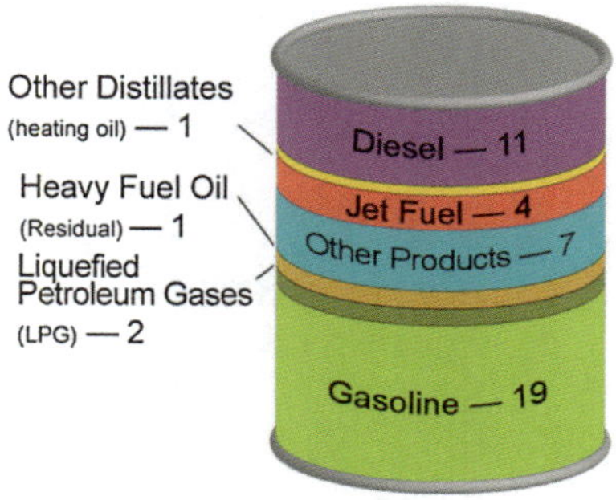

Fig. 15.2 Products (gallons) made from a 42-gal *barrel* of crude oil (http://www.eia.doe.gov/kids/energy.cfm?page=oil_home-basics)

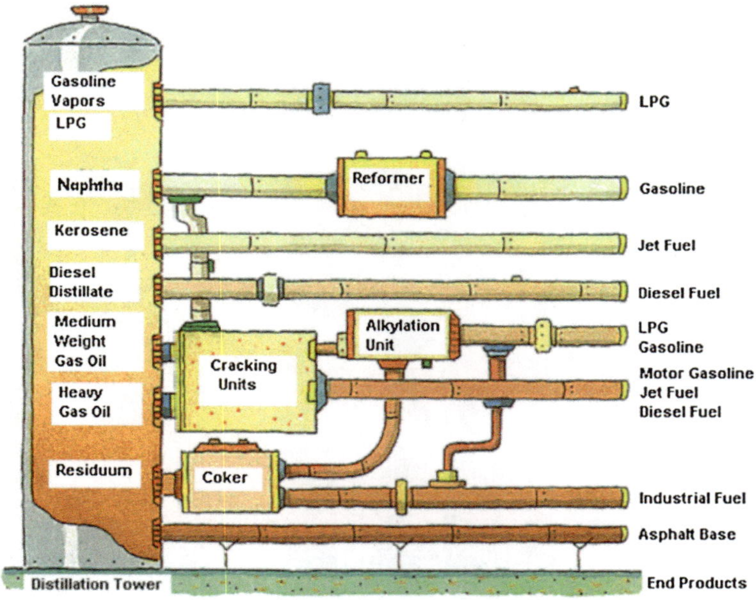

Fig. 15.3 Simplified schematic of the separation and conversion processes in a petrorefinery that result in a wide range of hydrocarbon products (http://www.eia.gov/kids/energy.cfm?page=oil_home-basics)

15.3 Biorefineries: Past and Present

The term "biorefinery" appears to have first been proposed in a journal article by Lynd et al. (1999) in an article discussing the concept of "biocommodity engineering". These authors outlined a vision for the evolution of biorefineries from facilities producing only a few products to those generating many outputs:

Table 15.1 Generalized comparison of petroleum refineries and biorefineries

	Petroleum refinery	Biorefinery	Comments
Feedstock	Crude oil	Lignocellulosic biomass (grasses, wood)	
		Sugar/starch fraction of biomass (maize, sugarcane)	
		Vegetable oil or tallow	
		Algal biomass	
		Municipal solid waste	
Feedstock variability	Moderate	High	Biomass feedstocks may range from wood to municipal solid waste to algal biomass
Unit operations: separation	Distillation, evaporation, and other thermal methods based on volatility differences at early and late stages of processing	Separation of biomass components possible with solvents but rarely done	Petroleum refineries use continuous processes
		Separation of products by distillation and other thermal methods	Biorefineries generally use batch processes
Unit operations: reaction	Continuous-flow heterogeneous catalytic or thermal reactions	Biochemical: soluble enzyme-catalyzed reactions and suspended cell fermentations	Petroleum refineries use continuous processes
		Thermochemical: thermal decomposition reactions and heterogeneous catalytic reactions	Biorefineries generally use batch processes
Number of unit operations per facility	High	Low	
Degree of process integration	High	Low	

(continued)

Table 15.1 continued

	Petroleum refinery	Biorefinery	Comments
Current product spectrum	Many (>100)	Few	Typical corn grain and sugarcane ethanol biorefinery products are ethanol, electricity, and distillers grains
Flexibility of product spectrum	High	Very low	
Size	10,000 to >80,000 m^3/d crude oil feedstock (Wikipedia 2013)	400–1,200 m^3/d product (Renewable Fuels Association 2013)	Processing capabilities vary widely for both types of refinery. Biorefinery values shown are for ethanol

> Biocommodity processes and products are often treated as though one or at best a few of these products would be manufactured in a single plant. Although this approach may be necessary initially to keep the scope of marketing, financing, and technology development manageable for first-of-a-kind plants, a multiproduct biorefinery configuration is likely to be more cost effective in the long term. Such an evolution would be similar to that experienced in the petroleum refining industry in which the initial focus on production of primarily kerosene with little revenue from the remaining fraction of oil ultimately gave way to integrated refineries that convert virtually all feedstock fractions into a wide range of valuable products (Lynd et al. 1999).

This implied definition of a biorefinery as producing multiple products from biomass was subsequently adopted by the biomass community, and sometimes extended to include food products (Ohara 2003). The term "integrated biorefinery" was subsequently introduced, with essentially the same meaning (US Department of Energy 2013). Despite this goal of multiple products, current biorefineries produce relatively few products: corn grain ethanol biorefineries often produce only ethanol, electricity, and distillers grains, while those producing commodity chemicals focus on a single product (e.g., lactic acid or 3-hydroxypropionic acid).

While the concept of converting a complex feedstock into numerous products is a valuable and important part of the refinery analogy, it is interesting to compare petro- and biorefineries on other grounds (Table 15.1). In particular, the variability of biomass feedstocks that can be considered for a biorefinery (from high oil content algal biomass to high carbohydrate content wood) is much larger than for a petrorefinery, meaning that biorefineries most likely will need to be designed for particular feedstock types. The size of the two types of biorefinery is also a significant difference; petrorefineries have become extremely large, regional operations, whereas current biorefineries are much smaller. The smaller size of biorefineries is partly a matter of their stage of development, but also reflects issues associated with the supply and cost of transportation of their relatively low energy density biomass feedstock to the refinery gate.

There are two primary biorefinery platforms for lignocellulosic biomass, based on the type of conversion process used:

- Biological: Biomass is pretreated using a variety of chemical and thermal methods (Blanch 2012; Chundawat et al. 2011) to reduce the particle size and make the carbohydrate polymers more accessible to the subsequent enzymatic depolymerization step to fermentable sugars. Yeasts and other microorganisms are then used to ferment the sugars to biofuels (e.g., ethanol, butanol) and other chemical products.
- Thermochemical: A process involving exposure to high temperatures, typically pyrolysis or gasification, is used to convert lignocellulosic biomass primarily into a liquid (pyrolysis oil) or synthesis gas (a mixture of carbon monoxide, hydrogen, and carbon dioxide), respectively (Digman et al. 2009). These are then processed further using catalyzed chemical reactions to form various products (Zhou et al. 2011).

15.4 Opportunities to Further Realize the Biorefinery Concept

15.4.1 General Considerations

The primary aspect of the biorefinery concept is the output of multiple products from a single facility. In 2004, the US Department of Energy produced reports identifying the non-fuel products that were determined to be the best targets for biorefineries on the basis of the pathways for production and the market sizes (US Department of Energy 2004). An updated version of this "Top 10" list was published by Bozell and Petersen (2010). The chemicals identified in these reports (Table 15.2) have been the focus of many research and development efforts toward the goal of biorefining.

One factor that makes biorefineries different than their petroleum counterparts is that oil is first fractionated and then each of those fractions is chemically converted to one or more products. In contrast, both biological and thermochemical biorefinery types convert biomass to an intermediate form (sugars, pyrolysis oil, or synthesis gas) before those compounds are fractionated and/or formed into the end products. This suggests that there are two main options for generating a suite of biorefinery products:

- Enable simultaneous production of products by incorporating a fractionation step prior to conversion
- Design biorefineries to yield different products sequentially with a flexible platform (e.g., change catalysts or microorganisms, operate at different conditions).

Table 15.2 "Top" chemicals for production from biomass

Conversion type	Biological conversion	Thermochemical: synthesis gas	Biological or thermochemical
Reference	US Department of Energy (2004)	US Department of Energy (2004)	Bozell and Petersen (2010)
Chemicals	1,4-diacids (succinic, fumaric and malic)	Hydrogen	Ethanol
	2,5-furan	Methanol	Furans
	Dicarboxylic acid		Glycerol and derivatives
	3-hydroxy propionic acid		Biohydrocarbons (isoprene)
	Aspartic acid		Lactic acid
	Glucaric acid		Succinic acid
	Glutamic acid		Hydroxypropionic acid/ aldehyde
	Itaconic acid		Levulinic acid
	Levulinic acid		Sorbitol
	3-hydroxybutyrolactone		Xylitol
	Glycerol		
	Sorbitol		
	Xylitol/Arabinitol		

The first opportunity for fractionation to obtain more products from biomass is independent of the type of biorefinery. Here, chemicals could be extracted directly from the lignocellulosic biomass. In one of the few examples of this concept, hexane has been used to extract a mixture of long-chain alcohols from sugar cane and switchgrass (Ravindranath et al. 2009). This mixture, called policosanol, has been shown to have cholesterol-lowering potency (McCarty 2002).

15.4.2 Thermochemical Biorefinery

In the case of pyrolysis oil, fractionation as well as chemical reactions can be used to form products. For example, Naik et al. (2010) evaluated supercritical CO_2 for this purpose and determined that certain high value products (furanoids, pyranoids, and bezenoids) could be separated from the pyrolysis oil. Upgrading reactions (with hydrogen) are used to lower the oxygen content of the molecules in the oil mixture (Zhang et al. 2005). Ultimately, pyrolysis oil must be further fractionated or blended into streams in a petrorefinery to obtain specific products.

Because synthesis gas has been produced for many years, catalysts and processes have been developed to convert this gas in one or more steps to a range of products from hydrocarbons to organic acids and aldehydes. However, a 2004 report from the US Department of Energy on suggested products to make from synthesis gas recommended only hydrogen and methanol because catalyst costs for the other products were too high (Table 15.2) (US Department of Energy 2004). Catalyst development for Fischer–Tropsch and other reactions is an ongoing research topic (Abelló and Montané 2011) and thus new options may emerge.

Depending upon the operating conditions of a thermochemical biorefinery, a significant amount of biochar may be produced (Manyà 2012). This carbonaceous material has been shown to have significant benefits for soil fertility (Spokas et al. 2012) and has also received considerable attention as a means of sequestering atmospheric carbon (Meyer et al. 2011).

15.4.3 Biochemical Biorefinery

While the metabolic capabilities of naturally occurring microorganisms has provided the ability to produce a range of fermentation products, the advent of metabolic engineering and synthetic biology technologies has dramatically increased the number of products that can be produced by microorganisms growing on sugars (Steen et al. 2010; Zhang et al. 2011). This includes not only previously produced fuel molecules such as alcohols and fatty acids, but also alkanes (Schirmer et al. 2010) and other hydrocarbons (Dugar and Stephanopoulos 2011; Jang et al. 2012). The same approaches have been used to engineer microorganisms for the production of non-fuel molecules from biomass (Curran and Alper 2012; Du et al.

2011; Jarboe et al. 2010). In most instances, metabolic engineering strategies have been implemented in the easily modified industrial strains of *Escherichia coli* and *Saccharomyces cerevisiae*, but research has also targeted known bacterial fermenters such as *Clostridium* sp. (Tracy et al. 2012).

With this increasing capacity to modify the metabolism of microorganisms and to form almost any metabolic product, biorefinery designers must decide between an organism optimized to produce the highest yield of only one molecule (and thus the sequential approach to achieving a multiple product biorefinery) or an organism that produces a suite of useful molecules but with lower concentrations and yields of any one of them. The type and cost of the separation processes involved will be critical in making this choice.

15.5 Future Perspectives

Within the next two years, several commercial-scale cellulosic biorefineries will begin production. This is an important stage in the development of this industry. Relative to current petrorefineries, the product diversity is much smaller. However, this is an expected starting point, and petrorefineries also began with a small product portfolio. It is interesting to note the rapid increase in refereed journal publications referencing not only "biorefinery" but also the terms "bio-based products" and especially "biorenewables" (Fig. 15.1).

In the future, several developments can be expected as biorefining matures:

- The implementation of new types of chemical conversions, such as hydrogenolysis (Ruppert et al. 2012) and aqueous-phase reforming (Huber and Dumesic 2006; Vispute and Huber 2009)
- Combinations of thermochemical and biological (chemical catalysts for sugars and polyols (Ruppert et al. 2012; Zhou et al. 2011), microbial fermentations of syngas (Henstra et al. 2007; Munasinghe and Khanal 2010)
- The development of metabolically engineered microorganisms that are robust in an industrial setting and have higher yields of the desired products
- The development of cost-effective processes to extract and purify high-value nutraceuticals from biorefinery feedstocks
- The development of continuous biorefinery processing over the current batch process-dominated approach
- Increased acceptability of bio-based products in the chemical marketplace.

References

Abelló S, Montané D (2011) Exploring iron-based multifunctional catalysts for Fischer–Tropsch synthesis: a review. ChemSusChem 4:1538–1556

Blanch HW (2012) Bioprocessing for biofuels. Curr Opin Biotechnol 23:390–395

Bozell JJ, Petersen GR (2010) Technology development for the production of biobased products from biorefinery carbohydrates-the US Department of Energy's "top 10" revisited. Green Chem 12:539–554

Chundawat SPS, Beckham GT, Himmel ME, Dale BE (2011) Deconstruction of lignocellulosic biomass to fuels and chemicals. Annu Rev Chem Biomol Eng 2:121–145

Curran KA, Alper HS (2012) Expanding the chemical palate of cells by combining systems biology and metabolic engineering. Metab Eng 14:289–297

Digman B, Joo HS, Kim D-S (2009) Recent progress in gasification/pyrolysis technologies for biomass conversion to energy. Environ Prog Sust Energy 28:47–51

Du J, Shao Z, Zhao H (2011) Engineering microbial factories for synthesis of value-added products. J Ind Microbiol Biotechnol 38:873–890

Dugar D, Stephanopoulos G (2011) Relative potential of biosynthetic pathways for biofuels and bio-based products. Nat Biotechnol 29:1074–1078

Henstra AM, Sipma J, Rinzema A, Stams AJM (2007) Microbiology of synthesis gas fermentation for biofuel production. Curr Opin Biotechnol 18:200–206

Huber G, Dumesic J (2006) An overview of aqueous-phase catalytic processes for production of hydrogen and alkanes in a biorefinery. Catal Today 111:119–132

Jang Y-S, Park JM, Choi S, Choi YJ, Seung DY, Cho JH, Lee SY (2012) Engineering of microorganisms for the production of biofuels and perspectives based on systems metabolic engineering approaches. Biotechnol Adv 30:989–1000

Jarboe LR, Zhang X, Wang X, Moore JC, Shanmugam KT, Ingram LO (2010) Metabolic engineering for production of biorenewable fuels and chemicals: contributions of synthetic biology. J Biomed Biotechnol 2010:1–18

Lynd L, Wyman C, Gerngross T (1999) Biocommodity engineering. Biotechnol Prog 15:777–793

Manyà JJ (2012) Pyrolysis for biochar purposes: a review to establish current knowledge gaps and research needs. Environ Sci Technol 46:7939–7954

McCarty MF (2002) Policosanol safely down-regulates HMG-CoA reductase–potential as a component of the esselstyn regimen. Med Hypotheses 59:268–279

Meyer S, Glaser B, Quicker P (2011) Technical, economical, and climate-related aspects of biochar production technologies: a literature review. Environ Sci Technol 45:9473–9483

Munasinghe PC, Khanal SK (2010) Biomass-derived syngas fermentation into biofuels: opportunities and challenges. Bioresour Technol 101:5013–5022

Naik S, Goud VV, Rout PK, Dalai AK (2010) Supercritical CO_2 fractionation of bio-oil produced from wheat-hemlock biomass. Bioresour Technol 101:7605–7613

National Research Council (2000) Biobased industrial products: research and commercialization priorities. The National Academies Press, Washington DC. http://www.nap.edu/catalog.php?record_id=5295

Ohara H (2003) Biorefinery. Appl Microbiol Biotechnol 62:474–477

Ravindranath SV, Uppugundla N, Lay JO, Clausen EC, Wilkins M, Ingraham RG, West C, Wu Y, Carrier DJ (2009) Policosanol, alpha-tocopherol, and moisture content as a function of timing of harvest of switchgrass (*Panicum virgatum* L.). J Agric Food Chem 57:3500–3505

Renewable Fuels Association (2013) Biorefinery locations. http://www.ethanolrfa.org/bio-refinery-locations/. Accessed 5 April 2013

Ruppert AM, Weinberg K, Palkovits R (2012) Hydrogenolysis goes bio: from carbohydrates and sugar alcohols to platform chemicals. Angew Chem Int Ed Engl 51:2564–2601

Schirmer A, Rude MA, Li X, Popova E, del Cardayre SB (2010) Microbial biosynthesis of alkanes. Science 329:559–562

Spokas KA, Cantrell KB, Novak JM, Archer DW, Ippolito JA, Collins HP, Boateng AA, Lima IM, Lamb MC, McAloon AJ, Lentz RD, Nichols KA (2012) Biochar: a synthesis of its agronomic impact beyond carbon sequestration. J Environ Qual 41:973–989

Steen EJ, Kang Y, Bokinsky G, Hu Z, Schirmer A, Mcclure A, del Cardayre SB, Keasling JD (2010) Microbial production of fatty-acid-derived fuels and chemicals from plant biomass. Nature 463:559–562

Suenaga H, Smith A (2011) Volatility dynamics and seasonality in energy prices: implications for crack-spread price risk. Energy J 32:27–58

Tracy BP, Jones SW, Fast AG, Indurthi DC, Papoutsakis ET (2012) Clostridia: the importance of their exceptional substrate and metabolite diversity for biofuel and biorefinery applications. Curr Op Biotechnol 23:364–381

US Department of Energy (2004) Top value added chemicals from biomass. Volume 1: results of screening for potential candidates from sugars and synthesis gas

US Department of Energy (2013) Integrated biorefineries. http://www1.eere.energy.gov/biomass/integrated_biorefineries.html. Accessed 5 April 2013

Vispute TP, Huber GW (2009) Production of hydrogen, alkanes and polyols by aqueous phase processing of wood-derived pyrolysis oils. Green Chem 11:1433–1445

Wikipedia (2013) List of oil refineries. http://en.wikipedia.org/wiki/List_of_oil_refineries. Accessed 5 April 2013

Zhang F, Rodriguez S, Keasling JD (2011) Metabolic engineering of microbial pathways for advanced biofuels production. Curr Opin Biotechnol 22:775–783

Zhang S, Yan Y, Li T, Ren Z (2005) Upgrading of liquid fuel from the pyrolysis of biomass. Biores Technol 96:545–550

Zhou C-H, Xia X, Lin C-X, Tong D-S, Beltramini J (2011) Catalytic conversion of lignocellulosic biomass to fine chemicals and fuels. Chem Soc Rev 40:5588–5617

Chapter 16
Catalytic Dehydration of Lignocellulosic Derived Xylose to Furfural

Basudeb Saha, Nathan S. Mosier and Mahdi M. Abu-Omar

Abstract In this chapter we present different biorefinary strategies for the production of Furfural, a top ten platform chemical for making next generation fine chemicals and liquid fuels. Several research articles have been published demonstrating the production of furfural using homogeneous and heterogeneous catalysts in single and biphasic solvent systems. This article summarizes the finding of the most recent research articles with critical discussion on the factors that control the yield and selectivity of furfural. Among several factors, special emphasis has been given on the improvement of partition coefficient of biphasic solvent systems and the effect of pore size of the heterogeneous catalyst in enhancing furfural yield and selectivity. Catalytic dehydration of xylose and its isomer form has been exemplified with Lewis and Brønsted acidic catalysts in understanding the mechanistic role of the individual acid sites in improving furfural yields and minimizing by-products formation.

Keywords Furfural • Biphasic solvent • Isomerization • Sustainable process • Liquid fuels

---

B. Saha · M. M. Abu-Omar (✉)
Brown Laboratory, Department of Chemistry and School of Chemical Engineering,
The Center for Catalytic Conversion of Biomass to Biofuels (C3Bio),
Purdue University, 560 Oval Drive, West Lafayette, IN 47907, USA
e-mail: mabuomar@purdue.edu

N. S. Mosier
School of Agricultural and Biological Engineering, The Center for Catalytic Conversion
of Biomass to Biofuels (C3Bio), and Laboratory of Renewable Resources Engineering,
Purdue University, 500 Central Drive, West Lafayette, IN 47907, USA

M. C. McCann et al. (eds.), *Plants and BioEnergy*, Advances in Plant Biology 4,
DOI: 10.1007/978-1-4614-9329-7_16,

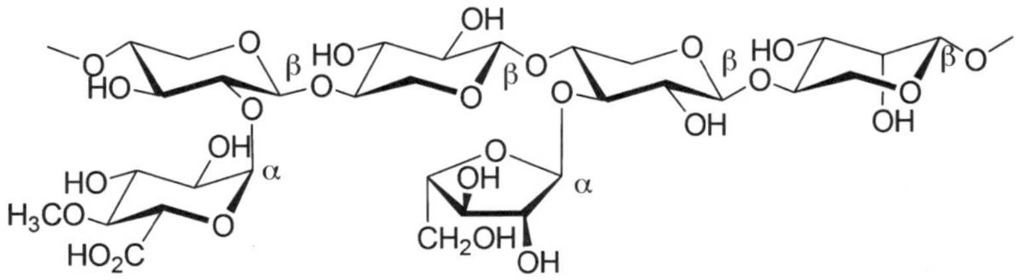

Fig. 16.1 Structure of xylan showing glycosidic bonds of xylose monomers

16.1 Introduction

Furfural is an important biomass derived platform chemical (Werpy and Peterson 2004), with an annual production volume of more than 200,000 tons (Kamm et al. 2006). The Quaker Oats Company commercialized a method for the production of furfural by treating hulls with dilute sulphuric acid as early as 1921 (Brownlee and Miner 1948). The general methodology of furfural production involves hydrolysis of xylan (Fig. 16.1), a polymer of xylose, which is present in lignocellulosic biomass, followed by catalytic dehydration of xylose with homogeneous or heterogeneous acidic materials. Xylose, a C_5 sugar unit of hemicellulose, is the second-most abundant component of biomass after cellulose (Dumitriu and Dekker 2005). Therefore, utilization of this abundant renewable C_5 sugars for the production of useful chemicals and fuels via environmentally and economically viable process is considered as a sustainable remediation to address the concern of diminishing petroleum reservoir, variability in fossil energy price and high dependence on petroleum feedstock (Dodds and Gross 2007).

The research trend of furfural production and exploration of its potential applications have received significant attention in recent years after publication of "Top Value Added Chemicals from Biomass" by Werpy and Peterson (2004). Furfural can be used as a precursor for several high value chemicals and biofuel. Candidates include 2-methylfuran (2-MF), 2-methyltetrahydrofuran (2-MeTHF), furfural alcohol (FA), ethyl lelulinate (EL), γ-valerolactone (gVL) and long chain hydrocarbons of diesel fraction (Corma et al. 2007; Dutta et al. 2012). Recently, furfural hydrogenation product, 2-MF, derived hydrocarbon blended gasoline has been tested for 90,000 km road trial with promising outcomes, which prompted initiation of commercial scale production of liquid hydrocarbon from 2-MF (Lange et al. 2012). A scheme for furfural production and its potential applications is shown in Fig. 16.2.

16.2 Furfural Production in Monophasic Solvent

The catalytic conversion of xylose, xylan and pre-treated biomass substrates to furfural in monophasic and biphasic solvent systems has been investigated by using homogeneous and heterogeneous catalytic materials containing Lewis and Brønsted

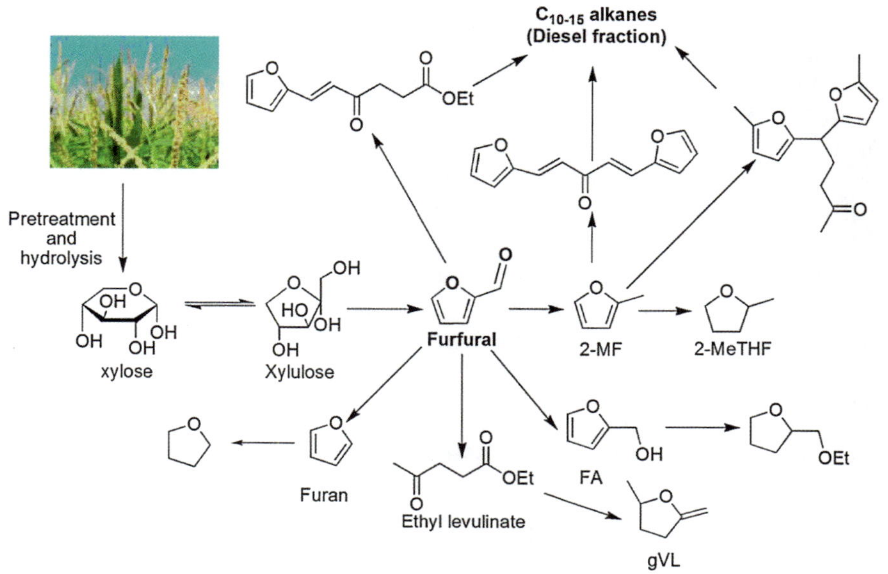

Fig. 16.2 Furfural production and its utilization routes for chemicals and liquid fuels

acidic sites. Like the Quaker Oats process, the conversion of xylose to furfural with mineral acid catalysts (HCl, H_2SO_4) in aqueous medium achieved a maximum of 50 % furfural yields (Sievers et al. 2009; Weingarten et al. 2010; Yemis and Mazza 2011). However, mineral acid catalyzed processes suffer from corrosiveness and environmental issues associated with the use of strong acids. In this context, Lewis acidic metal chloride catalysts are advantageous for furfural production via isomerisation of xylose to xylulose intermediate. Binder et al. have shown that $CrCl_3$ and $CrCl_2$ catalysts are effective for xylose conversion in N,N-dimethylacetamide (DMA)-LiCl solvent, giving a maximum of 56 % furfural yield in 4 h at 100 °C (Binder et al. 2010). Depolymerisation of Birch xylan is, however, a major challenge using this catalytic system, resulting in poor furfural yield (15 %). Although pre-treatment of xylan with HCl or 1-ethyl-3-methylimidazolium chloride ([EMIM]Cl) IL (ionic liquid) shows a significant improvement in xylan hydrolysis (77 %), a proportional increase in furfural yield was not observed in the subsequent dehydration step in DMA-LiCl. The same catalytic system in pure [EMIM]Cl solvent produced 63 % furfural from xylan in shorter time (3 min) when the reaction was carried out under microwave assisted hearing at 100 °C (Zhang and Zhao 2010). This method is also effective for intact biomass corn stalk, rice straw and pine wood due to high dissolution of biomass in imidazolium-based ionic liquids. Furfural yields from these biomass species are in the range of 23–31 % based on their pentose composition by weight (von Sivers and Zacchi 1995; Liu and Wyman 2005; Jin and Chen 2007). Potential drawback of the IL solvent is that ionic liquids are expensive and the separation of furfural from high boiling point ILs is energy intensive. Besides cost, ILs tend to deactivate by the water formed during the dehydration reaction. Xylose conversion has also been investigated in high boiling point organic solvents such as dimethylsulfoxide (DMSO) using heteropolyacids (Dias et al. 2005) and Nafion catalysis (Lam et al. 2011). Although these reusable catalysts give modest furfural yields in the range of 58–67 mol %, selectivity of the desired product is an issue due to the formation of undesired humin by-products via oligomerization between xylose and furfural (Dee and Bell 2011). Such humin formation has also been a challenge for acid-catalyzed formation of furfural from xylose in 1-butyl-3-methylimidazolium chloride ([BMIM]Cl) using Brønsted acids such as H_2SO_4 (Sievers et al. 2009). Besides humin formation in DMSO, the high boiling point of this solvent also possess a challenge for cost-effective separation of furfural with high purity, and therefore, these processes are economically unfavorable for larger scale commercial applications.

16.3 Furfural Production in Biphasic Media

Because of aforementioned disadvantages of monophasic solvent systems using high boiling point organic solvents or poor yield in pure aqueous medium, current research effort of furfural production is directed towards utilization of biphasic reaction systems in batch or continuous reactor. In case of biphasic system, aqueous or modified aqueous solution is used as the reactive phase. The organic layer of the biphasic system acts as an extracting solvent for continuous separation of furfural into the organic phase immediately as its being formed in the reactive phase. Thus, lower concentration of furfural in the aqueous phase limits its rehydration with water and thereby improves furfural yields (Qi et al. 2009). This method allows easy separation and reusability of the reactive aqueous phase containing both homogeneous and heterogeneous catalysts. The partition coefficient (R), which is the ratio of furfural in the organic phase to that in the aqueous phase, is an important parameter in determining the overall performance of the biphasic system. Higher partitioning of furfural into the organic layer improves effective extraction and hence increases product selectivity as well as yield. Besides the nature of the organic solvents in determining the partition coefficient, the presence of inorganic salt, e.g. NaCl, in the aqueous phase also increase the R values due to salting-out effect (Eisen and Joffe 1966; Tan and Aravinth 1999). vom Stein et al. (2011) developed a method for furfural production by using a biphasic solvent system comprising an aqueous solution of $FeCl_3 \cdot 6H_2O$, NaCl and xylose as a reactive phase and biomass derived 2-methyltetrahydrofuran (2-MeTHF) as the extracting organic phase. This method exhibited a maximum 71 % furfural yield with 98 % extraction capability into the organic phase. Besides pure xylose conversion, this biphasic system containing $FeCl_3$ catalyst has been shown to convert Beechwood biomass extracted non-purified xylose solution (30 wt%) to furfural at a production of rate of 3.5 g furfural/h (Fig. 16.3).

Similar to the iron system, the water-NaCl-THF biphasic medium is effective for $AlCl_3 \cdot 6H_2O$ catalyzed conversion of xylose, giving 75 % furfural yield in 5 min under microwave assisted heating (Yang et al. 2012a, b). The potential of the combined $AlCl_3$/water-NaCl-THF system has further demonstrated for

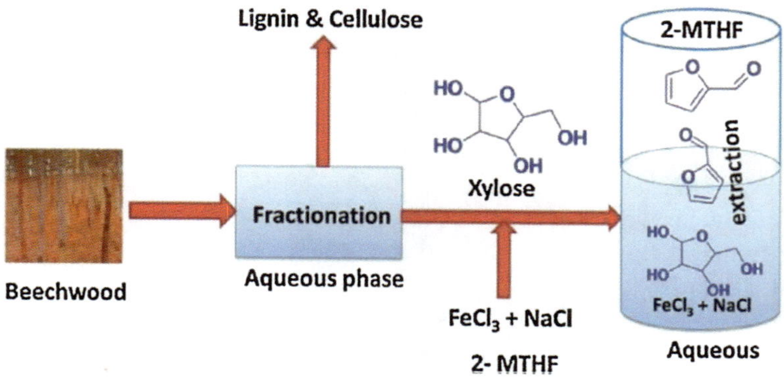

Fig. 16.3 Conversion of Beechwood biomass derived xylose to furfural in aqueous-2-MeTHF biphasic medium using iron catalyst

Table 16.1 Furfural yields from various sources of lignocellulosic biomass

Biomass	Temp (°C)	Time (min)	Furfural (%)	Xylose (%)
Corn stover	160	60	55	<1
Pinewood	160	60	38	<1
Switchgrass	160	60	56	1
Poplar	160	60	64	<1
Cellulose/xylan	160	60	66	0
Pinewood	180	30	61	<1

sustainable furfural production by converting lignocellulosic biomass (corn stover, pinewood, switchgrass, and poplar) (Yang et al. 2012a, b) to 55–66 mol% furfural (Table 16.1) based on literature values (Lu and Mosier 2007, 2008) of their pentose contents by weight.

Another strategy for converting the hemicellulose fraction of biomass to furfural utilizes a biomass derived solvent, 2-*sec*-butylphenol (SBP), as an extracting phase and an acidic aqueous layer as the reactive phase (Gürbüz et al. 2012). This biphasic system containing mineral acid catalyst produce high concentrations of furfural with maximum 78 % yield. The use of SBP solvent having high R values (90 and 50 with and without NaCl saturation in the aqueous phase) is advantageous because of (1) fast furfural extraction, (2) low amounts of mineral acids dissolve in SBP thereby eliminating energy intensive separation of mineral acid from the product layer, and (3) SBP can be derived from lignin. While the authors list SBP higher boiling point than the furfural product as an advantage, this feature can be interpreted as a drawback because it requires intensive energy to remove the product via distillation (boiling point of furfural 162 °C). Nevertheless, similar lignin derived alkylphenols, e.g., eugenol, can be envisaged as effective extracting solvents for furfural production. A recent study shows an efficient xylose and xylan conversion process with 72 % furfural yield using maleic acid catalyst in an aqueous medium at 200 °C (Kim et al. 2012). The synergistic effect of KCl and KI salts results in a higher furfural yield (88 %) and selectivity (95 %).

Besides homogeneous catalysts, several heterogeneous catalysts, including Zr-P, SiO_2-Al_2O_3, WO_x/ZrO_2, γ-Al_2O_3 and HY zeolite (Weingarten et al. 2011), Amberlyst-15 (Takagaki et al. 2010), hydrotalcite (Tuteja et al. 2012), tin-tungsten mixed oxide (Yamaguchi et al. 2011), Nafion 117 (Lam et al. 2011), and Sn-beta zeolite (Choudhary et al. 2011) have been employed for furfural production to take advantages of their easy separation and recyclability properties. These studies reveal that catalysts with higher Lewis acid sites are most active. However, catalyst pore confinement is found to have an adverse effect on furfural selectivity. Adsorption–desorption studies and decomposition experiments with furfural in aqueous solution have confirmed that HY zeolite causes furfural to irreversibly adsorb inside the zeolite pores and oligomarize to form humin side-products. Therefore, micropores containing catalysts may not be suitable for xylose dehydration due to strong adsorption of the product inside the catalyst pore.

16.4 Mechanistic Studies of Furfural Formation

Experimental evidence suggests that glucose dehydration to 5-hydrxymethylfurfural (HMF) occurs via the isomerization of glucose to fructose (Binder and Raines 2009; Roman-Leshkov et al. 2010; De et al. 2011; Pagán-Torres et al. 2012). The occurrence of similar isomerization for xylose to xylulose intermediate followed by dehydration to furfural was first established by comparing the xylose and xylulose conversion rates with a series of Lewis and Brønsted acidic catalysts (Choudhary et al. 2011). The results show that xylose does not react with Brønsted acidic catalysts such as Amberlyst-15 or HCl; however, when xylulose is the reactant, conversion is ~66 %, and furfural yield is 24 %. On the other hand, a catalytic system containing both Brønsted (HCl) and Lewis acidic sites (Sn-beta) can directly convert xylose to furfural with evidence of formation of xylulose and lyxose as intermediates. This result supports a reaction network in which xylulose dehydrates rapidly to furfural via Brønsted acid catalysis and xylose isomerizes to xylulose with a Lewis acid catalyst advocating for dual acidic sites of a catalyst. Formation of xylulose is a key step to furfural and requires either functional group rearrangement or a change in configuration on the C1 and C2 carbon atoms. Structural studies using X-ray absorption fine structure (EXAFS) spectroscopy reveals that Sn is substituted in pairs on opposite sides of six-member rings, i.e. uniform crystallographic location of Sn in the β crystal structure that leads to sites with uniform catalytic activity and high chemical selectivity (Fig. 16.4) (Bare et al. 2005). The results of Sn-beta zeolite catalyzed process indicate that the active site of the catalyst interacts with the carbonyl group of C1 and the adjacent hydroxyl group on C2. Kinetic studies of isomerization reactions indicate that certain acids and metals are able to transfer the hydrogen directly through a hydride shift between C–2 and C–1 (Collyer and Blow 1990). Involvement of

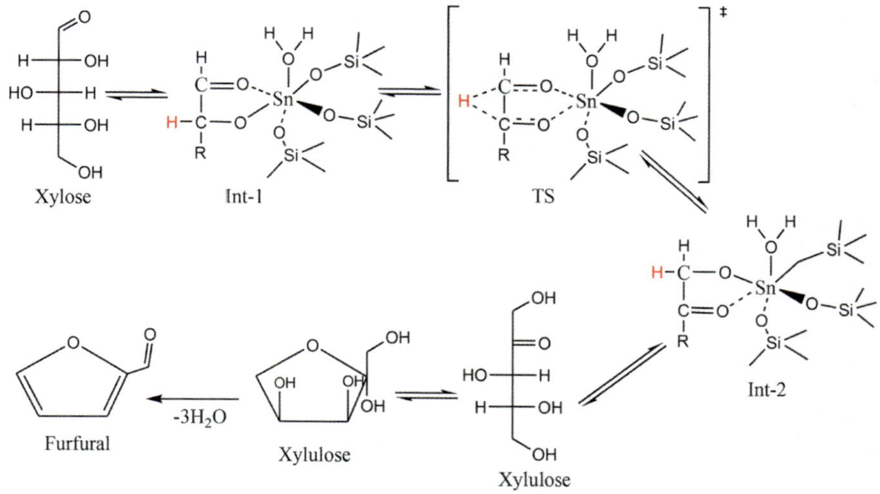

Fig. 16.4 Plausible reaction mechanism for xylose conversion to furfural *via* xylulose intermediate

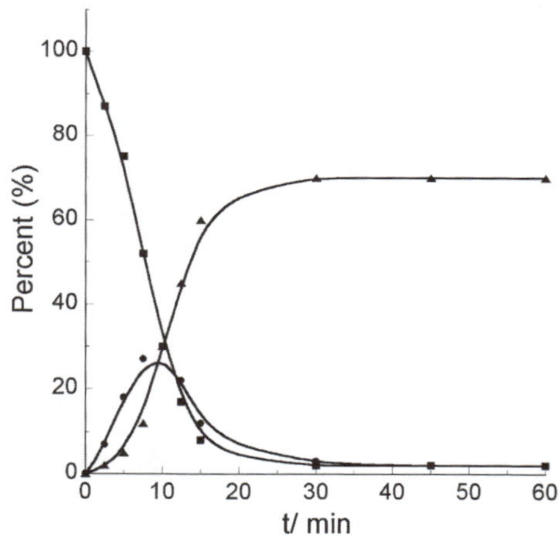

Fig. 16.5 Kinetic profile for xylose conversion with $AlCl_3$ catalyst in water-NaCl/THF solvent (*black square* xylose, *black circle* xylulose, *black up-pointing triangle* furfural)

1,2-hydride transfer step in the isomerisation of xylose to xylulose is also consistent with results from deuterium labeling experiments using $CrCl_3$ as catalyst (Binder et al. 2010). Lewis acidity in the catalyst is essential to polarize the carbonyl group in the ketone while coordinating both the alcohol and the ketone to facilitate a hydride shift between the two carbons (Corma et al. 2001).

$AlCl_3 \cdot 6H_2O$ catalyzed conversion of xylose in water-NaCl/THF biphasic solvent system shows direct evidence for xylulose formation as an intermediate species (Yang et al. 2012a, b). The kinetic profile (Fig. 16.5) of xylose conversion reveals xylulose formation in the early stages of the reaction, suggesting that isomerization of xylose to xylulose takes place during the course of the dehydration reaction. Under comparable reaction conditions, xylose conversion under conventional heating required longer reaction times (20 min) and afforded lower selectivity (64 %) for furfural along with cross-polymer humin by-product. Kinetic study of xylose conversion with maleic acid as catalyst (Kim et al. 2012) revealed that xylose conversion rates are lower in aqueous medium, which may be due to furfural acting as a Brønsted base and reacting with H_3O^+. Thus, total acid concentration of the aqueous medium decreases and consequently the conversion rate of xylose is retarded (Antan et al. 1991).

16.5 Conclusions

Furfural is a promising biomass derived platform chemical for the production of high value chemicals and liquid fuels that are currently obtained from petroleum feedstock. Development of effective catalytic and reaction systems to achieve high yield and selectivity of furfural from lignocellulosic biomass containing abundant xylose constitute a viable strategy for the modern biorefinery. Although several improvements have been made in laboratory scale processes to achieve high furfural yield and selectivity by developing effective catalytic system containing balanced Lewis and Brønsted acidic sites and biphasic reaction media, high production cost of furfural is still a challenge to its sustainable utilization for making other high value chemicals and fuels on commercial scale. A combination of both chemistry and engineering efforts to design more effective catalysts through clear understanding of the exact roles of Lewis and Brønsted acidic sites in minimization of by-products formation, improvement in furfural yield and selectivity, and cost-effective purification of the desired product(s) are necessary to lower furfural production cost.

Acknowledgments The authors acknowledge financial support from the Center for direct Catalytic Conversion of Biomass to Biofuels (C3Bio), an Energy Frontier Research Center funded by the U.S. Department of Energy, Office of Science, and Office of Basic Energy Sciences under Award Number DE-SC0000997.

References

Antan MJ, Leesomboon T, Mok WS, Richards GN (1991) Mechanism of formation of 2-furaldehyde from D-xylose. Carbohydr Res 217:71–85

Bare SR, Kelly SD, Sinkler W, Low JJ, Modica FS, Valencia S, Corma A, Nemeth LT (2005) Uniform catalytic site in Sn-β-Zeolite determined using x-ray absorption fine structure. J Am Chem Soc 127:12924–12932

Choudhary V, Pinar AB, Sandler SI, Vlachos DG, Lobo RF (2011) Xylose isomerization to xylulose and its dehydration to furfural in aqueous media. ACS Catal 1:1724–1728

Binder JB, Raines RT (2009) Simple chemical transformation of lignocellulosic biomass into furans for fuels and chemicals. J Am Chem Soc 131:1979–1985

Binder JB, Blank JJ, Cefali AV, Raines RT (2010) Synthesis of furfural from xylose and xylan. Chem Sus Chem 3:1268–1272

Brownlee HJ, Miner CS (1948) Industrial development of furfural. Ind Eng Chem 40:201–204
Collyer CA, Blow DM (1990) Observations of reaction intermediates and the mechanism of aldose-ketose interconversion by D-xylose isomerase. Proc Natl Acad Sci USA 87:1362–1366
Corma A, Nemeth LT, Renz M, Valencia S (2001) Sn-zeolite beta as a heterogeneous chemoselective catalyst for Baeyer–Villiger oxidations. Nature 412:423–425
Corma A, Ilorra S, Velty A (2007) Chemical routes for the transformation of biomass into chemicals. Chem Rev 107:2411–2502
De S, Dutta S, Saha B (2011) Microwave assisted conversion of carbohydrates and biopolymers to 5-hydroxymethylfurfural with aluminium chloride catalyst in water. Green Chem 13:2859–2868
Dias AS, Pillnger M, Valente AA (2005) Liquid phase dehydration of D-xylose in the presence of Keggin-type heteropolyacids. Appl Catal A Gen 285:126–131
Dee SJ, Bell AT (2011) A study of the acid-catalyzed hydrolysis of cellulose dissolved in ionic liquids and the factors influencing dehydration of glucose and the formation of humins. Chem Sus Chem 4:1166–1173
Dodds D, Gross R (2007) Chemicals from biomass. Science 318:1250–1251
Dumitriu S, Dekker M (2005) Polysaccharides: structural diversity and functional versatility, 2nd edn. Marcel Dekker, New York
Dutta S, De S, Saha B, Alam I (2012) Advanced in conversion of hemicellulosic biomass to furfural and upgrading to biofuels. Catal Sci Tech 2:2025–2036
Eisen EO, Joffe J (1966) Salt effects in liquid–liquid equilibria. J Chem Eng Data 11:480–484
Gürbüz EI, Wettstein SG, Dumesic JA (2012) Conversion of hemicellulose to furfural and levulinic acid using biphasic reactors with alkylphenol solvents. Chem Sus Chem 5:383–387
Jin S, Chen H (2007) Fractionation of fibrous fraction from steam-exploded rice straw. Process Biochem 42:188–192
Kamm B, Gruber PR, Kamm M (2006) Biorefineries—industrial processes and products. Wiley, Weinheim
Kim ES, Liu S, Abu-Omar MM, Mosier NS (2012) Selective conversion of biomass hemicellulose to furfural using maleic acid with microwave heating. Energy Fuels 26:1298–1304
Lam E, Majid E, Leung ACW, Cong JH, Mahmoud KA, Luong JHT (2011) Synthesis of furfural from xylose by heterogeneous and reusable nafion catalysts. Chem Sus Chem 4:535–541
Lange JP, Heide E, van der Buijtenen JV, Trice R (2012) Furfural—a promising platform for lignocellulosic biofuels. Chem Sus Chem 5:150–166
Liu C, Wyman EE (2005) Partial flow of compressed hot water through corn stover to enhance hemicelluloses sugar recovery and enzymatic digestibility of cellulose. Biores Technol 96:1978–1986
Lu YL, Mosier NS (2007) Biomimetic catalysis for hemicellulose hydrolysis in corn stover. Biotechnol Prog 23:116–123
Lu YL, Mosier NS (2008) Kinetic modeling analysis of maleic acid-catalyzed hemicellulose hydrolysis in corn stover. Biotechnol Bioeng 101:1170–1181
Pagán-Torres YJ, Wang T, Gallo JMR, Shanks BH, Dumesic JA (2012) Production of 5-hydroxymethylfurfural from glucose using a combination of Lewis and Brønsted acid catalysts in water in a biphasic reactor with an alkylphenol solvent. ACS Catal 2:930–934
Qi X, Watanabe M, Aida TM, Smith RL Jr (2009) Efficient process for conversion of fructose to 5-hydroxymethylfurfural with ionic liquids. Green Chem 11:1327–1331
Weingarten R, Tompsett GA, Conner WC Jr, Huber GW (2011) Design of solid acid catalysts for aqueous phase dehydration of carbohydrates: the role of Lewis and Brønsted acid sites. J Catal 279:174–182
Roman-Leshkov Y, Moliner M, Labinger JA, Davis ME (2010) Mechanism of glucose isomerization using a solid Lewis acid catalyst in water. Angew Chem Int Ed 49:8954–8957
Sievers C, Musin I, Marzialetti T, Valenzuela Olarte MB, Agrawal PK, Jones CW (2009) Acid-catalyzed conversion of sugars and furfurals in an ionic-liquid phase. Chem Sus Chem 2:665–671
Takagaki A, Ohara M, Nishimura S, Ebitani K (2010) One-pot formation of furfural from xylose via isomerization and successive dehydration reactions over heterogeneous acid and base catalysts. Chem Lett 39:838–840

Tan TC, Aravinth S (1999) Liquid-liquid equilibria of water/acetic acid/1-butanol system–effects of sodium (potassium) chloride and correlations. Fluid Phase Equilib 163:243–257

Tuteja J, Nishimura S, Ebitani K (2012) One-pot synthesis of furans from various saccharides using a combination of solid acid and base catalysts. Bull Chem Soc Jpn 85:275–281

vom Stein T, Grande PM, Leitner W, de Maria PD (2011) Iron-catalyzed furfural production in biobased biphasic systems: from pure sugars to direct use of crude xylose effluents as feedstock. Chem Sus Chem 4:1592–1594

von Sivers M, Zacchi G (1995) A techno-economical comparison of three processes for the production of ethanol from pine. Biores Technol 51:43–52

Weingarten R, Cho J, Conner WC Jr, Huber GW (2010) Kinetics of furfural production by dehydration of xylose in a biphasic reactor with microwave heating. Green Chem 12:1423–1429

Werpy T, Peterson G (2004) Top value added chemicals from biomass, vol. 1. Pacific Northwestern National Laboratory

Yamaguchi K, Sakurada T, Ogasawara Y, Mizuno N (2011) Tin-tungsten mixed oxide as efficient heterogeneous catalyst for conversion of saccharides to furan derivatives. Chem Lett 40:542–543

Yang Y, Hu CW, Abu-Omar MM (2012a) Conversion of carbohydrates and lignocellulosic biomass into 5-hydroxymethylfurfural using $AlCl_3$ $6H_2O$ catalyst in a biphasic solvent system. Green Chem 14:509–513

Yang Y, Hu C-W, Abu-Omar M-M (2012b) Synthesis of furfural from xylose, xylan, and biomass using $AlCl_3$ $6H_2O$ in biphasic media via xylose isomerization to xylulose. Chem Sus Chem 5:405–410

Yemis O, Mazza G (2011) Acid-catalyzed conversion of xylose, xylan and straw into furfural by microwave-assisted reaction. Biores Technol 102:7371–7378

Zhang Z, Zhao ZK (2010) Microwave-assisted conversion of lignocellulosic biomass into furans in ionic liquid. Biores Technol 101:1111–1114

Chapter 17
Catalytic Oxidation of Lignin for the Production of Low Molecular Weight Aromatics

Joseph J. Bozell

Abstract Lignin offers a number of attractive features as a starting material for chemical production, and the only large scale source of aromatic moieties in nature. It is highly abundant, comprising 15–25 wt% of lignocellulosic feedstocks, such as agricultural materials or forest resources, making it the second most available source of renewable carbon after cellulose.

17.1 Introduction

Lignin offers a number of attractive features as a starting material for chemical production, and the only large scale source of aromatic moieties in nature (Bozell et al. 2007). It is highly abundant, comprising 15–25 wt% of lignocellulosic feedstocks, such as agricultural materials or forest resources, making it the second most available source of renewable carbon after cellulose. However, lignin faces significant disadvantages as a chemical feedstock, with the primary drawback being a high level of structural heterogeneity that arises from two primary sources. First, the biosynthesis of lignin introduces heterogeneity as lignin is manufactured in the plant cell wall (Fig. 17.1) (Boerjan et al. 2003).

Lignin biosynthesis uses the three primary monolignols, p-hydroxycinnamyl alcohol, coniferyl alcohol and sinapyl alcohol, leading to the well-recognized p-hydroxyphenyl, guaiacyl and syringyl units, respectively, in the resulting lignin biopolymer. During biosynthesis, these monolignols are converted into highly delocalized phenoxy radicals that undergo radical–radical coupling and conversion to the lignin polymer. Softwoods are primarily constructed of guaiacyl units, while hardwoods contain significant amounts of both guaiacyl and syringyl units.

J. J. Bozell (✉)
Center for Renewable Carbon, Center for the Catalytic Conversion of Biomass (C3Bio), University of Tennessee, Knoxville, TN 37996, USA
e-mail: jbozell@utk.edu

M. C. McCann et al. (eds.), *Plants and BioEnergy*, Advances in Plant Biology 4,
DOI: 10.1007/978-1-4614-9329-7_17, © Springer Science+Business Media New York 2014

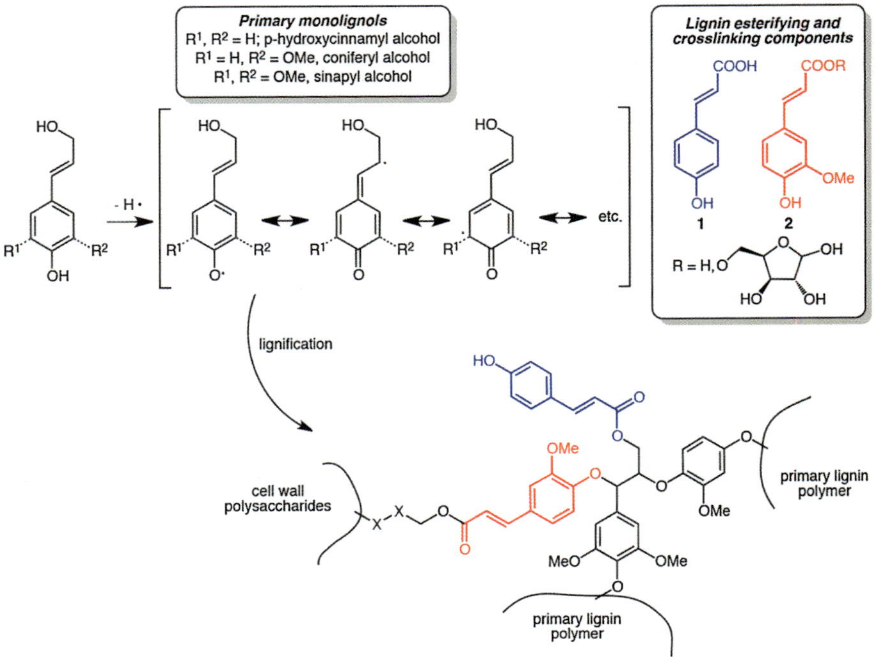

Fig. 17.1 Overview of lignin biosynthesis

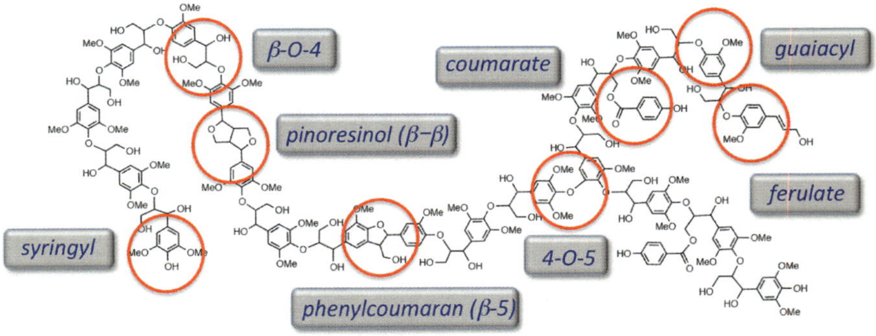

Fig. 17.2 Suggested structure of native poplar lignin

Herbaceous feedstocks (grasses) offer additional complexity, as their structure includes extensive crosslinking between lignin and hemicelluloses through ferulate ester linkages. Moreover, the ferulates themselves undergo dimerization and trimerization, and can be introduced into the bulk lignin. The result of these coupling reactions is the production of lignin as a complex aromatic biopolymer (Fig. 17.2, illustrating hardwood poplar) (Vanholme et al. 2010).

Figure 17.2 also identifies many of the well-recognized substructural units that comprise the lignin polymer as it is found in the plant. However, using lignin as a chemical feedstock requires its isolation from the lignocellulosic matrix, leading to the second source of lignin's structural heterogeneity. Any process used to isolate lignin inevitably induces structural changes in the native material through loss of some substructural units and introduction of new interunit linkages. As an example, switchgrass samples subjected to organosolv fractionation showed a marked change in the concentration of β-O-4 units as the severity of the fractionation increased (Bozell et al. 2011b). Quantitative ^{13}C analysis of the NMR spectral region between 88 and 77 ppm identifies lignin's β-O-4 units, which can be 50 % or greater of the interunit linkages in native lignin. After organosolv fractionation, NMR spectra indicate that it is possible to nearly eliminate all of these linkages at high severity (160 °C, 0.1 M acid catalyst). Other pretreatment processes result in similar changes. Dilute acid pretreatment can reduce β-O-4 units by 36 % (Samuel et al. 2010), and steam explosion can nearly eliminate them under proper conditions (Li et al. 2009). In parallel, lignin's structure realizes an increase in the number of free phenolic –OH groups as the β-O-4 units are cleaved.

Ongoing biorefinery development is providing access to high purity lignin as new, inexpensive sources of renewable carbon through technology such as organosolv fractionation (Bozell et al. 2011a). Despite these structural challenges, all lignin, regardless of source, comprises a network of electron rich aromatic rings. Accordingly, such systems would be anticipated to undergo a wide range of oxidation processes. Development of new oxidation catalysts would ideally employ environmentally benign terminal oxidants such as O_2 or HOOH, could be adapted to operation in aqueous media, and would demonstrate reactivity designed for the substructural units present in lignin as isolated by the biorefinery. This chapter provides a brief overview of work in both nonselective and selective lignin oxidation, and efforts within our group to develop selective catalytic oxidation processes for lignin and lignin models.

17.2 Nonselective Catalytic Lignin Oxidation

Lignin oxidation is a critical component of the kraft process used in the manufacture of pulp and paper. The kraft industry is the largest current producer of lignin, through the generation of about 120×10^6 tons of pulp in 2005, leading to the parallel production of about 72×10^6 t of lignin (Auhorn and Niemela 2007). Further, kraft cellulose pulp retains a small amount of recalcitrant lignin that must be removed chemically. Historically, however, the industry has not viewed this lignin as a potential source of chemicals, and instead uses a wide range of nonselective oxidation processes for its consumption. The simplest oxidation of lignin is its combustion as fuel for the kraft operation's chemical recovery boiler, a vital component of a mill's heat and energy balance (Fengel and Wegener 1984). In parallel, residual lignin in kraft cellulose not extracted during pulping is subjected to nonselective bleaching processes for its rapid oxidative deconstruction and removal. Industrially, ClO_2 is extensively used as a primary oxidant, but well recognized bleaching processes employing O_2, HOOH, O_3 or sequential combinations of these oxidants are also employed (Dence and Reeve 1996).

Although ClO_2 is an exceptionally efficient stoichiometric oxidant for lignin, continuing industrial interest in eliminating all chlorine-containing reagents from industrial bleaching has prompted evaluation of metal catalyzed approaches for the activation of O_2 or HOOH to minimize or replace ClO_2. Extensive work has been reported on nonselective catalytic oxidation of monomeric, dimeric and oligomeric lignin models (Collinson and Thielemans 2010). Oxidation methodology includes the use of O_2 with catalytic Co(II) or Mn(II) salts as one electron biomimetic oxidants of a series of lignin models under harsh conditions (Dicosimo and Szabo 1988), Cu(bis-*ortho*-phenanthroline) (Sippola and Krause 2005), or Co and Fe metalloporphyrins (Crestini et al. 2004; Cui and Dolphin 1995; Zhu and Ford 1993), in organic or aqueous medium for the oxidation of several lignin models using O_2, aerobic oxidation catalyzed by polyoxometalates (Evtuguin et al. 2000; Weinstock et al. 1997), or HOOH oxidation catalyzed by methyltrioxorhenium. (Crestini et al. 2006; Crestini et al. 2005). Biocatalytic oxidation of lignin models using O_2 and a variety of laccases or laccases in the presence of a mediator has also been widely examined as a means for removal of residual lignin from cellulose (Barreca et al. 2003; Crestini et al. 2003; Elegir et al. 2005; Lahtinen et al. 2009; Li et al. 1999; Rochefort et al. 2004), and has been compared to metal catalyzed systems (Bohlin et al. 2005). Electrocatalytic processes that mimic the action of laccase in laccase mediator systems have been examined for the conversion of lignin model monomers and dimers (Rochefort et al. 2002).

Expansion of these model studies to isolated lignin or cellulose pulp has also been carried out. An extensive study on lignin oxidation to substituted aromatic carboxylic acids used a mixed Co/Mn/Zr/Br catalyst in HOAc, modeling well-established industrial processes for the conversion of alkyl aromatic compounds to the corresponding carboxylic acid (e. g., the aerobic oxidation of *para*-xylene to terephthalic acid). Several types of lignin were examined with this catalyst system, and gave a maximum yield of 10.9 % organic products from hardwood organosolv lignin, as a mixture of vanillin, vanillic acid, syringaldehyde and syringic acid (Partenheimer 2009). Polyoxometalates and metalloporphyrins have also been examined with lignin in the development of chlorine-free bleaching processes (Perng et al. 1994; Voitl and von Rohr 2008; Weinstock et al. 1996). Electrochemical processes have been used for lignin oxidation. Electrolysis of commercial alkali lignin was carried out at Ru/V/Ti/O electrodes in ionic liquids to give 3–6 wt% of low molecular weight aromatic products as a complex mixture of more than 10 compounds (Reichert et al. 2012). Oxidation of kraft lignin using IrO_2 electrodes afforded vanillin and vanillic acid as the primary products, but no yields were reported (Tolba et al. 2010).

17.3 Oxidation of Lignin to Vanillin

Historically, interest in selective oxidation of lignin to discrete low molecular weight compounds began with efforts to produce vanillin from lignosulfonates isolated from the sulfite pulping process. A much smaller amount of lignin (about

5.7×10^6 t in 2005) is available from the sulfite process, but in contrast to the kraft process, sulfite pulping generally does not use the lignin as fuel, affording lignosulfonates as a separate process stream. The production of vanillin from lignin in the pulp and paper industry has a long history, with an initial observation of vanillin in lignin wastes appearing in 1875, and the first commercial scale production in the US starting in 1936 (Hocking 1997). The process has typically been carried out by alkaline oxidation of lignosulfonates isolated from sulfite pulping of wood at high temperature using air as the oxidant. Yields of lignin are quite low (5–10 %), but the pulp industry has employed the process for the purpose of generating additional revenue from pulping operations. Although 85 % of vanillin today is now produced from guaiacol, (da Silva et al. 2009) production from lignin is still carried out on a limited scale by industry, in particular, Borregard (Voitl and von Rohr 2010). The primary disadvantage to production of lignin from lignosulfonates is the cost associated with disposal of a large amount of residual lignin waste after vanillin recovery. These costs led to elimination of nearly all commercial lignin-based vanillin operations by the early 1990s (Hocking 1997).

Efforts have been made to improve this process through development of new oxidation catalysts. It has long been recognized that the yields of vanillin from sulfite liquors can be increased by the addition of metal catalysts such as Cu(II). Alkaline oxidation of sulfite waste liquor gave 22 % vanillin in the presence of $CuSO_4 \cdot 5H_2O$, and 13.5 % yield from sulfite waste liquor solids (Pearl 1942). Alternatively, alkaline nitrobenzene can be used as the oxidizing agent. Nitrobenzene oxidation of spruce wood meal gave a 23 % yield of vanillin (Creighton et al. 1941). However, the requirement for stoichiometric amounts of nitrobenzene in the oxidation significantly increases the cost of the process, making it unsuitable for industrial use. Development of a process catalytic in nitrobenzene was attempted, by combining nitrobenzene and Cu(II) catalyzed oxidations. However, the yield of vanillin using this process was never greater than 6 % (Bjorsvik 1999). As a result, the only remaining commercial process for conversion of lignin to vanillin still uses aerobic alkaline oxidation. Organosolv lignins have also been reported as starting materials for vanillin synthesis. Organosolv lignin from eucalyptus, sugarcane bagasse and softwood via the Acetosolv or Organocell process was converted to vanillin by oxidation in HOAc with O_2 catalyzed by $Co(OAc)_2$. The optimum yield reported for any of these systems was less than 6 % (Goncalves and Schuchardt 1999).

The much more abundant kraft lignin can also be converted to vanillin, although the yields remain uniformly low. Oxidation of black liquor from kraft pulping of *Pinus pilaster* with oxygen in alkaline medium at ~135 °C gave a maximum vanillin yield of 1.2 g/l, or about 0.9 g of vanillin/100 g of contained black liquor solids after 50 min. Comparative biocatalytic oxidation with *Acinetobacter anitratus* was carried out, but gave a tenfold lower production of vanillin than chemical oxidation (Mathias et al. 1995). Catalyzed and uncatalyzed oxidation of isolated kraft lignin from eucalyptus with oxygen, or oxygen and added Cu(II) or Co(II) catalysts gave <5 % yield of low molecular weight products as a mixture of materials (Villar et al. 2001). Engineering process analysis of this process to evaluate yield and kinetics of the oxidation has been reported (Araujo et al. 2010). Oxidation of kraft lignin under acidic conditions was carried out using O_2 as the oxidant in the presence of the polyoxometalate $H_3PMo_{12}O_{40}$ as a catalyst in MeOH at 170° to give a mixture of vanillin and methyl vanillate in combined yields of 7–8 %. However, the isolated material also contained significant amounts of oligomeric material, indicating that further purification of the reaction products would be necessary to obtain pure monomers (Voitl and von Rohr 2010, 2008). Again, nitrobenzene oxidation of kraft lignin improves vanillin production, with a 13 % vanillin yield reported from pine lignin (Mathias and Rodrigues 1995). Engineering studies to develop reactor systems for the continuous production and purification of vanillin from kraft lignin have been reported (da Silva et al. 2009).

17.4 Selective Catalytic Oxidation of Lignin

The current interest in developing biorefineries that integrate biobased chemical and biofuel production has dramatically expanded the availability of lignin beyond its traditional position within the pulp and paper industry. Accordingly, there has been a clear transition from simply removing lignin as an unwanted impurity to recognizing lignin as an important source of renewable carbon. The growing interest in lignin as a chemical feedstock is evidenced by several recent reports on catalytic processes for oxidative cleavage of lignin and lignin models from groups not traditionally associated with biomass conversion (Hanson et al. 2012; Nichols et al. 2010; Sergeev and Hartwig 2011; Son and Toste 2010; Zakzeski et al. 2010).

Our own research in this field dates from early efforts to selectively convert lignin to substituted anthraquinones as catalysts for cellulose pulp manufacture (Dimmel et al. 1999; Dimmel and Bozell 1991). More generally, our approaches target the primary unifying structural feature of lignin, its network of oxygenated aromatic rings, and in particular, the oxidation of *para*-substituted phenolics to benzoquinones and other low molecular weight aromatics. This approach is designed to take advantage of the structural units present in isolated biorefinery lignin, for example, the recognized increase in the concentration of free phenolic –OH groups that occurs when β-O-4 interunit linkages are cleaved. Many oxidative reactions exist for preparing benzoquinones from phenolics. However, the great majority of the reported oxidations are performed on phenolics that have no substituent *para* to the hydroxyl group of the phenol. In contrast, *every* phenolic unit in lignin contains a *para* substituent that must be selectively cleaved to realize a successful synthesis of benzoquinones or related simple aromatic compounds. We found that Co-Schiff base complexes catalyze this oxidation with O_2 as the terminal oxidant (Fig. 17.3, illustrated using syringyl alcohol as the substrate) (Bozell et al. 1995).

In the presence of oxygen and an external ligand (L; typically an aromatic base such as pyridine or imidazole) Co(salen) forms intermediate superoxo complex **1**, in which the normally high reactivity of oxygen is mediated. Subsequent reaction with a lignin model, such as syringyl alcohol, abstracts the phenolic hydrogen to generate an intermediate phenoxy radical that is ultimately converted to a benzoquinone,

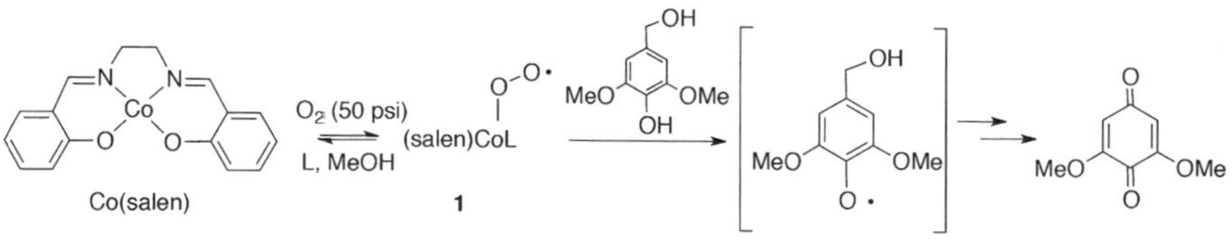

Fig. 17.3 Co-Schiff based catalyzed oxidation of *para*-substituted phenolics to quinones

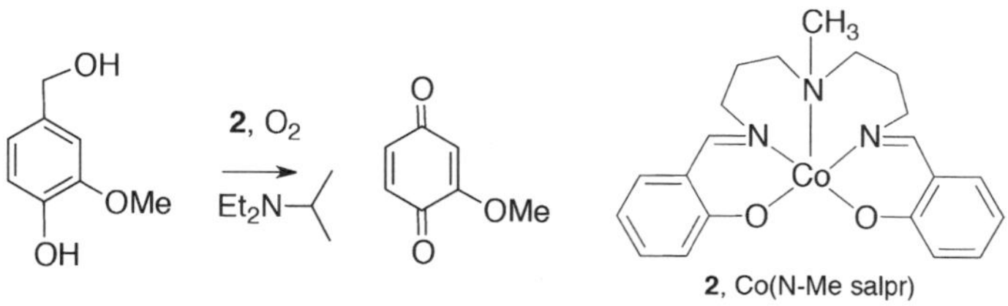

Fig. 17.4 Improved oxidation of guaiacyl models in the presence of a hindered base

specifically dimethoxybenzoquinone (DMBQ) for the process shown in Fig. 17.3. The yields of the reaction were highly dependent on the structure of the substrate. Syringyl models afforded quinone in 71–90 % yield, however, the corresponding guaiacyl models were much less reactive, giving monomethoxybenzoquinone (MMBQ) in 12–27 % yield. Depending on the substituent *para* to the phenolic hydroxyl group and the catalyst employed, benzylic oxidation leading to the production of substituted benzaldehydes could be observed in yields of 45–50 %.

Nonetheless, effective use of biorefinery lignin requires the ability to oxidize both syringyl and guaiacyl units in high yield, and the failure of this catalytic system with guaiacyl models was a disadvantage. These observations were attributed to an inability of the catalyst system to abstract a hydrogen atom from guaiacyl models to form the necessary phenoxy radical intermediate as shown in Fig. 17.3. Examination of Hammett σ^+ constants for structurally similar phenols revealed that the presence or absence of a single methoxy group on the aromatic ring could have a dramatic effect on the rate of hydrogen removal from the substrate. Subsequent work revealed that the yield of quinone from the guaiacyl model vanillyl alcohol is significantly improved by adding a sterically hindered base to oxidations catalyzed by Co(salen) or the alternative Co-Schiff base catalyst Co(N-Me salpr) (**2**, Fig. 17.4) (Cedeno and Bozell 2012). Oxidation of vanillyl alcohol to MMBQ proceeded in over 50 % yield upon addition of diisopropylethylamine, diisopropylamine or triethylamine. In contrast, oxidations in the absence of these bases displayed a maximum 21 % yield of quinone. Importantly, addition of this hindered base did not reduce the high yields of product observed for syringyl model oxidation.

Both NMR and UV–VIS measurements indicate that the hindered base does not coordinate to the Co catalysts and thus its role must involve other mechanistic pathways (Fig. 17.5).

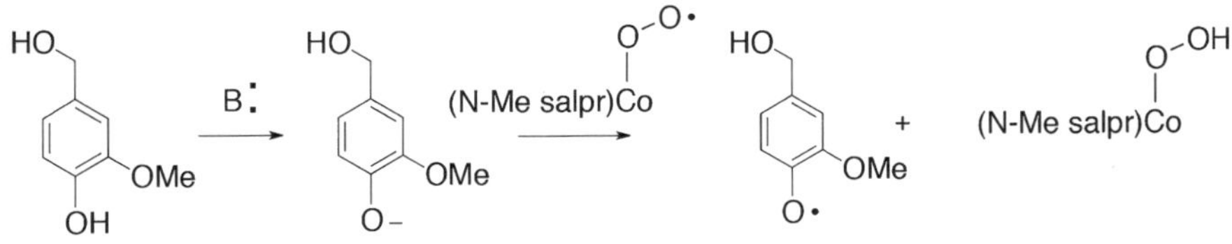

Fig. 17.5 Proposed mechanism for the oxidation of guaiacyl models in the presence of a hindered base

Our proposed mechanism postulates an initial deprotonation of the substrate by the sterically hindered base to form a more readily oxidized phenolate anion (Rappoport 2003). Transformation of the phenolate to the phenoxy radical occurs either with oxygen or the intermediate Co-superoxo complex, ultimately leading to quinone formation. This effect is currently limited, as addition of hindered bases to oxidations of other guaiacyl models such as isoeugenol, eugenol, or vanillin and vanillin acetal gave only low yields (0–10 %) of MMBQ. Further investigation into this alternative mechanism and computational evaluation of the intermediate Co complexes is currently underway.

Application of Co-Schiff base catalyzed oxidation to biorefinery lignin samples also induces conversion to quinones and structurally related aromatics. Treating mixtures of poplar and switchgrass lignin with Co(salen) and oxygen affords approximately 10 wt% yield of low molecular weight products, most of which is DMBQ. 2D-HMQC NMR analysis resolves these materials from the residual lignin present after the oxidation (Fig. 17.6).

Interestingly, oxidations of lignin are not improved by the addition of either an external basic ligand or a non-ligating hindered base, suggesting that the lignin itself may be acting in those capacities. Although the yields are still low, they are equivalent to similar lignin oxidation processes reported in the literature. Work is currently underway to optimize our catalyst system for the production of these materials.

Because of the tendency of Co(salen)/O_2 complexes to undergo deactivation during oxidation (Busch 1988), we also investigated other species that contain an oxygen centered free radical structurally analogous to Co-superoxo complexes and reported that stoichiometric NO_2 could be used for this conversion (Dimmel et al. 1996). Subsequent work revealed that *catalytic* NO_2 in the presence of O_2 also converted *para*-substituted phenols to benzoquinones (Bozell et al. 1998). In initial experiments, syringyl alcohol was treated with a stoichiometric amount of $NaNO_2$ and a small amount of concentrated HNO_3 or HCl (a convenient source of NO_2) in MeOH under argon at −20 °C to afford DMBQ in low yield. However, introduction of 1 atmosphere of O_2 to this reaction has a dramatic effect, allowing isolation of DMBQ in much higher yields (80–90 %) using only catalytic amounts of $NaNO_2$ to produce 80–90 % yields of DMBQ from syringyl alcohol. With $NaNO_2$ levels as low as 5 %, DMBQ was still formed in yields of 70–75 % (Fig. 17.7).

The mechanism of NO_2 oxidation displays similarities to that of Co-Schiff base/O_2 oxidations, but is complicated by the presence of additional oxides of nitrogen

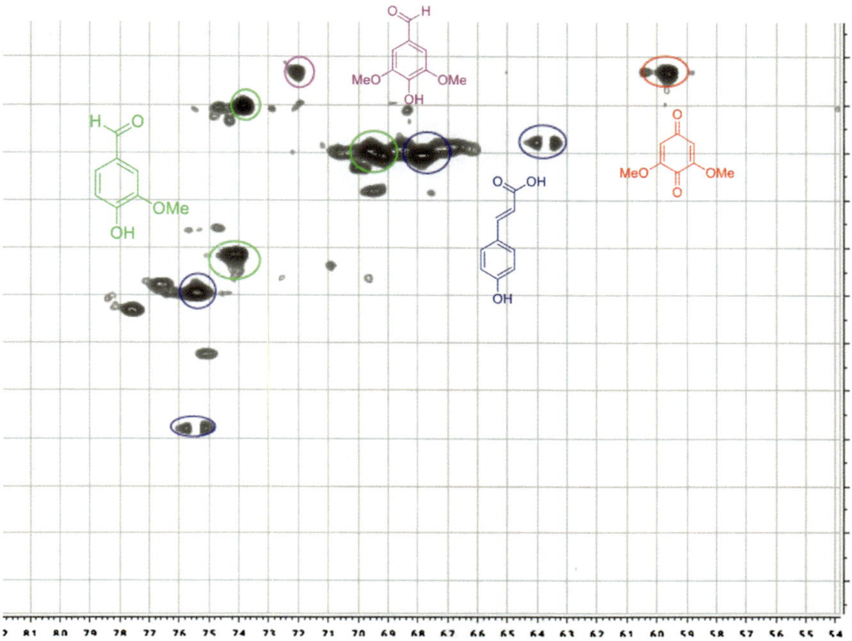

Fig. 17.6 Typical 2D HMQC spectrum of products from Co-Schiff base/O_2 oxidation

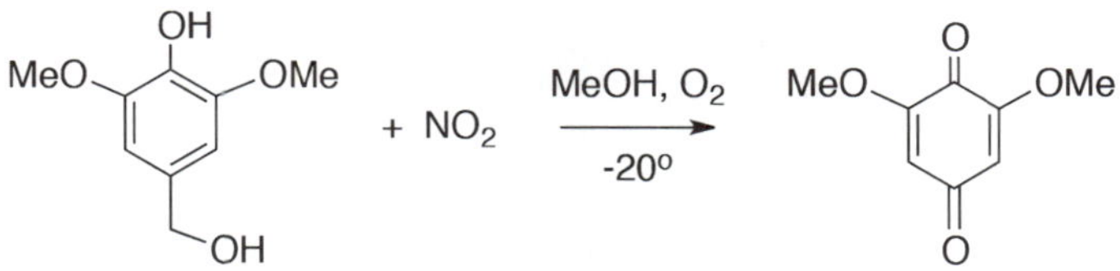

Fig. 17.7 Oxidation of lignin models using catalytic NO_2 in O_2

that can be formed under these conditions (Bosch et al. 1994). Kochi has shown that NO_2 oxidation of hydroquinone dialkyl ethers to quinones occurs via a radical cation that results from the reaction of the substrate with the NO_2 disproportionation product $NO^+NO_3^-$ (Rathore et al. 1994). Thus, substrates used in our study are assumed to undergo conversion to a radical cation upon reaction with NO^+. Formation of the quinone occurs via reaction of this cation with the nitrate counterion (Fig. 17.8).

This sequence forms HONO that is converted back to NO_2 via N_2O_3 formation and subsequent O_2 oxidation of N_2O_3 to N_2O_4, continuing the catalytic cycle. In parallel, we observe that direct reaction with NO_2 affords ring nitration rather than oxidation through formation of a phenoxy radical and subsequent trapping of that radical by a second molecule of NO_2. This mechanistic path is similar to that described for the reaction of phenols with Co-Schiff base complexes and O_2 (Nishinaga et al. 1981). We have also examined this process for the oxidation of organosolv lignin samples, and found that the yield of quinone is again fairly low, around 6–8 %.

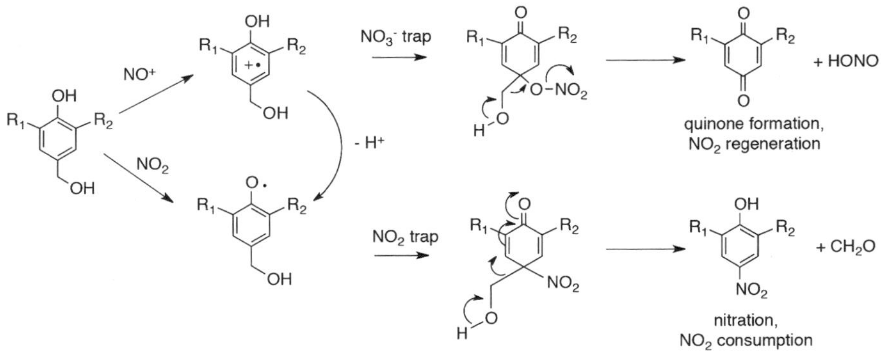

Fig. 17.8 Proposed mechanistic steps in catalytic NO_2 oxidation of *para*-substituted phenolics

17.5 Conclusions

The ability to convert lignin's heterogeneous structure to a single compound in high yield remains a grand challenge to the effective use of lignocellulosic feedstocks within the biorefinery. However, the transition of lignin from a byproduct largely sequestered within the pulp and paper industry to a material of general availability for biorefining has driven development of new catalytic oxidations designed to take advantage of the structural features resulting from its isolation. With the growing interest in lignin as a chemical feedstock, new catalytic processes offer the potential of improved conversion of this highly abundant material.

Acknowledgments This work was supported as part of the Center for Direct Catalytic Conversion of Biomass to Biofuels (C3Bio), an Energy Frontier Research Center funded by the US Department of Energy, Office of Science, Office of Basic Energy Sciences under Award Number DE-SC0000997, and US Department of Energy Office of Industrial Technologies.

References

Araujo JDP, Grande CA, Rodrigues AE (2010) Vanillin production from lignin oxidation in a batch reactor. Chem Eng Res Des 88:1024–1032

Auhorn WJ, Niemela K (2007) Process chemicals for the production of chemical pulp. Prof Papermaking 2:10–20

Barreca AM, Fabbrini M, Galli C, Gentili P, Ljunggren S (2003) Laccase/mediated oxidation of a lignin model for improved delignification procedures. J Mol Catal B 26:105–110

Bjorsvik HR (1999) Fine chemicals from lignosulfonates. 1. Synthesis of vanillin by oxidation of lignosulfonates. Org Proc Res Devel 3:330–340

Boerjan W, Ralph J, Baucher M (2003) Lignin biosynthesis. Ann Rev Plant Biol 54:519–546

Bohlin C, Persson P, Gorton L, Lundquist K, Jonsson LF (2005) Product profiles in enzymic and non-enzymic oxidations of the lignin model compound erythro-1-(3,4-dimethoxyphenyl)2-(2-methoxyphenoxy)-1,3-propanediol. J Mol Catal B 35:100–107

Bosch E, Rathore R, Kochi JK (1994) Novel catalysis of hydroquinone autoxidation with nitrogen-oxides. J Org Chem 59:2529–2536

Bozell JJ, Hames BR, Dimmel DR (1995) Cobalt-Schiff base complex-catalyzed oxidation of para-substituted phenolics—preparation of benzoquinones. J Org Chem 60:2398–2404
Bozell JJ, Hoberg JO, Dimmel DR (1998) Catalytic oxidation of para-substituted phenols with nitrogen dioxide and oxygen. Tetrahedron Lett 39:2261–2264
Bozell JJ, Holladay JE, Johnson D, White JF (2007) Top value added chemicals from biomass. Volume II—results of screening for potential candidates from Biorefinery Lignin. Pacific Northwest National Laboratory, Richland, WA
Bozell JJ, Black SK, Myers M, Cahill D, Miller WP, Park S (2011a) Solvent fractionation of renewable woody feedstocks: Organosolv generation of biorefinery process streams for the production of biobased chemicals. Biomass Bioenergy 35:4197–4208
Bozell JJ, O'Lenick CJ, Warwick S (2011b) Biomass fractionation for the biorefinery: heteronuclear multiple quantum coherence-nuclear magnetic resonance investigation of lignin isolated from solvent fractionation of switchgrass. J Agric Food Chem 59:9232–9242
Busch DH (1988) Synthetic dioxygen carriers for dioxygen transport. In: Martell AE, Sawyer DT (eds) Oxygen complexes and oxygen activation by transition metals. Plenum, New York
Cedeno D, Bozell JJ (2012) Catalytic oxidation of para-substituted phenols with cobalt-Schiff base complexes/O_2-selective conversion of syringyl and guaiacyl lignin models to benzoquinones. Tetrahedron Lett 53:2380–2383
Collinson SR, Thielemans W (2010) The catalytic oxidation of biomass to new materials focusing on starch, cellulose and lignin. Coord Chem Rev 254:1854–1870
Creighton RHJ, McCarthy JL, Hibbert H (1941) Aromatic aldehydes from spruce and maple woods. J Am Chem Soc 63:312
Crestini C, Jurasek L, Argyropoulos DS (2003) On the mechanism of the laccase-mediator system in the oxidation of lignin. Chem Eur J 9:5371–5378
Crestini C, Pastorini A, Tagliatesta P (2004) Metalloporphyrins immobilized on motmorillonite as biomimetic catalysts in the oxidation of lignin model compounds. J Mol Catal A 208:195–202
Crestini C, Pro P, Neri V, Saladino R (2005) Methyltrioxorhenium: a new catalyst for the activation of hydrogen peroxide to the oxidation of lignin and lignin model compounds. Biorg Med Chem 13:2569–2578
Crestini C, Caponi MC, Argyropoulos DS, Saladino R (2006) Immobilized methyltrioxo rhenium (MTO)/H_2O_2 systems for the oxidation of lignin and lignin model compounds. Biorg Med Chem 14:5292–5302
Cui F, Dolphin D (1995) Iron porphyrin catalyzed oxidation of lignin model compounds: Oxidation of phenylpropane and phenylpropene model compounds. Can J Chem 73:2153–2157
da Silva EAB, Zabkova M, Araujo JD, Cateto CA, Barreiro MF, Belgacem MN, Rodriques AE (2009) An integrated process to produce vanillin and lignin-based polyurethanes from kraft lignin. Chem Eng Res Des 87:1276–1292
Dence CW, Reeve DW (1996) Pulp bleaching—principles and practice. TAPPI, Atlanta
Dicosimo R, Szabo HC (1988) Oxidation of lignin model compounds using single-electron-transfer catalysts. J Org Chem 53:1673–1679
Dimmel DR, Bozell JJ (1991) Pulping catalysts from lignin. Tappi J 74:239–241
Dimmel DR, Karim MR, Savidakis MC, Bozell JJ (1996) Pulping catalysts from lignin. 5. Nitrogen dioxide oxidation of lignin models to benzoquinones. J Wood Chem Technol 16:169–189
Dimmel DR, Althen E, Savidakis M, Courchene C, Bozell JJ (1999) New quinone-based pulping catalysts. Tappi J 82:83–89
Elegir G, Daina S, Zoia L, Bestetti G, Orlandi M (2005) Laccase mediator system: oxidation of recalcitrant lignin model structures present in residual kraft lignin. Enzyme Microb Technol 37:340–346
Evtuguin DV, Daniel AID, Silvestre AJD, Amado FML, Neto CP (2000) Lignin aerobic oxidation promoted by molybdovanadophosphate polyanion $PMo_7V_5O_{40}$ $^{(8-)}$. Study on the oxidative cleavage of beta-O-4 aryl ether structures using model compounds. J Mol Catal A 154:217–224

Fengel D, Wegener G (1984) Wood: Chemistry, ultrastructure, reactions. Walter de Gruyter, Berlin

Goncalves AR, Schuchardt U (1999) Oxidation of organosolv lignins in acetic acid—Influence of oxygen pressure. Appl Biochem Biotechnol 77–9:127–132

Hanson SK, Wu R, Silks LAP (2012) C–C or C–O bond cleavage in a phenolic lignin model compound: selectivity depends on vanadium catalyst. Angew Chem Int Ed 51:3410–3413

Hocking MB (1997) Vanillin: synthetic flavoring from spent sulfite liquor. J Chem Educ 74:1055–1059

Lahtinen M, Kruus K, Boer H, Kemell M, Andberg M, Viikari L, Sipila J (2009) The effect of lignin model compound structure on the rate of oxidation catalyzed by two different fungal laccases. J Mol Catal B 57:204–210

Li KC, Xu F, Eriksson KEL (1999) Comparison of fungal laccases and redox mediators in oxidation of a nonphenolic lignin model compound. Appl Environ Microbiol 65:2654–2660

Li JB, Gellerstedt G, Toven K (2009) Steam explosion lignins; their extraction, structure and potential as feedstock for biodiesel and chemicals. Bioresour Technol 100:2556–2561

Mathias AL, Rodrigues AE (1995) Production of vanillin by oxidation of pine kraft lignins with oxygen. Holzforschung 49:273–278

Mathias AL, Lopretti MI, Rodrigues AE (1995) Chemical and biological oxidation of pinus-pinaster lignin for the production of vanillin. J Chem Technol Biotechnol 64:225–234

Nichols JM, Bishop LM, Bergman RG, Ellman JA (2010) Catalytic C–O bond cleavage of 2-aryloxy-1-arylethanols and its application to the depolymerization of lignin-related polymers. J Am Chem Soc 132:12554–12555

Nishinaga A, Tomita H, Nishizawa K, Matsuura T, Ooi S, Hirotsu K (1981) Regioselective formation of peroxyquinolatocobalt(III) complexes in the oxygenation of 2,6-di-tert-butylphenols with cobalt(II) Schiff-base complexes. J Chem Soc Dalton, pp 1504–1514

Partenheimer W (2009) The aerobic oxidative cleavage of lignin to produce hydroxyaromatic benzaldehydes and carboxylic acids via metal/bromide catalysts in acetic acid/water mixtures. Adv Synth Catal 351:456–466

Pearl IA (1942) Vanillin from lignin materials. J Am Chem Soc 64:1429–1431

Perng YS, Oloman CW, Watson PA, James BR (1994) Catalytic oxygen bleaching of wood pulp with metal porphyrin and phthalocyanine complexes. Tappi J 77:119–125

Rappoport Z (2003) The chemistry of phenols. Wiley, Hoboken

Rathore R, Bosch E, Kochi JK (1994) Selective nitration versus oxidative dealkylation of hydroquinone ethers with nitrogen-dioxide. Tetrahedron 50:6727–6758

Reichert E, Wintringer R, Volmer DA, Hempelmann R (2012) Electro-catalytic oxidative cleavage of lignin in a protic ionic liquid. Phys Chem Chem Phys 14:5214–5221

Rochefort D, Bourbonnais R, Leech D, Paice MG (2002) Oxidation of lignin model compounds by organic and transition metal-based electron transfer mediators. Chem Commun pp 1182–1183

Rochefort D, Leech D, Bourbonnais R (2004) Electron transfer mediator systems for bleaching of paper pulp. Green Chem 6:14–24

Samuel R, Pu YQ, Raman B, Ragauskas AJ (2010) Structural characterization and comparison of switchgrass ball-milled lignin before and after dilute acid pretreatment. Appl Biochem Biotechnol 162:62–74

Sergeev AG, Hartwig JF (2011) Selective, nickel-catalyzed hydrogenolysis of aryl ethers. Science 332:439–443

Sippola VO, Krause AOI (2005) Bis(o-phenanthroline)copper-catalysed oxidation of lignin model compounds for oxygen bleaching of pulp. Catal Today 100:237–242

Son S, Toste FD (2010) Non-oxidative vanadium-catalyzed C–O bond cleavage: application to degradation of lignin model compounds. Angew Chem Int Ed 49:3791–3794

Tolba R, Tian M, Wen JL, Jiang ZH, Chen AC (2010) Electrochemical oxidation of lignin at IrO_2-based oxide electrodes. J Electroanal Chem 649:9–15

Vanholme R, Demedts B, Morreel K, Ralph J, Boerjan W (2010) Lignin biosynthesis and structure. Plant Physiol 153:895–905

Villar JC, Caperos A, Garcia-Ochoa F (2001) Oxidation of hardwood kraft-lignin to phenolic derivatives with oxygen as oxidant. Wood Sci Technol 35:245–255

Voitl T, von Rohr PR (2008) Oxidation of lignin using aqueous polyoxometalates in the presence of alcohols. ChemSusChem 1:763–769

Voitl T, von Rohr PR (2010) Demonstration of a process for the conversion of kraft lignin into vanillin and methyl vanillate by acidic oxidation in aqueous methanol. Ind Eng Chem Res 49:520–525

Weinstock IA, Atalla RH, Reiner RS, Moen MA, Hammel KE, Houtman CJ, Hill CL (1996) A new environmentally benign technology and approach to bleaching kraft pulp. Polyoxometalates for selective delignification and waste mineralization. New J Chem 20:269–275

Weinstock IA, Atalla RH, Reiner RS, Moen MA, Hammel KE, Houtman CJ, Hill CL, Harrup MK (1997) A new environmentally benign technology for transforming wood pulp into paper—Engineering polyoxometalates as catalysts for multiple processes. J Mol Catal A 116:59–84

Zakzeski J, Jongerius AL, Weckhuysen BM (2010) Transition metal catalyzed oxidation of Alcell lignin, soda lignin, and lignin model compounds in ionic liquids. Green Chem 12:1225–1236

Zhu WM, Ford WT (1993) Oxidation of lignin model compounds in water with dioxygen and hydrogen-peroxide catalyzed by metallophthalocyanines. J Mol Catal 78:367–378

Printed by Printforce, the Netherlands